深入理解
Spring Cloud
与实战

方剑 / 编著

电子工业出版社
Publishing House of Electronics Industry
北京·BEIJING

内 容 简 介

本书共分 10 章，主要介绍 Spring Cloud 各个核心组件的设计原理，以及目前流行的 Spring Cloud Alibaba 和 Netflix 组件，并且剖析 Spring Cloud 对流处理、批处理，以及目前业界流行的 Serverless 的支持。在介绍各部分内容时，本书将理论与实践相结合，对每个核心知识点都给出了具体的案例应用，以帮助读者掌握核心组件的设计理念。

本书适合对 Spring Cloud 感兴趣并且想透彻理解 Spring Cloud 的读者阅读，也适合正在进行微服务选型的开发者阅读。

未经许可，不得以任何方式复制或抄袭本书之部分或全部内容。
版权所有，侵权必究。

图书在版编目（CIP）数据

深入理解 Spring Cloud 与实战 / 方剑编著.—北京：电子工业出版社，2021.1
ISBN 978-7-121-39973-2

Ⅰ．①深… Ⅱ．①方… Ⅲ．①互联网络—网络服务器 Ⅳ．①TP368.5

中国版本图书馆 CIP 数据核字（2020）第 226698 号

责任编辑：李利健
印　　刷：北京天宇星印刷厂
装　　订：北京天宇星印刷厂
出版发行：电子工业出版社
　　　　　北京市海淀区万寿路 173 信箱　　　　邮编：100036
开　　本：787×980　1/16　　印张：28.25　　字数：663 千字
版　　次：2021 年 1 月第 1 版
印　　次：2025 年 1 月第 8 次印刷
定　　价：106.00 元

凡所购买电子工业出版社图书有缺损问题，请向购买书店调换。若书店售缺，请与本社发行部联系，联系及邮购电话：（010）88254888，88258888。
质量投诉请发邮件至 zlts@phei.com.cn，盗版侵权举报请发邮件至 dbqq@phei.com.cn。
本书咨询联系方式：010-51260888-819，faq@phei.com.cn。

推荐序 1

阿里巴巴是国内比较早采用微服务架构的公司，在 2011 年就对外开源了 Dubbo RPC 框架。阿里巴巴内部也在 Dubbo 的基础上逐渐摸索并演进出更多配套的技术组件，如配置管理、微服务网关、链路跟踪、限流降级、容灾演练等，形成了一套完整而又独立的内部技术栈。彼时，在外面的开源体系下，始终没有一套完整的微服务开发框架来很好地支持微服务不同技术组件的松耦合集成。Spring Cloud 正是为此而生。基于"习惯优于配置"的设计理念，Spring Cloud 定义了微服务架构"全家桶"内各个组件的接口标准，让开发者可以更容易地组装自己的微服务。由于和 Spring 的设计理念一脉相承，Spring Cloud 具备了强大的社区影响力，并成为当下流行的 Java 微服务框架。

2018 年的时候，我正负责阿里云一款名为企业级分布式应用服务（EDAS）的云产品，主要为企业提供微服务架构的一站式产品化方案。当时走访多个客户并进行交流后发现，不少客户对我们的产品"既爱又恨"：一方面，他们认可阿里云在微服务上的实践经验，EDAS 能帮助他们更快地构建微服务；另一方面，他们对业务被阿里云这套技术架构绑定有很强的担忧。几经考虑，产品要解决技术绑定问题，最佳途径无疑是支持开源、开放。因此，我们决定除了重新启动 Dubbo 开源，原本只用于商业化的微服务组件也一并开源，比如用于服务注册/发现和配置推送的 Nacos、用于限流降级保护的 Sentinel、用于分布式事务的 Seata 等。为了让这些组件能够更容易地互相集成，我们又推出了 Spring Cloud Alibaba，提供对 Spring Cloud 标准的完整实现适配，将这些微服务组件和开源的现有体系无缝融合。当时方剑（洛夜）就是 Spring Cloud Alibaba 开源项目组的负责人和核心开发成员，有不少 Spring Cloud 组件都是方剑设计和编写的，并且他参与了整个项目的开源过程，以及 EDAS 对 Spring Cloud 框架的支持工作，事无巨细都有了解。

　　当听到他说在写这本关于 Spring Cloud 的书时，我想，无论是对 Spring Cloud 框架设计原理的理解，还是对 Spring Cloud 各个组件的实现的理解，他应该是国内最资深的人之一。在读完本书后，我觉得值得向 Spring Cloud 的开发者和技术爱好者推荐本书，书中技术点分析翔实，层次清晰，完全可用于 Spring Cloud 微服务实战指导。建议所有打算将 Spring Cloud 用于开发的朋友仔细阅读本书。

司徒放（姬风）

阿里云智能资深技术专家

阿里云智能云原生应用 PaaS 与 Serverless 产品线负责人

推荐序 2

作为近年来流行的微服务框架，Spring Cloud 俨然成了现代微服务架构的事实标准。从 Spring Framework 到 Spring Boot，再到 Spring Cloud，几乎每一次迭代，Spring 社区总是能重新定义一种全新的架构形态，并且推动其成为事实标准。目前市面上的 Spring Cloud 书籍主要以介绍 Spring Cloud Netflix 技术栈为主，但随着 Netflix OSS 进入维护状态，让人不免对 Spring Cloud 的未来产生了担心，这时候方剑站出来创建了 Spring Cloud Alibaba，把集阿里巴巴微服务"十年之功"的开源体系，如 Apache Dubbo、Nacos、Sentinel、RocketMQ 等和 Spring Cloud 完美融合在一起。为此，我深感钦佩。因此，我认为由他来撰写这样一本书实在是再合适不过的。

本书内容紧跟技术前沿，深入浅出，核心知识点都配有相应的实操案列，基于 Spring Boot 2.x，结合 Spring Cloud Alibaba 所有的核心组件，将 Spring Cloud 的实现原理抽丝剥茧地为大家一一梳理。书中包括的 Spring Cloud Data Flow 和 Spring Cloud Function 相关内容是市面上其他同类书中较少涉及的，这两部分内容展示了 Spring 社区未来对大数据和 Serverless 两个重要领域的布局，也表明了作者对 Spring 技术发展趋势的判断。建议想了解 Spring Cloud，以及对其内部实现感兴趣的人阅读本书。

作为方剑的同事，我被他对技术的极致追求所折服。在平时的工作交流中，他对 Spring

Cloud 的洞见经常让我受益。方剑应该是国内与 Spring 社区接触最紧密的人之一，他和 Josh Long、Andy Clement、Spencer Gibb 等"大牛"都有深入的沟通和频繁的交流，可以说对 Spring 完整生态的理解相当深入。同时，作为 Spring Cloud Alibaba 创始成员，他带领 Spring Cloud Alibaba 成为国内首个在 Spring 社区"毕业"的中国项目。他对技术的洞察力和领导力让我对本书充满了信心。

"行路难，行路难，多歧路，今安在？长风破浪会有时，直挂云帆济沧海。"希望本书就是那条船，能够帮助更多的读者受益，抵达梦想的彼岸。希望更多的人了解 Spring Cloud 之后，参与到 Spring 开源社区的建设中，为社区贡献自己的一份力量。

张乎兴
阿里云智能高级技术专家，
Apache Member，Apache Dubbo/Tomcat PMC Member

推荐序 3

2017 年,我与方剑在国内开源社区相识。那时,他给人的印象是一名乐于分享并且积极上进的青年。后来,得知他入职阿里巴巴中间件部门,并一同参与了 Spring Cloud 与阿里巴巴及阿里云的技术整合项目,也就是 Spring Cloud Alibaba。

作为一名资深开发人员,方剑的职业素养高,具备专业、严谨、热情的态度和强烈的责任心。记得 2018 年 11 月,我们一起参加 Spring One Tour 2018 北京峰会。会后,方剑给 Spring Cloud Leader Spencer 提出了不少关于 Spring Cloud 实现上的建议。有趣的是,方剑刚走出会议大厅就接到来自社区贡献者军哥(nepxion.com 项目发起人)的电话,反馈 Spring Cloud Alibaba 孵化版本中的问题。经过排查,问题的确存在。数次电话沟通中,方剑面露歉意,并承诺尽快修复问题。我想,大多数工程师都经历过这种情况。晚间,我们来到 Josh Long(Spring Developer Advocate)下榻的酒店,方剑向对方演示 Spring Cloud Alibaba RocketMQ 在本地和阿里云环境的操作,时间不知不觉已到凌晨四点。后来,Josh Long 还录制了相关的视频分享在 YouTube 上。

作为 Spring Cloud Alibaba 项目的负责人,方剑除了贡献代码,更多的是把精力投入社区运营、维护,以及与 Spring 官方的沟通上。经过一年多的不懈努力,2019 月 7 月 24 日,Spring 官方宣布 Spring Cloud Alibaba"毕业",仓库迁移至 Alibaba Github OSS 下。这意味着

Spring Cloud Alibaba 是国内首个进入 Spring 社区的开源项目。

作为一名技术布道者，方剑身体力行，2018 年至 2019 年走遍了全国 IT "重镇"，分享 Spring Cloud 和 Spring Cloud Alibaba 等技术。几乎每场分享过后，方剑都将自己的技术心得和体会以博文的方式发布，文章的技术深度和广度是有目共睹的。同时，方剑意识到单凭技术文章很难系统性地探讨技术细节，由此萌生了写书的念头，他还与我讨论过写书的一些技巧和注意事项。

在他写作期间，Spring Cloud 的内核在不断变化，这也给图书的编写增添了难度，不得不与时俱进，及时进行调整。不过，我个人认为对这本书的等待是值得的。尽管方剑是 Spring Cloud Alibaba 的核心开发人员，但本书并没有专题讨论 Spring Cloud Alibaba，也不是单纯地指导读者如何使用 Spring Cloud，而是从宏观和微观的视角深入讨论 Spring Cloud 架构的发展，以及各个组件之间的特性和联系。所以，本书不仅适合 Spring Cloud 初探者，也可以供资深从业人员参考。

小马哥（mercyblitz）
Apache Dubbo PMC，Spring Cloud Alibaba 项目架构师

推荐序 4

我自己有过写书的想法，但从来没有想过给其他人的图书写序。洛夜（方剑）找我为他的新书写序时，我一开始是推辞的，因为我认为阿里巴巴有很多"牛人"，他们写序会更有影响力，对读者也更有帮助。但是他说："你应该写，毕竟是你给我的机会。"我突然醒悟了，我觉得我的确应该写，不是因为我给了洛夜机会，而是洛夜给了我一个机会缅怀我们一起为 Spring Cloud Alibaba 奋斗的日子。我们需要记录一下 Spring Cloud Alibaba 的成长史，也是我们奉献给 Spring Cloud Alibaba 的奋斗史，这的确是作为亲历者才有的最真挚、最纯粹的情感，别人无法取代。

Spring Cloud Alibaba 的奋斗史有几点值得我们记住：

（1）艰难抉择。一边是营收的 KPI，一边是从零开始做开源。我们小组 5 个人反复讨论了多次，最终决定开发 Spring Cloud Alibaba，理由很简单：我们有更好的经过生产验证且更完整的分布式能力，我们希望更多的人受益。

（2）独自前行。决定要做之后，困难如期而至。因为所有的兄弟团队的第一要务就是营收，开源在他们看来是可有可无的。怎么办？我们只好独自前行，其中的难度可想而知。我们首先必须深入了解每个组件的实现细节，不断看文档，反复做调试，"吃透"每个功能的设计原理，同时必须深刻理解 Spring Cloud 的核心标准，做一个完美的适配。写完之后，再邀请兄弟团队调试代码。就这样一步一步地完成核心模块的适配，再逐个研究、追加新的模块，最终实现 Spring Cloud Alibaba 的第一版开源。

（3）遭遇危机。2019 年年初，Spring Cloud Netflix 宣布停止更新，它的停更再次给我们敲响了警钟，商业化公司纯粹靠情怀来支撑开源是无法长久的，但这也证明了我们对开源与商业化进行融合的前景判断是正确的，必须坚持下去。

（4）激流勇进。随着 Spring Cloud Alibaba 的逐渐成熟，各个模块都有了专职人员支持，而且影响力也越来越大。

回忆完我们的奋斗史后，再聊一下洛夜，一句话总结：他是一个纯粹的技术男。纯粹的技术男有且仅有两个爱好：写最难的代码，打最"暴力"的游戏。这是属于我们的"Logo"。还记得洛夜刚加入阿里巴巴的时候，话极少，喜欢写代码，一写就写到很晚。后来在做 Spring Cloud Alibaba 项目期间，他更是到了废寝忘食的地步，每天工作到凌晨一两点是家常便饭，眼睛总是红的。可以说，没有他的熬夜，就没有今天的 Spring Cloud Alibaba。

由于他是一个纯粹的技术男，导致他只能输出干货。对于他的演讲，我也反馈过很多次，因为干货太多、信息量太大，所以很容易让人犯困。我非常希望他演讲时能讲几个笑话，但是他一直没有做到。这本《深入理解 Spring Cloud 与实战》凝结了他所有的心血，还是一如既往地全是干货。可以说这是你学习 Spring Cloud Alibaba 的必备良书，有此一本足矣。

彭文杰
阿里云智能高级技术专家，微课
网前 CTO，卫生部考试系统前总架构师

前　言

据 O'Reilly 在 2020 年 7 月 15 日公布的企业微服务市场现状的调研数据来看，有 77％的组织采用了微服务，其中，92％的组织成功地使用了微服务。Spring Cloud 和 Apache Dubbo 作为非常流行的两款微服务开发框架，深受开发者们关注。笔者编写本书的目的是让读者透彻理解 Spring Cloud 这套开发框架。

2018 年 7 月，笔者有幸参与了 Spring Cloud Alibaba 开源项目的建设，并成为该项目的创始人之一。在随后的日子里，笔者经历了 Spring Cloud Alibaba 进入 Spring Cloud 孵化器、毕业、Spring Cloud 官网上线 Spring Cloud Alibaba、Spring Initializr 上线 Spring Cloud Alibaba 组件、GitHub Star 数超过 15000 个、Used By 数超过 7000 个、企业使用数超过 1000 个等诸多具有里程碑意义的事件。在整个过程中，笔者有幸与 Spring Cloud 的创始人 Spencer Gibb 交流过 Spring Cloud 的一些设计理念，还通过与 Spring 首席布道师 Josh Long 一起制作 *Bootiful Podcast* 视频来介绍 Spring Cloud Alibaba。这些都是非常难忘的经历。

在建设 Spring Cloud Alibaba 的过程中，笔者对 Spring Cloud 整个生态的知识点进行了深入学习，发现 Spring Cloud 生态体系其实非常庞大，大家平时可能更多地关注于服务注册/发现、配置管理、熔断器和网关这 4 大方面，但实际上 Spring Cloud 还包括消息、消息总线、任务调度、应用部署、Serverless、CI/CD 等诸多领域的知识。

本书共分 10 章，主要介绍 Spring Cloud 各个核心组件的设计原理，以及目前流行的 Spring Cloud Alibaba 和 Netflix 组件，并且剖析 Spring Cloud 对流处理、批处理，以及目前业界流行的 Serverless 的支持。在介绍各部分内容时，本书将理论与实践相结合，对每个核心知识点都给出了具体的案例应用，以帮助读者掌握核心组件的设计理念。

笔者一开始是比较抗拒写这本书的，因为市面上已有不少与 Spring Cloud 相关的书籍，再编写一本可能也没有太大意义。正是有了 Spring Cloud Alibaba 的这段经历，以及对最新 Spring Cloud 版本的理解（本书所用的版本是 Spring Cloud Hoxton.RELEASE），所以才想把对 Spring Cloud 的理解写出来与大家分享。

由于笔者平时工作非常忙，写书过程中又遇到了其他事情，只能每天早起、晚睡挤出时间来编写，所以导致本书历时近一年时间才完成。

因笔者能力有限，书中难免有错漏之处，恳请读者批评、指正。对本书的意见和建议，读者可以通过电子邮件发送给笔者（电子邮箱为 fangjian0423@apache.org），或者在本书对应的 GitHub 代码仓库上提交 issue。本书所有的示例代码均已发布到 GitHub 官网的"fangjian0423/deep-in-spring-cloud-samples"页面下，每个项目的说明文档（README）都注明了对应的章节。

本书能顺利出版，首先要感谢笔者的老板和同事，是他们让笔者有机会参与 Spring Cloud Alibaba 开源项目的建设。其次要感谢笔者的妻子，是她的鼓励让笔者有动力编写这本书，而且她在笔者写作期间分担了不少家庭事务，让笔者有更多的时间投入写作。最后要感谢电子工业出版社的编辑李利健和她的同事们对本书提出的修改建议。

作　者

目 录

第 1 章 Spring Cloud 生态 1
- 1.1 Spring Cloud 基础：Spring Boot 概述 1
- 1.2 Spring Boot 核心特性 3
  - 1.2.1 Web 服务器：WebServer 3
  - 1.2.2 条件注解：@ConditionalOnXX 7
  - 1.2.3 工厂加载机制 10
  - 1.2.4 配置加载机制 12
  - 1.2.5 Spring Boot Actuator 15
- 1.3 Spring Cloud 概述 17
  - 1.3.1 Spring Cloud 诞生背景 17
  - 1.3.2 Netflix OSS 19
  - 1.3.3 Spring Cloud 项目 20
  - 1.3.4 Spring Cloud 版本 21
  - 1.3.5 Spring Cloud 最新动态 23

第 2 章 服务注册与服务发现 26

2.1 微服务架构演进 26
2.2 使用 Alibaba Nacos 体验第一个 Spring Cloud 微服务应用 29
2.2.1 下载并启动 Nacos Server 30
2.2.2 启动 Nacos Discovery Provider 进行服务注册 32
2.2.3 启动 Nacos Discovery Consumer 进行服务发现 33
2.3 使用 Netflix Eureka 替换 Alibaba Nacos 注册中心 36
2.3.1 启动 Eureka Server 37
2.3.2 启动 Eureka Discovery Provider 进行服务注册 38
2.3.3 启动 Nacos Discovery Consumer 进行服务发现 39
2.4 Spring Cloud 统一服务注册/发现编程模型 40
2.4.1 DiscoveryClient 和 ReactiveDiscoveryClient 40
2.4.2 ServiceInstance 和 Registration 44
2.4.3 ServiceRegistry 48
2.4.4 ServiceRegistryEndpoint 50
2.5 双注册双订阅模式 51
2.5.1 双注册双订阅模式分析 51
2.5.2 案例：使用双注册双订阅模式将 Eureka 注册中心迁移到 Nacos 注册中心 ... 54

第 3 章 负载均衡与服务调用 57

3.1 负载均衡原理 57
3.2 Spring Cloud LoadBalancer 负载均衡组件 59
3.3 Netflix Ribbon 负载均衡 74
3.3.1 RibbonLoadBalancerClient 74
3.3.2 RibbonServer 和 Server 79
3.3.3 ServerIntrospector 82
3.3.4 ILoadBalancer 84
3.3.5 ServerList 87
3.3.6 ServerListUpdater 90
3.3.7 ServerStats 92
3.3.8 Netflix Ribbon 配置项总结 94

		3.3.9 Ribbon 缓存时间 97
	3.4	Dubbo LoadBalance 负载均衡 98
	3.5	OpenFeign：声明式 Rest 客户端 100
		3.5.1 OpenFeign 概述 100
		3.5.2 OpenFeign 对 JAX-RS 的支持 102
		3.5.3 OpenFeign 底层执行原理 103
	3.6	Dubbo Spring Cloud：服务调用的新选择 105
	3.7	再谈路由和负载均衡 111
	3.8	案例：应用流量控制 112
		3.8.1 流量控制应用的业务场景 113
		3.8.2 使用 Netflix Ribbon 完成应用灰度发布 115

第 4 章 配置管理 119

	4.1	配置中心背景概述 119
	4.2	Spring/Spring Boot 与配置 120
	4.3	Spring Cloud 与配置 122
		4.3.1 使用 Alibaba Nacos 体验配置的获取及动态刷新 123
		4.3.2 从 Spring Cloud 配置中心获取配置的原理 125
		4.3.3 Spring Cloud 配置动态刷新 135
	4.4	Spring Cloud Config Server/Client 152
		4.4.1 Spring Cloud Config Server 152
		4.4.2 Spring Cloud Config Client 165
		4.4.3 Spring Cloud Config Client 与 Service Discovery 整合 171
		4.4.4 Spring Cloud Config 配置动态刷新 174
	4.5	再谈配置动态刷新 175
	4.6	案例：Spring Cloud 应用流量控制策略动态生效 180

第 5 章 熔断器 184

	5.1	熔断器模式概述 184
	5.2	手动实现一个断路器 186
		5.2.1 定义 State 枚举和 Counter 计数器类 186

5.2.2　定义 CircuitBreaker 类 188
　　5.2.3　使用 CircuitBreaker 进行场景测试 191
5.3　Spring Cloud Circuit Breaker 的技术演进 193
5.4　Alibaba Sentinel 199
　　5.4.1　Sentinel 核心概述 199
　　5.4.2　Spring Cloud Alibaba Sentinel 204
　　5.4.3　Sentinel VS OpenFeign 和 RestTemplate 207
　　5.4.4　Sentinel 限流与 Dashboard 211
　　5.4.5　Sentinel 的高级特性 215
5.5　Netflix Hystrix 219
　　5.5.1　Hystrix 核心概述 219
　　5.5.2　Spring Cloud Netflix Hystrix 225
　　5.5.3　Hystrix 限流与 Dashboard 227
　　5.5.4　Hystrix 的高级特性 231
5.6　Resilience4j 235
　　5.6.1　Resilience4j 体验 236
　　5.6.2　Spring Cloud Resilience4j 241
　　5.6.3　Resilience4j 的高级特性 244
5.7　案例：使用 Sentinel 保护应用，防止服务雪崩 248

第 6 章　Spring 生态消息驱动 253

6.1　消息中间件概述 254
6.2　Spring 与消息 257
　　6.2.1　消息编程模型的统一 257
　　6.2.2　消息的发送和订阅 258
　　6.2.3　WebSocket 265
　　6.2.4　案例：使用 spring-messaging 处理 WebSocket 268
6.3　Spring Integration 271
　　6.3.1　Spring Integration 核心组件概述 272
　　6.3.2　Spring Integration 核心组件的使用 275
6.4　Spring Cloud Stream 277
　　6.4.1　使用 Spring Cloud Stream 发送和接收消息 277

	6.4.2 理解 Binder 和 Binding	280
	6.4.3 深入理解 Spring Cloud Stream	282
	6.4.4 Spring Cloud Stream 的高级特性	284

第 7 章 消息总线 ... 295

- 7.1 消息总线概述 ... 295
- 7.2 深入理解 Spring Cloud Bus ... 296
  - 7.2.1 Spring Cloud Bus 的使用 ... 297
  - 7.2.2 Spring Cloud Bus 的原理 ... 300
  - 7.2.3 Spring Cloud Bus 事件 ... 302
  - 7.2.4 Spring Cloud Bus 源码分析 ... 305
- 7.3 案例：使用 Spring Cloud Bus 完成多节点配置动态刷新 ... 309

第 8 章 Spring Cloud Data Flow ... 311

- 8.1 批处理/流处理概述 ... 312
- 8.2 流处理案例：信用卡反欺诈系统 ... 315
- 8.3 批处理案例：统计 GitHub 仓库的各项指标数据 ... 323
- 8.4 Spring Cloud Data Flow 批处理任务组合 ... 331
- 8.5 Spring Cloud Data Flow Shell ... 335
- 8.6 Spring Cloud Skipper ... 337
- 8.7 Spring Cloud Deployer ... 341
  - 8.7.1 TaskLauncher 接口 ... 342
  - 8.7.2 AppDeployer 接口 ... 344
  - 8.7.3 LocalAppDeployer ... 348
- 8.8 Spring Cloud Task ... 349
  - 8.8.1 体验 Spring Cloud Task ... 349
  - 8.8.2 深入理解 Spring Cloud Task ... 351
  - 8.8.3 Spring Cloud Task Batch ... 354
- 8.9 Spring Batch ... 358
  - 8.9.1 Spring Batch 核心组件 ... 358
  - 8.9.2 案例：使用 Spring Batch 完成便利店每日账单统计 ... 361

第 9 章 网关 366

9.1 API 网关概述 366
9.2 Netflix Zuul 368
9.3 非阻塞式的 Spring Cloud Gateway 371
9.4 Route 路由信息 376
9.5 Predicate 机制 377
9.5.1 PredicateDefinition 和 AsyncPredicate 377
9.5.2 RoutePredicateFactory 378
9.5.3 内置 RoutePredicateFactory 381
9.6 Filter 机制 382
9.6.1 FilterDefinition 和 GatewayFilter 382
9.6.2 GlobalFilter 386
9.6.3 内置 GatewayFilterFactory 387
9.6.4 网关内置的 GlobalFilter 390
9.7 整合注册中心和配置中心 391
9.8 GatewayControllerEndpoint 397
9.9 案例：使用 Spring Cloud Gateway 进行路由转发 398

第 10 章 Spring Cloud 与 Serverless 401

10.1 Serverless 401
10.2 Java Function 403
10.3 Spring Cloud Function 405
10.4 Spring Cloud Function 与 Spring 生态的整合 411
10.4.1 Spring Cloud Function 与 Spring Web/WebFlux 412
10.4.2 Spring Cloud Function 与 Spring Cloud Stream 420
10.4.3 Spring Cloud Function 与 Spring Cloud Task 427
10.5 案例：使用 GCP Cloud Functions 体验 Spring Cloud Function 430

第 1 章 Spring Cloud 生态

Spring Cloud 是一个构建在 Spring Boot 基础上的微服务开发框架,也是涉及服务注册/发现、分布式配置、API 网关、熔断器、分布式消息、消息总线、负载均衡、链路追踪等多个领域的一个整体解决方案。深入理解 Spring Cloud 的前提是理解 Spring Boot 的设计。

本章将介绍 Spring Boot 的核心特性,这些核心特性涉及 Web 服务器、条件注解、工厂加载机制、配置加载机制等内容。

1.1 Spring Cloud 基础:Spring Boot 概述

本章要介绍的 Spring Cloud 生态基于 Spring Boot。因此,必须先熟悉 Spring Boot 的原理。

Spring Boot 的诞生源自 Spring 团队内部的一个想法:Improve Containerless Web Application Architectures(改进无容器的 Web 应用结构)。也就是说,Mike Youngstrom(AWS 软件工程

师）认为基于 Servlet 的开发太烦琐了，会给 Spring 开发人员增加新的学习负担，因为使用原生的 Servlet 容器开发模式会面临以下问题：

- 需要 web.xml 或其他 Servlet 容器需要的配置文件，配置文件中的内容在不同的 Servlet 容器里有不同的格式。
- 需要构造 WAR 包并部署到 Servlet 容器中。
- 不同的 Servlet 容器有不同的配置方式，比如端口和线程池的方式都需要熟悉。
- ClassLoader 继承机制复杂。
- 监控和管理功能需要在应用外部进行配置。
- 单独配置日志功能。
- 理解 ApplicationContext 内的各种配置及继承关系。

Mike Youngstrom 认为越简单的开发框架越受欢迎，所以想要使用这种无容器（这里的无容器指的是 Servlet 容器）的开发方式。使用无容器开发方式可以带来以下好处：

- 提供一个统一的组件模型，不需要知道 Servlet 组件模型。
- 统一所有的配置格式，这样开发者只需要学习 Spring 配置模型，这套配置模型会生效于应用，以及对应的第三方组件（比如数据库、Redis）。
- 从 main 函数执行将会简化应用的启动和关闭。
- 一个更简单的纯 Java 类加载层次结构。
- 简单的开发工具。不需要复杂的 IDE 来构造 WAR 包，并将其部署到 Servlet 容器中，只需执行应用程序主类即可。

之后的一些 Support embedded servlet containers（支持内置 Servlet 容器）、Add Grails like BootStrap functionality（添加类似 Grails 的 Bootstrap 功能）等都被关联到这个 Issue 中。这些 Issue 都对 Spring Core 模块里的代码进行了修改。

Spring Boot 创始人 Phil Webb 针对这些现象评论：我们决定启动一个名为 Spring Boot 的新项目来解决这些问题，而不是在 Spring Core 模块里修改。

Mike Youngstrom 提出的无容器开发方式带来的好处目前在 Spring Boot 中都已经有了对应的功能。

- 内置 Servlet 容器，提供统一的容器抽象。这个 Servlet 容器抽象有 Tomcat、Jetty、Undertow 和 Netty 实现类。
- 对于所有组件的配置项，可以使用相同的格式在 application.properties 和 application.yml

中配置，不再需要为单独的组件创建不同的配置方式。例如，端口号不再需要在 Servlet 容器的配置文件中配置，只需在 application.properties 中配置 server.port 项；日志的级别不再需要在日志框架的配置文件里配置，只需在 application.properties 中配置 logging.level.root 等相关配置项。
- 提供了 SpringApplication 用于直接运行 Spring 应用，无须关注 ApplicationContext 的构造。
- 只存在一个 ClassLoader，不会像单独的 Servlet 容器那样拥有独立的 ClassLoader。
- 内置 Servlet 容器，直接执行 main 方法即可，无须构造 WAR 包。

简单了解了 Spring Boot 诞生的背景之后，接下来剖析 Spring Boot 核心特性的内容。

1.2 Spring Boot 核心特性

1.2.1 Web 服务器：WebServer

内置 Servlet 容器已经成为过去式了，新版本的 Spring Boot 将此特性称为 WebServer。

早期的 Java Web 应用都需要构造成 WAR 包（WEB-INF 目录下有个 web.xml 文件，大家一定不陌生），然后部署到 Servlet 容器（比如 Tomcat、Undertow、Jetty）。

如果是 Spring MVC 应用，需要在 web.xml 中配置 DispatcherServlet 这个 Servlet 和对应的 url-pattern，url-pattern 默认会拦截所有的请求。DispatcherServlet 拦截请求后，再通过内部的 HandlerMapping 根据 URL 信息去匹配 Controller，最终找到匹配到的 Controller，然后使用 Controller 内部的方法对请求进行处理。

WebServer 表示一个可配置的 Web 服务器（比如 TomcatWebServer、JettyWebServer、NettyWebServer、UndertowServletWebServer、UndertowWebServer），可以通过 WebServer 接口对外提供的 start 方法启动服务器，用 stop 方法停止服务器。有了 WebServer 后，我们不再需要关心外部的 Web 服务器、web.xml 文件、各种 Servlet 和 Filter 的配置等因素，只需要编写代码并把项目打包成 JAR 文件后直接运行即可，这非常适合云原生架构中的可独立部署特性。

WebServer 接口的定义如下：

```
public interface WebServer {
```

```java
/**
 * 启动 WebServer，如果 WebServer 已经启动，调用该方法不会有影响
 * @throws WebServerException 如果 Server 无法启动，则抛出该异常
 */
void start() throws WebServerException;

/**
 * 停止 WebServer，如果 WebServer 已经停止，调用该方法不会有影响
 * @throws WebServerException 如果 Server 无法停止，则抛出该异常
 */
void stop() throws WebServerException;

/**
 * 返回 Server 监听的端口号
 * @return 端口号（如果没有监听，则返回 -1）
 */
int getPort();
}
```

对于 WebServer，可通过 ReactiveWebServerFactory 或 ServletWebServerFactory 工厂接口去创建。这两个工厂接口对应的代码如下：

```java
@FunctionalInterface
public interface ServletWebServerFactory {
    WebServer getWebServer(ServletContextInitializer... initializers);
}
@FunctionalInterface
public interface ReactiveWebServerFactory {
    WebServer getWebServer(HttpHandler httpHandler);
}
```

ConfigurableWebServerFactory 接口继承自 WebServerFactory 接口（这是一个空接口）和 ErrorPageRegistry 接口（错误页注册表，内部维护错误页信息，提供 addErrorPages 方法以添加错误页到 WebServer 中），ConfigurableWebServerFactory 接口表示这是一个可配置的 WebServerFactory 接口，定义如下：

```java
public interface ConfigurableWebServerFactory
        extends WebServerFactory, ErrorPageRegistry {

    /**
     * 设置 WebServer 要监听的端口。如果没有设置,默认使用 8080。
     * 设置为-1,表示禁止自动启动(启动应用上下文,但不监听任何端口)
     */
    void setPort(int port);

    /**
     * 设置要绑定的地址
     */
    void setAddress(InetAddress address);

    /**
     * 设置出现异常时要显示的错误页(Error Page)
     */
    void setErrorPages(Set<? extends ErrorPage> errorPages);

    /**
     * SSL 相关配置
     */
    void setSsl(Ssl ssl);

    /**
     * 设置获取 SSL 的 SslStoreProvider
     */
    void setSslStoreProvider(SslStoreProvider sslStoreProvider);

    /**
     * 设置 HTTP2 相关配置
     */
    void setHttp2(Http2 http2);

    /**
     * 设置 HTTP compression 相关配置
```

```
 */
void setCompression(Compression compression);

/**
 * 设置 Server HEADER 的值
 */
void setServerHeader(String serverHeader);
}
```

ConfigurableWebServerFactory 提供了 WebServer 常用的配置信息，它的子接口表示各个 WebServer 实现类独有的配置，比如 ConfigurableJettyWebServerFactory 对应 Jetty 独有的配置，ConfigurableTomcatWebServerFactory 对应 Tomcat 独有的配置。真正创建 WebServer 的工厂类（如 TomcatServletWebServerFactory）通过继承和实现接口的方式实现了 ConfigurableWebServerFactory 和 ServletWebServerFactory 这两个接口。

Spring Boot 对于 WebServer 概念新增了一些事件，比如 WebServerInitializedEvent 事件，表示 ApplicationContext 刷新过后且 WebServer 处于 ready 状态下会触发的事件。

 提示：Spring Cloud 服务注册的时机就是在 WebServerInitializedEvent 事件被触发的时候。

我们常用的 spring-boot-starter-web 模块默认使用的是 Tomcat（Pom 里存在 spring-boot-starter-tomcat 依赖），如果要使用 Undertow 或者 Jetty，需要在 spring-boot-starter-web 依赖中排除 spring-boot-starter-tomcat 依赖，然后加上 spring-boot-starter-undertow（对应 Undertow）或 spring-boot-starter-jetty（对应 Jetty）依赖。

```
<!-- 以使用 Jetty 为例 -->
<dependency>
    <groupId>org.springframework.boot</groupId>
    <artifactId>spring-boot-starter-web</artifactId>
    <exclusions>
        <exclusion>
            <groupId>org.springframework.boot</groupId>
            <artifactId>spring-boot-starter-tomcat</artifactId>
        </exclusion>
    </exclusions>
```

```xml
</dependency>

<dependency>
    <groupId>org.springframework.boot</groupId>
    <artifactId>spring-boot-starter-jetty</artifactId>
</dependency>
```

若要使用 NettyServer，则需要使用 `spring-boot-starter-webflux`（内部的 `spring-boot-starter-reactor-netty` 触发 ReactiveWebServerFactoryConfiguration#EmbeddedNetty 自动化配置生效）代替 `spring-boot-starter-web`。

1.2.2 条件注解：@ConditionalOnXX

Spring Boot 有一个很重要的模块——`spring-boot-autoconfigure`，该模块内部包含了很多第三方依赖的 AutoConfiguration(自动化配置类)，比如 `KafkaAutoConfiguration`、`GsonAutoConfiguration`、`ThymeleafAutoConfiguration`、`WebMvcAutoConfiguration` 等。这些 AutoConfiguration 只会在特定的情况下才会生效，这里的特定情况其实就是条件注解。

Spring Boot 提供的条件注解如表 1-1 所示。

表 1-1

条件注解	作 用	条件注解解析类
ConditionalOnBean	ApplicationContext 存在某些 Bean 时条件才成立	OnBeanCondition
ConditionalOnClass	classpath 中存在某些 Class 时条件才成立	OnClassCondition
OnCloudPlatformCondition	在某些云平台（Kubernetes、Heroku、Cloud Foundry）下条件才成立	OnCloudPlatformCondition
ConditionalOnExpression	SPEL 表达式成立时条件才成立	OnExpressionCondition
ConditionalOnJava	JDK 某些版本条件才成立	OnJavaCondition
ConditionalOnJndi	JNDI 路径存在时条件才成立	OnJndiCondition
ConditionalOnMissingBean	ApplicationContext 不存在某些 Bean 时条件才成立	OnBeanCondition
ConditionalOnMissingClass	classpath 中不存在某些 Class 时条件才成立	OnClassCondition
ConditionalOnNotWebApplication	在非 Web 环境下条件才成立	OnWebApplicationCondition
ConditionalOnProperty	Environment 中存在某些配置项时条件才成立	OnPropertyCondition

续表

条件注解	作用	条件注解解析类
ConditionalOnResource	存在某些资源时条件才成立	OnResourceCondition
ConditionalOnSingleCandidate	ApplicationContext 存在且只有一个 Bean 时条件才成立	OnBeanCondition
ConditionalOnWebApplication	在 Web 环境下条件才成立	OnWebApplicationCondition

表 1-1 中，条件注解仅仅只是一个注解，真正的判断逻辑在这些条件注解的解析类内部。解析类内部会根据注解的属性来判断是否满足条件，比如，OnJavaCondition 条件注解解析类对应的是@ConditionOnJava 条件注解，其内部会判断当前应用的 JDK 版本是否正确。内部处理逻辑的代码如下：

```java
@Order(Ordered.HIGHEST_PRECEDENCE + 20)
class OnJavaCondition extends SpringBootCondition {

    // 根据 JavaVersion 获取当前 jdk 的版本。JavaVersion 是 SpringBoot 封装的一个类
    // 其内部会根据某些类的某些方法是否存在来判断 JDK 版本
    // 比如 Optional 的 empty 方法存在，表示这是 JDK 1.8
    // Optional 的 stream 方法存在，表示这是 JDK 9
    private static final JavaVersion JVM_VERSION = JavaVersion.getJavaVersion();

    @Override
    public ConditionOutcome getMatchOutcome(ConditionContext context,
AnnotatedTypeMetadata metadata) {
        // 得到 ConditionalOnJava 注解里的属性
        Map<String, Object> attributes =
metadata.getAnnotationAttributes(ConditionalOnJava.class.getName());
        Range range = (Range) attributes.get("range");
        JavaVersion version = (JavaVersion) attributes.get("value");
        // 判断是否满足条件
        return getMatchOutcome(range, JVM_VERSION, version);
    }

    protected ConditionOutcome getMatchOutcome(Range range, JavaVersion
runningVersion, JavaVersion version) {
        boolean match = isWithin(runningVersion, range, version);
        String expected = String.format((range != Range.EQUAL_OR_NEWER) ? "(older
than %s)" : "(%s or newer)", version);
```

```
        ConditionMessage message = ConditionMessage.forCondition(ConditionalOnJava.class,
expected)
                .foundExactly(runningVersion);
        return new ConditionOutcome(match, message);
    }

    // 判断 Java 版本是否满足条件
    private boolean isWithin(JavaVersion runningVersion, Range range, JavaVersion
version) {
        if (range == Range.EQUAL_OR_NEWER) {
            return runningVersion.isEqualOrNewerThan(version);
        }
        if (range == Range.OLDER_THAN) {
            return runningVersion.isOlderThan(version);
        }
        throw new IllegalStateException("Unknown range " + range);
    }
}
```

理解了这些条件注解后，Spring Boot 在哪里使用这些条件注解来判断是否需要加载自动化配置类呢？Spring Boot 通过 spring-context 模块中提供的 ConditionEvaluator 完成这个动作。

ConditionEvaluator 在 AnnotatedBeanDefinitionReader、ClassPathScanningCandidate-ComponentProvider、ConfigurationClassBeanDefinitionReader、ConfigurationClassParser、AnnotatedBeanDefinitionReader 这些类扫描组件、配置类解析组件扫描、解析组件的时候来判断是否需要跳过（skip）某些配置类，具体的解析逻辑这里不再展开介绍。

Spring Cloud 内部定义的一些新的条件注解如表 1-2 所示。

表 1-2

条件注解	作 用	条件注解解析类
ConditionalOnBlockingDiscoveryEnabled	被@ConditionalOnProperty 修饰，存在 spring.cloud.discovery.blocking.enabled 配置项且为 true 时，条件成立（不配置默认为 true）。在 DiscoveryClient 场景下使用	—

续表

条件注解	作　用	条件注解解析类
ConditionalOnReactiveDiscoveryEnabled	被@ConditionalOnProperty 修饰，存在 spring.cloud.discovery.reactive.enabled 配置项且为 true 时，条件成立（不配置时默认为 true）。在 Reactive-DiscoveryClient 场景下使用	—
ConditionalOnDiscoveryHealthIndicatorEnabled	被@ConditionalOnProperty 修饰，存在 spring.cloud.discovery.client.health-indicator.enabled 配置项且为 true 时，条件成立（不配置默认为 true）。主要用于 HealthIndicator 相关的自动化配置	—

1.2.3　工厂加载机制

1.2.2 节介绍了条件注解作用于自动化配置类。那么这些自动化配置类是从哪里被加载的呢？这是通过工厂加载机制（factory loading mechanism）来实现的，这个机制会从 META-INF/spring.factories 文件中加载自动化配置类。下面是 spring-boot-autoconfigure 模块里 META-INF/spring.factories 文件的一部分内容：

```
# 应用初始化器
org.springframework.context.ApplicationContextInitializer=\
org.springframework.boot.autoconfigure.SharedMetadataReaderFactoryContextInitializer,\
org.springframework.boot.autoconfigure.logging.ConditionEvaluationReportLoggingListener

# 应用监听器
org.springframework.context.ApplicationListener=\
org.springframework.boot.autoconfigure.BackgroundPreinitializer

# 自动注入的 ImportListener
org.springframework.boot.autoconfigure.AutoConfigurationImportListener=\
org.springframework.boot.autoconfigure.condition.ConditionEvaluationReportAutoConfigurationImportListener
```

```
# 自动注入的 Filter
org.springframework.boot.autoconfigure.AutoConfigurationImportFilter=\
org.springframework.boot.autoconfigure.condition.OnBeanCondition,\
...

# 自动装配类
org.springframework.boot.autoconfigure.EnableAutoConfiguration=\
org.springframework.boot.autoconfigure.admin.SpringApplicationAdminJmxAutoConfiguration,\
org.springframework.boot.autoconfigure.aop.AopAutoConfiguration,\
org.springframework.boot.autoconfigure.amqp.RabbitAutoConfiguration,\
...

# 错误分析
org.springframework.boot.diagnostics.FailureAnalyzer=\
org.springframework.boot.autoconfigure.diagnostics.analyzer.NoSuchBeanDefinitionFailureAnalyzer,\
...

# 模板提供者
org.springframework.boot.autoconfigure.template.TemplateAvailabilityProvider=\
org.springframework.boot.autoconfigure.freemarker.FreeMarkerTemplateAvailabilityProvider,\
org.springframework.boot.autoconfigure.mustache.MustacheTemplateAvailabilityProvider,\
...
```

spring.factories 是一个 properties 格式的文件。key 是一个类的全称，比如，"org.springframework.boot.autoconfigure.EnableAutoConfiguration"，value 是用 ","分割的自动化配置类的全称列表。

启动 Spring Boot 应用的@SpringBootApplication 注解内部被@EnableAutoConfiguration 注解修饰，@EnableAutoConfiguration 注解会导入 AutoConfigurationImportSelector 这个 ImportSelector。AutoConfigurationImportSelector 内部的 selectImports 要导入的配置类是通过 SpringFactoriesLoader 获取的。

```java
protected List<String> getCandidateConfigurations(AnnotationMetadata metadata,
AnnotationAttributes attributes) {
    List<String> configurations =
SpringFactoriesLoader.loadFactoryNames(getSpringFactoriesLoaderFactoryClass(),
getBeanClassLoader());
    Assert.notEmpty(configurations,
            "No auto configuration classes found in META-INF/spring.factories. If you " + "are using a custom packaging, make sure that file is correct.");
    return configurations;
}
protected Class<?> getSpringFactoriesLoaderFactoryClass() {
    return EnableAutoConfiguration.class;
}
```

在 SpringFactoriesLoader 的加载过程中，选择的 key（对应 spring.factories 文件中的 key）是 EnableAutoConfiguration 这个类对应的类全名。

Spring Cloud 内部也使用了工厂加载机制并扩展了一些 key。比如，org.springframework.cloud.bootstrap.BootstrapConfiguration 用于在 Bootstrap 过程中加载对应的配置类。

1.2.4 配置加载机制

Spring Boot 把配置文件的加载封装成了 PropertySourceLoader 接口，该接口的定义如下：

```java
public interface PropertySourceLoader {
    // 支持的文件后缀
    String[] getFileExtensions();
    // 把资源 Resource 加载成 PropertySource
    PropertySource<?> load(String name, Resource resource, String profile)
            throws IOException;
}
```

Spring Boot 对于该接口只有两种实现：

- PropertiesPropertySourceLoader：加载 properties 或 xml 文件。
- YamlPropertySourceLoader：加载 yaml 或 yml 文件。

提示:resources/application.properties 或 resources/application.yaml 配置文件就是被这两种 PropertySourceLoader 所加载的。

SpringApplication 内部维护着一个 `ApplicationListener` 集合属性,用于监听 `ApplicationEvent`。默认情况下,该属性会被工厂加载机制所加载(加载的 key 为 org.springframework.context.ApplicationListener):

```
public SpringApplication(ResourceLoader resourceLoader, Class<?>... primarySources) {
    ...
    setListeners((Collection) getSpringFactoriesInstances(ApplicationListener.class));
    ...
}
```

spring-boot 模块里的 META-INF/spring.factories 中存在 ConfigFileApplicationListener:

```
org.springframework.context.ApplicationListener=\
...
org.springframework.boot.context.config.ConfigFileApplicationListener,\
...
```

ConfigFileApplicationListener 是 Spring Boot 配置加载的核心类,它实现了 `EnvironmentPostProcessor` 接口。`EnvironmentPostProcessor` 接口是配置环境信息 Environment 的 PostProcessor,可以对应用的 Environment 进行修改。

由于 ConfigFileApplicationListener 实现了 ApplicationListener 接口,会监听 Spring 的事件。其中,对 ApplicationEnvironmentPreparedEvent 进行了监听,会调用 onApplicationEnvironmentPreparedEvent 方法:

```
@Override
public void onApplicationEvent(ApplicationEvent event) {
    if (event instanceof ApplicationEnvironmentPreparedEvent) {
        onApplicationEnvironmentPreparedEvent((ApplicationEnvironmentPreparedEvent) event);
    }
```

```java
    if (event instanceof ApplicationPreparedEvent) {
        onApplicationPreparedEvent(event);
    }
}

private void onApplicationEnvironmentPreparedEvent(ApplicationEnvironmentPreparedEvent event) {
    // 使用工厂加载机制获取 EnvironmentPostProcessor 集合
    List<EnvironmentPostProcessor> postProcessors = loadPostProcessors();
    // ConfigFileApplicationListener 也实现了 EnvironmentPostProcessor 接口,被添加到该集合中
    postProcessors.add(this);
    // 排序
    AnnotationAwareOrderComparator.sort(postProcessors);
    // 遍历 EnvironmentPostProcessor 集合并进行处理
    for (EnvironmentPostProcessor postProcessor : postProcessors) {
        postProcessor.postProcessEnvironment(event.getEnvironment(), event.getSpringApplication());
    }
}
```

ConfigFileApplicationListener 的 postProcessEnvironment 方法内部构造了 Loader，并调用 load 方法进行配置文件的加载：

```java
@Override
public void postProcessEnvironment(ConfigurableEnvironment environment, SpringApplication application) {
    addPropertySources(environment, application.getResourceLoader());
}
protected void addPropertySources(ConfigurableEnvironment environment, ResourceLoader resourceLoader) {
    RandomValuePropertySource.addToEnvironment(environment);
    new Loader(environment, resourceLoader).load();
}
```

Loader 内部有不少细节，比如，配置文件的文件名是根据 spring.config.name 配置项来决定

的，不设置时默认为 application。默认配置文件的加载路径为 classpath:/、classpath:/config/、file:./ 和 file:./config/，这个加载路径可以通过 spring.config.location 配置项来修改。spring.profiles.active 用于指定生效的 profile 等，这里不再具体展开介绍。

 提示：Spring Cloud 在加载过程中把 spring.config.name 配置设置成了 bootstrap，所以加载的文件名是 bootstrap.properties 或 bootstrap.yml。

1.2.5 Spring Boot Actuator

Spring Boot 提供了不少 Production-ready 的特性，这些特性需要引入模块 spring-boot-starter-actuator 才能自动生效。

```
<dependencies>
    <dependency>
        <groupId>org.springframework.boot</groupId>
        <artifactId>spring-boot-starter-actuator</artifactId>
    </dependency>
</dependencies>
```

其中，Endpoint 是比较核心的功能，其作用是让我们监控应用，以及跟应用之间的交互。比如，Health Endpoint、HttpTrace Endpoint 用于应用的健康检查和 HTTP 历史链路（应用监控），Loggers Endpoint 用于动态改变应用的日志级别（应用交互）。

表 1-3 列出了常用的一些 Endpoint。

表 1-3

Endpoint id	描述	默认是否开启
beans	显示 Spring ApplicationContext 中所有的 Bean	默认开启
conditions	显示条件注解是否满足条件的匹配信息	默认开启
configprops	显示所有@ConfigurationProperties 对应的配置信息	默认开启
env	Spring 配置信息，可以看到加载了哪些 PropertySource	默认开启
health	显示应用的健康信息，可以用来做应用的健康检查	默认开启
httptrace	显示 HTTP 链路信息	默认开启，需要一个 HttpTraceRepository bean
info	显示应用信息	默认开启
loggers	显示或修改应用的日志配置	默认开启

续表

Endpoint id	描 述	默认是否开启
mappings	显示 @RequestMapping 的映射信息	默认开启
shutdown	可以让应用优雅地关闭	默认关闭

 提示：在 Spring Boot 2.x 以后的版本中，Endpoint 对应的路径是/actuator/endpointid；对于 Spring Boot 1.x，路径是/endpointid。

Spring Cloud 也创建了一些新的 Endpoint，如表 1-4 所示。

表 1-4

Endpoint id	描 述	默认是否开启
features	显示 Spring Cloud 提供的 feature 列表	默认关闭
service-registry	显示或修改服务实例的状态	默认关闭

Spring Cloud 母公司 Pivotal 旗下的 Cloud Foundry 产品基于 Endpoint 做了很多图形化的界面，例如，Health Endpoint（健康检查）对应的界面如图 1-1 所示，Loggers Endpoint（修改日志级别）对应的界面如图 1-2 所示，HttpTrace Endpoint（显示 HTTP 链路）对应的界面如图 1-3 所示。

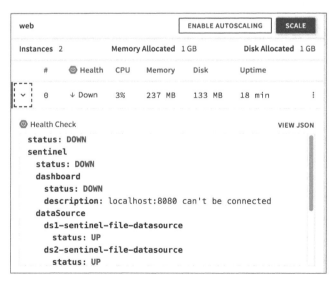

图 1-1

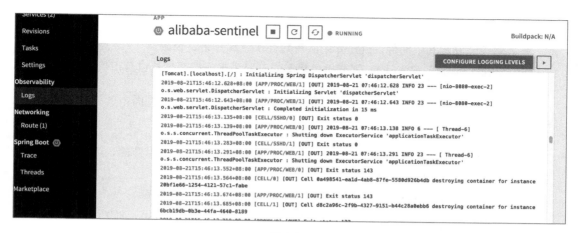

图 1-2

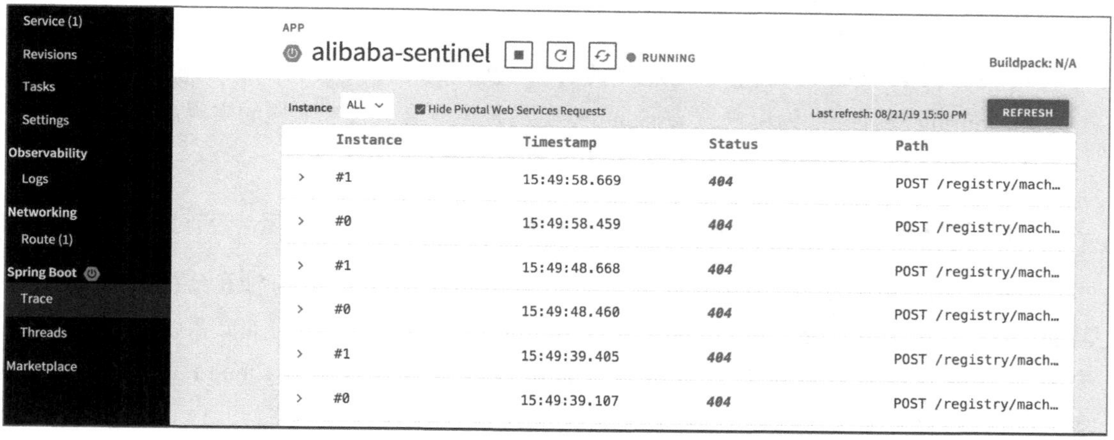

图 1-3

对 Spring Boot 的核心概念有了一定的了解之后，我们开始进入 Spring Cloud 世界。

1.3 Spring Cloud 概述

1.3.1 Spring Cloud 诞生背景

笔者认为 Spring Cloud 的诞生有以下三个原因：

- Spring 大家族中的大部分项目都是客户端项目，他们缺少服务器端项目和分布式项目的经验。
- Netflix 开源其中间件，Spring 团队与其合作进军分布式领域。
- Spring 母公司（Pivotal）需要提供更多的云服务来保持或增长自身市值。

Spring 自诞生以来，已经衍生出很多子项目。Spring 官网的项目页面中就列出了以下项目：

- Spring Boot：快速构建微服务应用。
- Spring Framework：为依赖注入、事务管理、Web 应用、数据访问、消息等提供核心支持。Spring Framework 内部包含了多个子模块。
- Spring Data：为数据访问提供统一的基于 Spring 编程模型的框架。
- Spring Integration：Spring 对 EIP（Enterprise Integration Patterns）的集成。
- Spring Batch：为批处理提供了统一的模型。
- Spring Security：权限、安全相关模型。
- Spring AMQP：封装基于 AMQP 的消息解决方案。
- Spring Session：为用户的 session（会话）提供一套 API 和对应的实现。
- Spring for Android：为 Android 开发提供的一套 Spring 框架。

我们可以发现，这些项目大部分都是客户端项目，适合开发单体应用，没有涉及分布式领域。

Pivotal 公司缺少分布式项目的经验，而且 Netflix 开源的组件正好涉及分布式领域的核心功能，双方一拍即合，合作诞生的 Spring Cloud 项目用于为快速构建分布式系统提供统一的开发模式。Spring Cloud 诞生以后，其母公司 Pivotal 的商业化 PaaS 产品 Cloud Foundry 上就提供了 Spring Cloud 的商业化服务（配置管理、服务发现和熔断器）。

国内的阿里云也有类似的 PaaS 产品，如企业级分布式应用服务（Enterprise Distributed Application Service，简称 EDAS）提供了应用生命周期的托管服务，以及与 Spring Cloud 相关的服务治理功能；微服务引擎（Micro Service Engine）提供了注册/配置中心的托管服务，以及与 Spring Cloud 相关的服务治理功能，如图 1-4 所示。

接下来，我们来了解与 Spring 合作的美国公司 Netflix。Netflix 的中文名是网飞（也称奈飞），是一家会员订阅制的流媒体播放平台。此流媒体播放平台的技术都是 Netflix 自己开发的，我们熟悉的 Spring Cloud Netflix 对应的一些组件，比如注册中心 Eureka、熔断工具 Hystrix、网关 Zuul 都是 Netflix 对外开源的产品。

图 1-4

1.3.2 Netflix OSS

Netflix OSS（Netflix Open Source Software）即 Netflix 开源软件，是 Netflix 开源的软件集合。这些软件覆盖了分布式开发的部分领域，它包含的组件如表 1-5 所示。

表 1-5

组件	描述
Eureka	服务注册中心，完成服务的注册和发现
Feign	声明式 REST 客户端，会基于接口自动生成动态代理
Ribbon	客户端负载均衡
Archaius	配置管理 API，提供动态类型化属性、线程安全配置操作、轮询框架、回调机制
Hystrix	Circuit Breaker 熔断器
Zuul	网关

Spring Cloud 与 Netflix 合作产生的 Spring Cloud Netflix 项目就是 Spring 体系与 Netflix 中间件集成的项目。

显然，Netflix OSS 并不能满足分布式应用领域的所有功能。Spring Cloud 在 Netflix 没有涉及的领域也都有相应的解决方案。

1.3.3　Spring Cloud 项目

Spring Cloud 早期与 Netflix 合作，其内部的很多接口借鉴了 Netflix 开源软件的实现。不过 Spring Cloud 的目的是为了做生态，不与 Netflix 进行强绑定。除了 Spring Cloud Netflix，Spring Cloud 后续也提供了 Spring Cloud Zookeeper 项目和 Spring Cloud Consul 项目，其中，前者是 Spring Cloud 与 Zookeeper（服务注册/发现与配置管理）集成的项目，后者是 Spring Cloud 与 Consul（服务注册/发现、配置管理、消息）集成的项目。

Netflix 开源的软件只涉及分布式应用开发中的一部分内容，Spring Cloud 生态还提供了其他领域的组件。比如 Spring Cloud Stream 组件统一消息 API，屏蔽底层消息中间件实现细节；Spring Cloud Sleuth 组件用于链路追踪；Spring Cloud Open Service Broker 实现 Open Service Broker API 等。表 1-6 列出了 Spring Cloud 生态的功能。

表 1-6

Spring Cloud 特性	实现组件
服务注册/发现	Alibaba Nacos、Netflix Eureka、Apache Zookeeper、HashiCorp Consul
分布式配置	Alibaba Nacos、Spring Cloud Config Server/Client、Apache Zookeeper、HashiCorp Consul
API 网关	Spring Cloud Gateway、Netflix Zuul
熔断器	Alibaba Sentinel、Netflix Hystrix、Resilience4j
分布式消息	SCS RocketMQ Binder、SCS Kafka Binder、SCS Rabbit Binder(SCS = Spring Cloud Stream)
消息总线	SCB RocketMQ、SCB Kafka、SCB AMQP(SCB = Spring Cloud Bus)
链路追踪	Spring Cloud Sleuth
服务调用	RestTemplate、OpenFeign、Dubbo Spring Cloud
负载均衡	Netflix Ribbon、Spring Cloud LoadBalancer、Dubbo LoadBalancer
快速构建 Service Broker 应用	Spring Cloud Open Service Broker
简化云平台服务连接串的获取	Spring Cloud Connectors
消费者驱动的契约测试（Consumer-Driven Contracts）解决方案	Spring Cloud Contrcat
Serverless	Spring Cloud Function
流处理和批处理	Spring Cloud Data Flow
分布式应用的安全认证	Spring Cloud Security
Spring Cloud 与 Kubernetes 的集成	Spring Cloud Kubernetes
任务调度	Spring Cloud Task

1.3.4　Spring Cloud 版本

Spring Cloud 在 2016 年发布了首个版本，到目前为止，发布的版本已经将近 50 个。若使用 Spring Cloud，需要引入 `org.springframework.cloud:spring-cloud-dependencies` 这个 BOM，在 maven 中以 import 的方式引入这个依赖：

```xml
<dependencyManagement>
    <dependencies>
        <dependency>
            <groupId>org.springframework.cloud</groupId>
            <artifactId>spring-cloud-dependencies</artifactId>
            <version>Hoxton.RELEASE</version>
            <type>pom</type>
            <scope>import</scope>
        </dependency>
    </dependencies>
</dependencyManagement>
```

BOM 引入后，不再需要单独为模块指定版本，比如，引入 `spring-cloud-starter-openfeign` 模块不需要指定版本号：

```xml
<dependency>
    <groupId>org.springframework.cloud</groupId>
    <artifactId>spring-cloud-starter-openfeign</artifactId>
</dependency>
```

Spring Cloud 版本号使用伦敦地铁站命名，第一个版本以 A 开头，第二个版本以 B 开头，以此类推。目前的 Spring Cloud 版本已经到了 Hoxton。

Spring snapshot 仓库会发布每个版本的 snapshot 版本，比如 Greenwich.BUILD-SNAPSHOT、Hoxton.BUILD-SNAPSHOT 等。如果要使用 snapshot 版本，需要在 maven 中加上 repository 配置，定义如下：

```xml
<repositories>
    <repository>
        <id>spring-snapshot</id>
        <name>Spring Snapshot</name>
        <url>https://repo.spring.io/snapshot</url>
```

```xml
        <snapshots>
            <enabled>true</enabled>
        </snapshots>
    </repository>
</repositories>
```

Spring milestone 仓库会发布每个版本的 milestone 版本和 RC 版本。比如，Hoxton.RELEASE 发布之前会发布 Hoxton.M1、Hoxton.M2、Hoxton.M3、Hoxton.RC1 版本。如果要使用 milestone 版本，需要在 maven 中加上 repository 配置：

```xml
<repositories>
    <repository>
      <id>spring-milestones</id>
      <name>Spring Milestones</name>
      <url>https://repo.spring.io/milestone</url>
    </repository>
</repositories>
```

正式版本会以.RELEASE 结尾，比如 Greenwich.RELEASE、Hoxton.RELEASE 等，会在中央仓库上发布。

发布 Release 版本以后会发布 SR 版本（SR 表示修正版或更新版），会修复 Release 版本里的一些问题。比如 Greenwich.SR1、Greenwich.SR2 是 Greenwich.RELEASE 的两个修正版本。

Spring Cloud 基于 Spring Boot 构建，每个版本也会有对应的 Spring Boot 版本。Spring Cloud 与 Spring Boot 的版本关系如表 1-7 所示。

表 1-7

版 本 号	发 布 时 间	依赖的 Spring Boot 版本
Angel.SR5	2016-01	Spring Boot 1.2.x
Brixton.RELEASE	2016-05	Spring Boot 1.3.x
Camden.RELEASE	2016-09	Spring Boot 1.4.x
Dalston.RELEASE	2017-04	Spring Boot 1.5.x
Edgware.RELEASE	2017-11	Spring Boot 1.5.x
Finchley.RELEASE	2018-06	Spring Boot 2.0.x
Greenwich.RELEASE	2019-01	Spring Boot 2.1.x
Hoxton.RELEASE	2019-11	Spring Boot 2.2.x

Spring Cloud Alibaba 在 Spring Cloud 孵化器内 "毕业" 时进行了仓库的迁移，如要使用 Spring Cloud Alibaba 对应的依赖，需要引入 com.alibaba.cloud:spring-cloud-alibaba-dependencies BOM：

```xml
<dependencyManagement>
    <dependencies>
        <dependency>
            <groupId>com.alibaba.cloud</groupId>
            <artifactId>spring-cloud-alibaba-dependencies</artifactId>
            <version>2.2.0.RELEASE</version>
            <type>pom</type>
            <scope>import</scope>
        </dependency>
    </dependencies>
</dependencyManagement>
```

Spring Cloud Alibaba 与 Spring Cloud 和 Spring Boot 的版本关系如表 1-8 所示。

表 1-8

Spring Cloud Alibaba 版本	Spring Cloud 版本	Spring Cloud Boot 版本
2.2.x.RELEASE	Spring Cloud Hoxton	Spring Boot 2.2.x
2.1.x.RELEASE	Spring Cloud Greenwich	Spring Boot 2.1.x
2.0.x.RELEASE	Spring Cloud Finchley	Spring Boot 2.0.x
1.5.x.RELEASE	Spring Cloud Edgware	Spring Boot 1.5.x

1.3.5 Spring Cloud 最新动态

自 2015 年诞生以来到本书编写之时，Spring Cloud 的生态发生了非常大的变化。Spring Cloud Netflix 大部分组件不再继续更新，其 GitHub 仓库甚至已经删除了这些维护组件。Spring Cloud Netflix 已经不再是 Spring Cloud 推荐的实现，其组件可以被其他 Spring Cloud 生态组件所替代，比如 Spring Cloud Gateway 代替 Netflix Zuul，Spring Cloud LoadBalancer 代替 Netflix Ribbon，Spring Cloud Circuit Breaker Sentinel 代替 Hystrix。

Spring Cloud Netflix 大部分组件不再继续更新的原因是 Netflix OSS 宣布 Hystrix、Eureka 2.x、Ribbon 和 Archaius 进入维护模式。这意味着不会添加新的特性，只会修改一些比较小的 Bug。

在 2019 年巴塞罗那 Spring I/O 大会和奥斯汀 Spring One Platform 大会上，Spring 团队都有一个主题为"How to live in a post Spring Cloud Netflix world"的环节来讲解 Spring Cloud Netflix 进入维护模式后使用其他 Spring Cloud 实现代替 Netflix 的解决方案。

Spring Cloud 团队在 2019 年 12 月通过 Spring Blog 对外宣布后续的产品路线图，下一个大版本 Ilford 会做以下两件事情：

- 删除处于维护模式中的项目（Spring Cloud Netflix 的绝大多数项目）。
- 简化 Spring Cloud 发版列车，云厂商相关的项目会被移出 Spring Cloud 仓库，并单独维护（Spring Cloud Alibaba 是第一个移出 Spring Cloud 仓库的云厂商项目）。

2018 年，阿里巴巴开源了 Spring Cloud Alibaba，这是一个整合阿里巴巴开源中间件与 Spring Cloud 生态的开源项目，其内部组件如表 1-9 所示。

表 1-9

组件	描述
Nacos	服务注册中心、服务配置中心
Dubbo	扩展 Spring Cloud 客户端，可以调用 Dubbo 服务
Seata	分布式事务解决方案
Spring Cloud Stream RocketMQ Binder	基于 RocketMQ 的 Spring Cloud Stream 实现，完成消息的处理
Sentinel	Circuit Breaker 熔断器，同时还有限流和系统保护的功能
Spring Cloud Bus RocketMQ	基于 RocketMQ 的消息总线实现

Spring Cloud Alibaba 自 2018 年 7 月开源依赖后，获得的关注度非常高。截至本书编写时，Spring Cloud Alibaba 在 GitHub 上的 star 数已经超过 15000 个，used by 超过 7000 个，至少 1000 家公司在生产环境使用 Spring Cloud Alibaba。

目前 Spring Cloud Alibaba 已经入驻 Spring 官网，官网上也有 Spring Cloud Alibaba 项目的介绍。Spring Cloud Alibaba 内的 Sentinel、RocketMQ Binder 也已经在 Spring Cloud 官方的 Spring Cloud Circuit Breaker、Spring Cloud Stream 的推荐实现列表上。

目前 Spring Cloud 实现的各个功能对比如表 1-10 所示。其中，处于维护模式或者不再继续开发的组件有 Archaius、Eureka 2.0、Hystrix 和 Ribbon。Spring Cloud 官方提供的 Spring Environment、Service Registry/Discovery、Spring Cloud Circuit Breaker、Spring Cloud Stream、Spring Cloud Bus、Spring Cloud LoadBalancer 都对各个领域的编程模型进行了统一。

表 1-10

	Spring Cloud Netflix	Spring Cloud 官方	Spring Cloud Alibaba	Spring Cloud Consul	Spring Cloud Kubernetes	Spring Cloud Zookeeper
分布式配置	Archaius（已不维护）	Spring Environment（编程模型统一）SCC Client/Server	Nacos	Consul	Config Map	Zookeeper
服务注册/发现	Eureka 1.x Eureka 2.x（已不维护）	Service Registry（编程模型统一）Service Discovery（编程模型统一）	Nacos	Consul	Api Server	Zookeeper
服务熔断	Hystrix（已不维护）	Spring Cloud Circuit Breaker（编程模型统一）	Sentinel	-	-	-
服务调用	Feign	OpenFeign RestTemplate	Dubbo RPC	-	-	-
服务路由	Zuul	Spring Cloud Gateway	Dubbo+Servlet	-	-	-
分布式消息	-	Spring Cloud Stream（编程模型统一）SCS RabbitMQ/Kafka	SCS RocketMQ	SCS Consul	-	-
消息总线	-	Spring Cloud Bus（编程模型统一）	SCB RocketMQ	SCB Consul	-	-
负载均衡	Ribbon（已不维护）	Spring Cloud LoadBalancer（编程模型统一）	Dubbo LB	-	-	-
分布式事务	-	-	Seata	-	-	-

第 2 章
服务注册与服务发现

本章将介绍单体应用如何演进到微服务，在此基础上引出 Spring Cloud 微服务开发框架。然后分别使用 Nacos 和 Eureka 体验 Spring Cloud 应用的开发，在这个过程中，对 Spring Cloud 服务注册和服务发现的编程模型进行分析和讲解。由于 Spring Cloud 统一了这套编程模型，所以在代码编写上无论是使用 Nacos 还是 Eureka，基本上无任何区别。接着介绍双注册双订阅模式在 Spring Cloud 上的应用，加深对 Spring Cloud 编程模型的理解，最后介绍一个注册中心的迁移案例来加深理解双注册双订阅模式的作用。

2.1 微服务架构演进

在单体应用时代，所有的功能（代码）耦合在一起，并且部署到同一个进程中。如图 2-1 所示，一个单体的在线商场应用由用户模块（UserService）、账号模块（AccountService）、商品模块（GoodsService）、订单模块（OrderService）等多个模块组成。

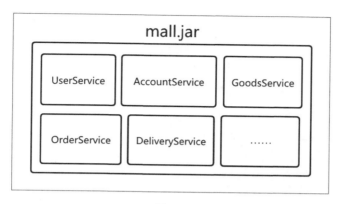

图 2-1

随着业务的增加，单体应用面临的问题也会越来越多，例如：

- 当部分模块出现问题的时候，会影响整个应用。比如，图 2-1 所示的应用在某次发布前，其订单模块修改了代码，导致 CPU 的使用率飙升到 100%，这时整个应用会不可用，也会影响其他模块的功能。
- 应用功能越多，部署成本越高。比如，某次发布只有订单模块新增了新的功能，但发布时必须发布整个应用，如果发布过程没有无损下线，就会导致部分请求报错，从而引起业务受损。
- 技术栈受限，要求所有的开发者必须使用相同的开发语言。

这种情况就像搭积木一样，积木搭得越高，整体结构就会变得越不稳定，往上面加新的积木（添加功能）或拿掉一些积木（删除功能）都非常危险，随时可能导致整个积木结构崩溃。这时出现了微服务架构，即把一个单体应用拆分成为各个子应用，各个子应用在单独的进程中。如图 2-2 所示，这是 Martin Fowler（ThoughtWorks 首席科学家，当今世界软件开发领域最具影响力的大师之一）在其博客上介绍的单体应用与微服务应用之间的区别（单体应用会把所有的功能放到一个进程里，扩容时相当于复制单体应用到多个服务器上。微服务架构会把每个功能单独放到一个进程里，扩容时根据需要对不同的微服务进行不同的操作）。

微服务架构的优点如下：

- 每个微服务可以独立开发、独立运行、独立部署，可以使用任意一种开发语言。
- 每个微服务之间是独立的，如果某个服务宕机，只会影响当前服务，不会对整个业务系统产生影响。
- 不同团队维护不同的微服务，职责单一。

- 可以针对不同的微服务做不同的扩/缩容策略，不会造成资源浪费。

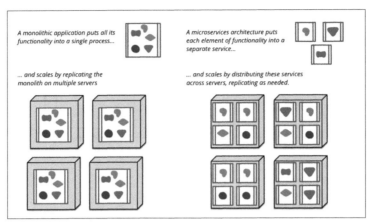

图 2-2

使用微服务架构虽然有以上优点，但也会有一些缺点。其中一个缺点就是微服务节点信息的维护，图 2-2 中的 5 个微服务分别拥有 5、3、1、3 和 4 个节点，这些服务的节点信息会随着时间随时变化，我们怎么感知这些节点信息的变化呢？如图 2-3 所示，UserService 服务依赖 OrderService 服务，而 OrderService 有 6 个实例节点，这 6 个节点的 IP 需要通过注册中心维护。

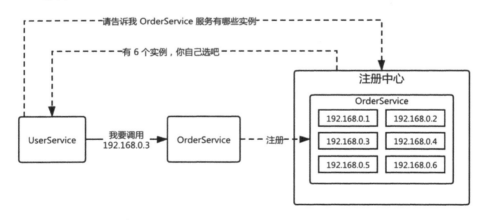

图 2-3

Spring Cloud 和 Apache Dubbo 内部服务节点信息的维护使用服务注册/服务发现机制实现。Spring Cloud 服务调用的过程如图 2-4 所示，多个 Provider（服务提供者）首先会在注册中心注册自身的实例信息，Consumer（服务消费者）去注册中心发现服务，得到服务实例列表，最终选择其中某个实例发起一次服务调用。Apache Dubbo 的架构如图 2-5 所示，它也使用依赖注册中心的服务注册/服务发现机制。Provider 向注册中心注册自身的实例信息，Consumer 发起服务调用的时候首先会去注册中心订阅 Provider 服务对应的实例列表，然后选择某个实例发起一次服务调用，调用过程会被 Monitor 监听模块监听并统计。

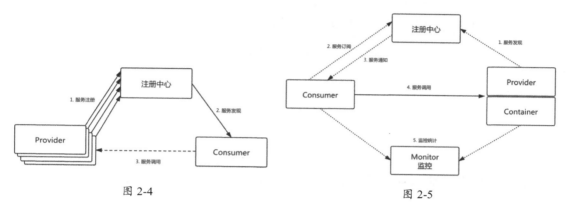

图 2-4　　　　　　　　　　　　　图 2-5

2.2　使用 Alibaba Nacos 体验第一个 Spring Cloud 微服务应用

　　Nacos 是阿里巴巴集团在 2018 年 7 月开源的一个易于构建云原生应用的动态服务发现、配置管理和服务管理平台，其在 2019 年 4 月发布了 1.0.0 GA 版本，可大规模投入生产环境使用。随着 Spring Cloud Alibaba 融入 Spring Cloud 生态，在 Spring Cloud 的注册中心选型上可以使用 Nacos 作为注册中心。

　　Nacos 的架构如图 2-6 所示。其中，Provider 应用注册服务到 Nacos 注册中心，Consumer 从 Nacos 注册中心订阅服务。Nacos 提供的 Nacos Console 可以完成各种运维操作。核心模块 Nacos Server 由以下 4 部分组成：

- Open API 暴露各种操作，客户端可以调用 Open API 完成各项事宜。
- Config Service 和 Naming Service 分别对应配置功能及服务注册/发现功能。
- Nacos Core 是整个 Nacos Server 的核心，上层的 Config Service 和 Naming Service 底层都依赖 Nacos Core。
- Nacos Core 底层提供一致性协议算法。

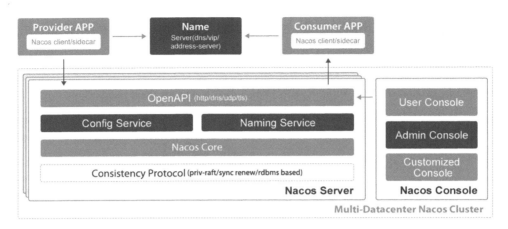

图 2-6

2.2.1 下载并启动 Nacos Server

进入 Nacos 的 GitHub 主页后打开 releases 页,就会显示所有版本的 Nacos 信息,如图 2-7 所示。

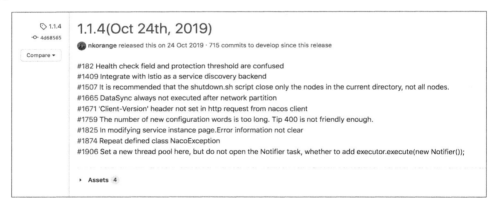

图 2-7

下面以 Nacos 1.1.4 为例进行说明,下载完成后进行解压,解压后的 Nacos 目录结构如下:

```
tree -L 1 ./
./
├── LICENSE
├── NOTICE
```

```
├── bin
├── conf
└── target
```

其中:

- LICENSE 表示开源协议,Nacos 使用的是 Apache-2.0 协议。
- NOTICE 表示注意点,包含 Nacos 开发团队所使用的开源协议、使用了第三方的一些类库等信息。
- bin 目录中是一些可执行文件,比如,startup.sh/shutdown.sh 表示 UNIX 系统下的启动/关闭脚本,startup.cmd/shutdown.cmd 表示 Windows 系统下的启动/关闭脚本。
- conf 目录存放配置文件,包含 Nacos 日志级别、数据库相关配置、集群等相关配置、启动端口等。
- target 目录存放 Nacos 生成的一些可执行文件,比如 nacos-server.jar。startup.sh 文件内部就会执行 nacos-server.jar。

下载完 Nacos 之后,以单机部署的模式进行启动:

```
sh bin/startup.sh -m standalone
```

启动成功之后,使用浏览器访问 http://localhost:8848(默认端口是 8848,可以通过 conf/application.properties 配置文件修改 server.port 配置项改变端口号)。然后使用默认的账号和密码(nacos/nacos)进入控制台(默认情况下,如果要使用 Nacos 上的生产环境,建议将内置的 Derby 内存数据库更改成 MySQL 数据库),如图 2-8 所示。

图 2-8

2.2.2 启动 Nacos Discovery Provider 进行服务注册

在 Java 工程脚手架上选择 Nacos Service Discovery 和 Spring Web 模块，并创建 spring-cloud-alibaba-nacos-provider 项目。

 提示：Java 工程脚手架是适合中国开发者的 Spring Initializr 代码框架生成器，自带 Demo Code，省去搜索引擎检索等复杂动作；有完善的工具链、免费的 IDEA 插件，方便直接在 IDE 中生成，更适合国内用户的网络环境。

新建 NacosProvider 启动类：

```java
@SpringBootApplication
public class NacosProvider {

    public static void main(String[] args) {
        SpringApplication.run(NacosProvider.class, args);   // ①
    }

    @RestController
    class EchoController {     // ②

        @GetMapping("/echo")    // ③
        public String echo(HttpServletRequest request) {
            return "echo: " + request.getParameter("name");    //④
        }

    }

}
```

上述代码中：

① 配合 @SpringBoot 注解启动一个 Spring Boot 应用。

② 使用 @RestController 定义一个 EchoController。

③ /echo 是 EchoController 对外暴露的访问路径。

④ /echo 返回 "echo: #{name 参数}"。

application.properties 配置文件如下：

```
# 要注册的服务名
spring.application.name=my-provider
# 应用要启动的端口，也是要注册的端口号
server.port=8080
# Nacos Server 地址信息
spring.cloud.nacos.discovery.server-addr=localhost:8848
```

启动 NacosProvider 后在应用控制台可以发现 nacos-provider 服务成功注册到 Nacos 的日志：

```
nacos registry, nacos-provider ip:8080 register finished
```

在 Nacos 控制台上也看到了服务注册成功的提示，如图 2-9 所示。

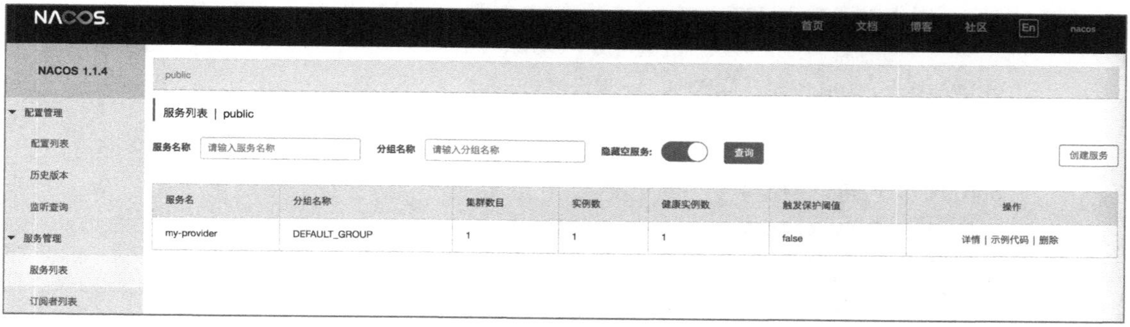

图 2-9

2.2.3　启动 Nacos Discovery Consumer 进行服务发现

同样，在 Java 工程脚手架上选择 Nacos Service Discovery 和 Spring Web 模块，创建 spring-cloud-alibaba-nacos-consumer 项目。

新建 NacosConsumer 启动类：

```
@SpringBootApplication
@EnableDiscoveryClient(autoRegister = false)    // ①
public class NacosConsumer {

    public static void main(String[] args) {
```

```java
        SpringApplication.run(NacosConsumer.class, args);
    }

    @Bean       // ②
    public RestTemplate restTemplate() {
        return new RestTemplate();
    }
}

@RestController
class HelloController {

    @Autowired      // ③
    private DiscoveryClient discoveryClient;

    @Autowired      // ④
    private RestTemplate restTemplate;

    private String serviceName = "my-provider";

    @GetMapping("/info")
    public String info() {
        List<ServiceInstance> serviceInstances =
discoveryClient.getInstances(serviceName); // ⑤
        StringBuilder sb = new StringBuilder();
        sb.append("All services: " + discoveryClient.getServices() + "<br/>");
        sb.append("my-provider instance list: <br/>");
        serviceInstances.forEach( instance -> {        // ⑥
            sb.append("[ serviceId: " + instance.getServiceId() +
                ", host: " + instance.getHost() +
                ", port: " + instance.getPort() + " ]");
            sb.append("<br/>");
        });
        return sb.toString();
    }

    @GetMapping("/hello")
    public String hello() {
```

```
            List<ServiceInstance> serviceInstances =
discoveryClient.getInstances(serviceName);
            ServiceInstance serviceInstance = serviceInstances.stream()
                .findAny().orElseThrow(() ->
                    new IllegalStateException("no " + serviceName + " instance available"));   // ⑦
            return restTemplate.getForObject(    // ⑧
                "http://" + serviceInstance.getHost() + ":" + serviceInstance.getPort() +
                    "/echo?name=nacos", String.class);
        }
    }
}
```

上述代码中：

① 这是一个 Consumer 应用，设置不自动注册到注册中心。

② 构造一个 RestTemplate，后续用于访问 HTTP 服务。

③ 在 HelloController 中自动注入 DiscoveryClient，这是 Spring Cloud Commons 模块提供的一个服务发现接口，Spring Cloud Alibaba Nacos Discovery 模块内部会初始化一个它的实现类——NacosDiscoveryClient，用于后续的服务发现操作。

④ 在 HelloController 中注入前面构造的 RestTemplate。

⑤ 使用 DiscoveryClient 获取 my-provider 服务对应的所有实例。

⑥ 使用 lambda 遍历获取的所有实例，去获取各个实例里的 host 和 port 信息。

⑦ 从所有的服务实例列表中任意获取一个，如果没有服务实例，则抛出 IllegalStateException 异常。

⑧ 使用 RestTemplate 调用服务实例对应的节点信息中的/echo 方法。

application.properties 配置文件如下：

```
# 应用名
spring.application.name=nacos-consumer
```

```
# 应用要启动的端口
server.port=8081
# Nacos Server 地址信息
spring.cloud.nacos.discovery.server-addr=localhost:8848
```

启动 NacosConsumer 后访问 http://ip:port/8081/info，返回内容如下：

```
All services: [my-provider]
my-provider instance list:
[ serviceId: my-provider, host: 30.5.121.216, port: 8080 ]
```

访问 http://ip:port/8081/hello 后，返回内容如下：

```
echo: nacos
```

2.3 使用 Netflix Eureka 替换 Alibaba Nacos 注册中心

Eureka 是 Netflix 开源的一款注册中心中间件，其架构如图 2-10 所示。

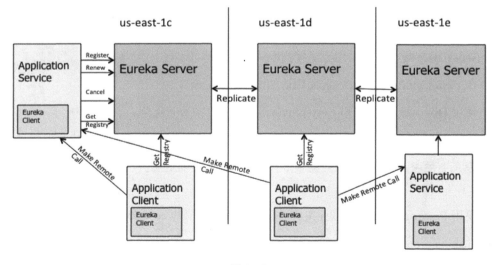

图 2-10

建议将 Eureka Server 集群分别部署在不同的可用区，达到高可用的效果。us-east-1c、us-east-1d

及 us-east-1e 分别表示 us-east 区下的 3 个不同可用区。使用 Eureka Client 客户端启动该应用后会发请求给 Eureka Server 进行注册,后续发送"心跳"进行续约(renew)。客户端也可以发起取消(cancel)进行服务下线、获取(get)服务信息等请求。Eureka Server 集群内的各个实例会进行数据复制,确保数据的一致性。

在此不推荐继续使用 Netflix Eureka 作为注册中心,其原因有以下 3 点:

- 在 2019 年巴塞罗那 Spring I/O 大会和奥斯汀 Spring One Platform 大会上,Spring 团队都有一个主题为"How to live in a post Spring Cloud Netflix world"的环节来讲解 Spring Cloud Netflix 进入维护模式后使用其他 Spring Cloud 实现代替 Netflix 的解决方案。
- Eureka 2.0 已经停止开发。Eureka 1.x 架构存在问题,比如,客户端采用 pull 模式拉取服务数据时,导致实时性不足和无谓的拉取性能消耗的问题;Eureka 集群的每一个实例都可以接受客户端的写请求,并且各个实例会进行数据复制,从而导致一些性能问题。
- 其他注册中心如 Alibaba Nacos、HaShicorp Consul 等一直在发展,差距会越来越大。

2.3.1 启动 Eureka Server

Eureka Server 与 Spring 体系进行了非常好的整合,一个 Spring Boot 应用引入对应的依赖后就代表一个 Eureka Server。在 Java 工程脚手架上选择 Eureka Server,并创建 spring-cloud-netflix-eureka-server 项目。

新建 EurekaServerApplication 启动类:

```
@SpringBootApplication
@EnableEurekaServer   //使用 @EnableEurekaServer 注解开启 Eureka Server
public class EurekaServerApplication {

    public static void main(String[] args) {
        SpringApplication.run(EurekaServerApplication.class, args);
    }

}
```

application.properties 配置文件如下:

```
# 应用名
spring.application.name=eureka-server
# 应用要启动的端口
server.port=8761
# 不向注册中心注册实例，因为该实例本身就是注册中心
eureka.client.register-with-eureka=false
# 不从 Eureka Server 中获取实例信息
eureka.client.fetch-registry=false
```

启动 EurekaServerApplication 后，进入 http://localhost:8761，确认 Eureka Server 启动成功。

2.3.2 启动 Eureka Discovery Provider 进行服务注册

在 Java 工程脚手架上选择 Eureka Discovery Client 和 Spring Web 模块，并创建 spring-cloud-netflix-eureka-provider 项目。

新建 EurekaProvider 启动类。EurekaProvider 内部的代码与 NacosProvider 的代码完全一样，在此不再重复解释。

EurekaProvider 的 application.properties 配置文件内容如下：

```
# 要注册的服务名
spring.application.name=my-provider
# 应用要启动的端口，也是要注册的端口号
server.port=8080
# Eureka Server 地址信息
eureka.client.service-url.defaultZone=http://localhost:8761/eureka
```

启动 EurekaProvider 后在控制台可以发现成功注册到 Eureka 的日志：

```
DiscoveryClient_MY-PROVIDER/ip:my-provider:8080 - registration status: 204
...
The response status is 200
```

在 Eureka 控制台上也可以看到服务注册成功的信息，如图 2-11 所示。

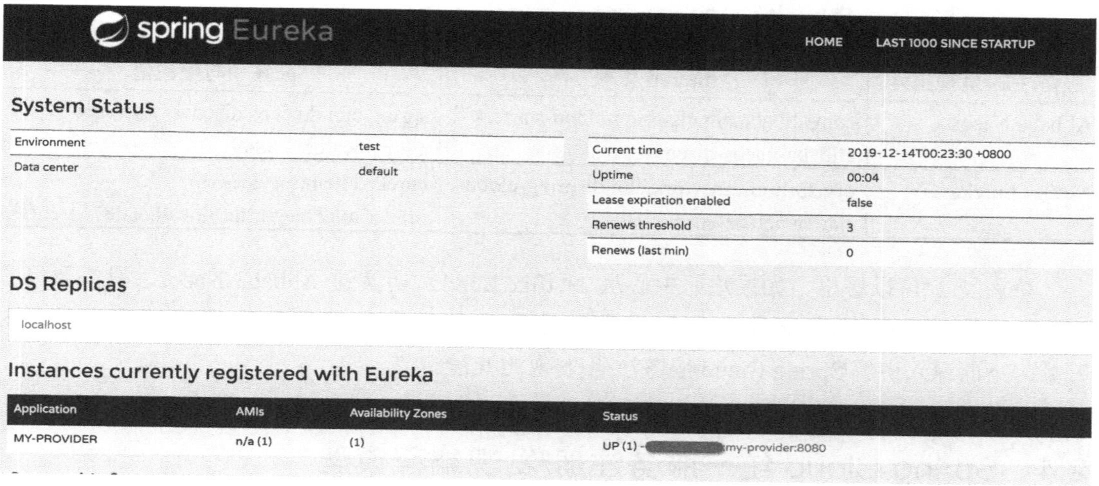

图 2-11

2.3.3 启动 Nacos Discovery Consumer 进行服务发现

同样，在 Java 工程脚手架上选择 Eureka Discovery Client 和 Spring Web 模块，创建 spring-cloud-alibaba-eureka-consumer 项目。

新建 EurekaConsumer 启动类的代码跟 NacosConsumer 相同，在此不再重复解释。

EurekaConsumer 的 application.properties 配置文件内容如下：

```
# 应用名
spring.application.name=eureka-consumer
# 应用要启动的端口
server.port=8081
# Eureka Server 地址信息
eureka.client.service-url.defaultZone=http://localhost:8761/eureka
```

启动 EurekaConsumer 后访问 http://ip:port/8081/info 和 http://ip:port/8081/hello，返回的内容也是一样的。

对比 NacosProvider 和 EurekaProvider，以及 NacosConsumer 和 EurekaConsumer，可以发现它们的代码没有任何差异，两者唯一的区别在 pom 依赖和配置内容上，如表 2-1 所示。

表 2-1

服务注册/发现组件	maven 依赖	配置项与配置值
Alibaba Nacos	com.alibaba.cloud:spring-cloud-starter-alibaba-nacos-discovery	spring.cloud.nacos.discovery.server-addr=localhost:8848
Netflix Eureka	org.springframework.cloud:spring-cloud-starter-netflix-eureka-client	eureka.client.service-url.defaultZone=http://localhost:8761/eureka

从表 2-1 可以看出，如果注册中心从 Netflix Eureka 切换到 Alibaba Nacos，只需要修改 maven 依赖和配置即可，无须修改任何代码。这是 Spring Cloud 统一服务注册/发现编程模型的好处。下面深入讲解 Spring Cloud 服务注册/发现的编程模型。

2.4　Spring Cloud 统一服务注册/发现编程模型

Spring Cloud 统一了服务注册和服务发现编程模型，其代码在 `spring-cloud-commons` 模块里，相关接口的说明如表 2-2 所示。

表 2-2

接　　口	作　　用
org.springframework.cloud.client.discovery.DiscoveryClient	代表服务发现常见的读取操作
org.springframework.cloud.client.discovery.EnableDiscoveryClient	使用该注解表示开启服务发现功能
org.springframework.cloud.client.discovery.ReactiveDiscoveryClient	基于响应式的代表服务发现常见的读取操作
org.springframework.cloud.client.serviceregistry.ServiceRegistry	注册与注销服务的操作封装
org.springframework.cloud.client.ServiceInstance	代表服务的一个实例

统一编程模型有以下两个优点。

- 无须关注底层服务注册/发现的实现细节，只需了解上层统一的抽象。
- 更换注册中心非常简单，只需修改 maven 依赖和对应的注册中心配置信息（比如注册中心地址、namespace、group 等）。

2.4.1　DiscoveryClient 和 ReactiveDiscoveryClient

DiscoveryClient 和 ReactiveDiscoveryClient 代表 Consumer 从注册中心发现 Provider 的服务发现操作，接口定义如下：

```java
public interface DiscoveryClient extends Ordered {

    /**
     * 默认的优先级，多个 DiscoveryClient 存在的情况下以优先级排序
     */
    int DEFAULT_ORDER = 0;

    /**
     * 具体服务发现组件的描述信息，在 HealthIndicator 中会被用到
     * @return 描述信息
     */
    String description();

    /**
     * 根据服务名查询所有的服务实例
     * @param serviceId 服务名
     * @return 服务实例集合
     */
    List<ServiceInstance> getInstances(String serviceId);

    /**
     * @return 返回注册中心所有的服务名
     */
    List<String> getServices();

    /**
     * 具体的服务发现组件的优先级，默认为 0
     * @return 优先级
     */
    @Override
    default int getOrder() {
        return DEFAULT_ORDER;
    }

}
```

2.2 节和 2.3 节介绍的 Alibaba Nacos 和 Netflix Eureka 服务注册/发现实例中都是直接注入 DiscoveryClient 完成服务实例的获取。它们对应的 DiscoveryClient 是依赖内部自动化配置类产生的 NacosDiscoveryClient 和 EurekaDiscoveryClient。

ReactiveDiscoveryClient 是在 Spring Cloud Hoxton M3 中首次加入的基于响应式的服务发现接口。

```java
public interface ReactiveDiscoveryClient extends Ordered {

    int DEFAULT_ORDER = 0;

    String description();

    Flux<ServiceInstance> getInstances(String serviceId);

    Flux<String> getServices();

    @Override
    default int getOrder() {
        return DEFAULT_ORDER;
    }
}
```

ReactiveDiscoveryClient 接口中的方法与 DiscoveryClient 几乎相同，只是把 List 类型转换成了 Reactor 里的 Flux 类型。

在 Spring WebFlux 场景下，ReactiveDiscoveryClient 会生效，这个 NacosReactiveConsumer 启动类内部的逻辑就是使用 ReactiveDiscoveryClient 的一个场景。

```java
@SpringBootApplication
@EnableDiscoveryClient(autoRegister = false)
public class NacosReactiveConsumer {

    public static void main(String[] args) {
        SpringApplication.run(NacosReactiveConsumer.class, args);
```

```
}

@RestController
class HelloController {

    @Autowired
    private ReactiveDiscoveryClient reactiveDiscoveryClient; // ①

    @Autowired
    private DiscoveryClient discoveryClient;

    private String serviceName = "my-provider";

    @GetMapping("/services")
    public Flux<String> info() {
        return reactiveDiscoveryClient.getServices(); // ②
    }

    @GetMapping("/instances")
    public Flux<String> instance() {
        return reactiveDiscoveryClient.getInstances(serviceName).map(instance -> // ③
            "[ serviceId: " + instance.getServiceId() +
                ", host: " + instance.getHost() +
                ", port: " + instance.getPort() + " ]");
    }

    @GetMapping("/hello")
    public Mono<String> hello() {
        List<ServiceInstance> serviceInstances = discoveryClient.getInstances(serviceName);
        ServiceInstance serviceInstance = serviceInstances.stream()
            .findAny().orElseThrow(() ->
                new IllegalStateException("no " + serviceName + " instance available"));
        return WebClient.create("http://" +      // ④
            serviceInstance.getHost() + ":" + serviceInstance.getPort())
```

```
                .get().uri("/echo?name=nacos").retrieve().bodyToMono(String.class);
        }

    }

}
```

上述代码中：

① 注入 ReactiveDiscoveryClient，在 WebFlux 场景下，NacosReactiveDiscoveryClient-Configuration 会自动构造 NacosReactiveDiscoveryClient。

② 使用 ReactiveDiscoveryClient#getServices 方法得到 Flux<String>。注意，这里是一个 Reactor 的 Flux 类型。

③ 得到 Flux<ServiceInstance>，然后使用 map 方法进行映射。

④ 使用 WebClient 这种 reactive 客户端去调用服务。

2.4.2 ServiceInstance 和 Registration

服务信息在每个注册中心的内部数据存储结构都有各自的模型。如图 2-12 所示，Zookeeper 注册中心内部的数据模型是一种树状结构，这棵树由不同的节点组成。顶层是 "/" 节点，"/" 节点有两个子节点，分别是 /app1 和 /app2，所有注册的实例信息在 Zookeeper 里都是一个节点，这个节点在 Zookeeper 内部被称为 ZNode。

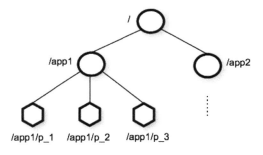

图 2-12

ZNode 在 Zookeeper Curator 客户端中的对应类是 org.apache.curator.x.discovery.ServiceInstance，该类内部包含 name、id、address、port、sslPort、payload 等属性。

如图 2-13 所示，这是 Eureka 注册中心内部的数据模型。所有注册的实例信息在 Eureka 内都是一个 InstanceInfo 对象实例，该类内部包含 instanceId、appName、status、port、asgName、dataCenterInfo、metadata 等属性。

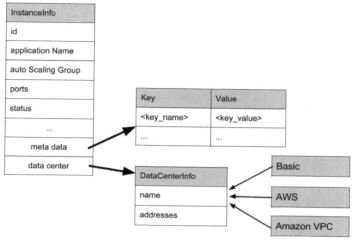

图 2-13

如图 2-14 所示，这是 Nacos 注册中心内部的数据模型。Nacos 服务信息由 3 元组 namespace（命名空间用于数据隔离）、group（分组）和 serviceName（服务名）组成。所有注册的实例信息在 Nacos 内都是一个 com.alibaba.nacos.api.naming.pojo.Instance 对象实例，该类内部拥有 instanceId、ip、port、weight、healthy、enabled、serviceName、cluseterName 等属性。

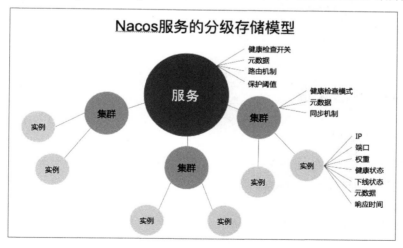

图 2-14

Spring Cloud 提供的 ServiceInstance 和 Registration 的作用是抽象实例在各种注册中心的数据模型。无论是 Zookeeper 的 ServiceInstance、Eureka 的 InstanceInfo，还是 Nacos 的 Instance，在 Spring Cloud 体系内会被统一抽象为 ServiceInstance 和 Registration。其中，ServiceInstance 表示客户端从注册中心获取到的实例数据结构，Registration 表示客户端注册到注册中心的实例数据结构，其表现如图 2-15 所示。

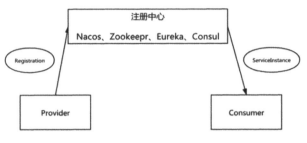

图 2-15

ServiceInstance 接口定义如下：

```java
public interface ServiceInstance {

    /**
     * @return 实例 ID，可以不实现。默认返回 null
     */
    default String getInstanceId() {
        return null;
    }

    /**
     * @return 注册的服务 ID
     */
    String getServiceId();

    /**
     * @return 服务实例的 hostname
     */
    String getHost();

    /**
```

```
 * @return 服务实例的端口
 */
int getPort();

/**
 * @return 是否使用 HTTPS
 */
boolean isSecure();

/**
 * @return 服务实例的 URI 地址
 */
URI getUri();

/**
 * @return 服务实例的 key/value（键值对）形式的 metadata 信息
 */
Map<String, String> getMetadata();

/**
 * @return The scheme of the service instance.
 */
default String getScheme() {
    return null;
}
}
```

使用 DiscoveryClient 可以基于服务名获取到这个服务下的所有 ServiceInstance 集合。比如，my-provider 服务在注册中心可能会存在两个 ServiceInstance，即 service-instance1: 192.168.1.1:8080 和 service-instance2: 192.168.1.2:8080。

Registration 接口定义如下：

```
public interface Registration extends ServiceInstance {

}
```

可以看到，这个接口继承了 ServiceInstance，并且没有额外的方法定义。因为在从注册中心获取实例信息和把实例信息注册到注册中心这两个过程中，实例信息的存储结构完全可以相同。笔者预测将来这个接口可能会新增一些方法。那么，哪个接口使用 Registration 注册服务呢？那就是下面要介绍的 ServiceRegistry。

2.4.3 ServiceRegistry

ServiceRegistry 用于服务信息的注册（register）和注销（deregister），其接口定义如下：

```
public interface ServiceRegistry<R extends Registration> {

    /**
     * 基于实例信息将其注册到注册中心
     * @param registration 实例信息
     */
    void register(R registration);

    /**
     * 基于实例信息将其从注册中心注销
     * @param registration 实例信息
     */
    void deregister(R registration);

    /**
     * 关闭 ServiceRegistry，这是一个生命周期方法
     */
    void close();

    /**
     * 设置服务实例的状态
     * @param registration 服务实例
     * @param status 状态
     * @see org.springframework.cloud.client.serviceregistry.endpoint.ServiceRegistryEndpoint
     */
    void setStatus(R registration, String status);
```

```
/**
 * 获取服务实例的状态
 * @param registration 服务实例
 * @param <T> The type of the status.
 * @return 服务实例当前状态
 * @see org.springframework.cloud.client.serviceregistry.endpoint.ServiceRegistryEndpoint
 */
<T> T getStatus(R registration);
}
```

如图 2-16 所示，这是服务注册和注销过程的一个简单说明。

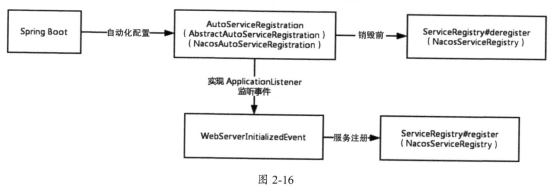

图 2-16

AutoServiceRegistration 是一个空接口，表示自动完成服务注册过程。接口具体的实现类（如 NacosAutoServiceRegistration）会在 NacosServiceRegistryAutoConfiguration 自动化配置类中被自动构造。

AbstractAutoServiceRegistration 抽象类实现了 AutoServiceRegistration 接口，同时也实现了 ApplicationListener 接口并监听 WebServerInitializedEvent。收到事件后，使用 ServiceRegistry 完成服务注册。应用程序关闭时触发 @PreDestroy 注解使用 ServiceRegistry 完成服务注销，这就是 Spring Cloud 服务信息的注册和注销时机。

 提示：Spring Cloud 2.2.0.RELEASE 之后的版本通过 WebServerInitializedEvent 事件的监听完成服务注册已被声明为过期方法。Spring Cloud 把注册时机的决定权交给了各个注册中心实现。

2.4.4 ServiceRegistryEndpoint

ServiceRegistryEndpoint 是 Spring Cloud 服务注册/发现功能对外暴露的 Endpoint，其 ID 是 service-registry，用于获取和设置服务实例的状态。获取和设置的动作由 ServiceRegistry 的 getStatus 和 setStatus 方法完成。

在 NacosProvider 项目中加上 org.springframework.boot:spring-boot-starter-actuator 依赖，配置 management.endpoints.web.exposure.include=* 后进行启动，启动后执行如下脚本：

```
curl --header "Content-Type:application/json" -XPOST http://localhost:8080/actuator/service-registry?status=down
```

如图 2-17 所示，执行后可看到在 Nacos 控制台的"操作"列下的"下线"按钮变成了"上线"按钮。

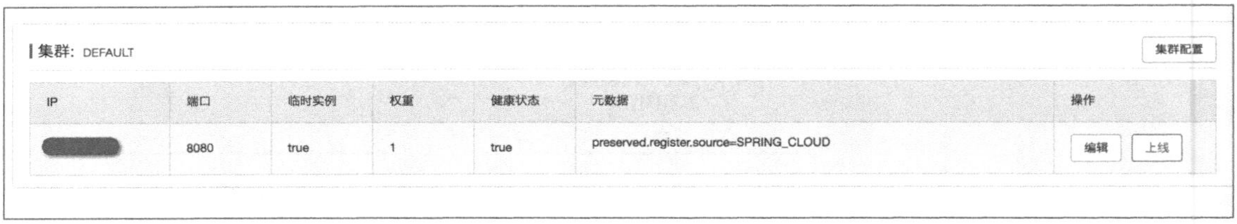

图 2-17

ServiceRegistryEndpoint 非常有用，可以完成应用的无损下线。应用的无损下线是什么意思呢？比如某个应用有 10 个实例，某天需要发布新版本，发布新版本意味着需要先下线部分实例（下线旧版本），然后使用新的部署包发布新版本，以此类推，直到所有的旧版本都被替换成新版本。无损下线的意思是在旧版本下线的过程中，服务调用不发生任何错误。

Spring Cloud 默认的服务实例更新机制会每隔 30s 去注册中心获取服务对应的实例列表信息来覆盖内存里的实例信息。如果一些实例在某个时间点完成下线操作，但是调用这个服务的客户端因为还没达到 30s 刷新时间，其内存中的服务实例列表中存在已经下线的实例，调用这些已经下线的实例的服务接口会发生连接超时异常（这个实例服务已经下线，对应的端口未开启，客户端访问会超时）。这就需要无损下线来避免异常的发生。

使用 ServiceRegistryEndpoint 让应用无损下线的思路如下：

① 调用 ServiceRegistryEndpoint，将需要下线的实例的服务下线。

② 服务下线之后，等待客户端达到 30s 刷新时间，通过刷新实例列表信息来删除已下线

的实例（这时候只是服务下线，实例并未下线）。

③ 实例下线（服务发现已经不再添加该实例，可以放心下线实例）。

调用 ServiceRegistryEndpoint 的过程如下：

① 在依赖文件中加上 org.springframework.boot:spring-boot-actuator 组件。

② 在配置文件中加上 management.endpoints.web.exposure.include = service-registry。

③ 使用 curl --header "Content-Type:application/json" -XPOST http://host:port/actuator/service-registry -d '{"status": "DOWN"}' 完成服务下线（如果是误操作，需要重新上线，只需将 DOWN 改成 UP 即可）。

2.5 双注册双订阅模式

双注册双订阅表示一个 Provider 应用可以将自身的实例信息注册到多个注册中心上，一个 Consumer 应用可以订阅到多个注册中心上的服务实例信息。

如图 2-18 所示，Provider 可以把自身的服务实例信息注册到 Nacos 和 Eureka 集群上，Consumer 发起服务订阅的时候可以从 Nacos 和 Eureka 上订阅服务。

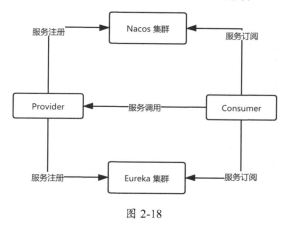

图 2-18

2.5.1 双注册双订阅模式分析

Spring Cloud 自身的编程模型是支持双注册双订阅模式的。

在服务注册侧,我们在 2.4.3 节分析过 Spring Cloud 各个注册中心都有 AutoServiceRegistration 的实现类,比如,NacosAutoServiceRegistration 和 EurekaAutoServiceRegistration 实现在类内部完成服务的注册。这些 AutoServiceRegistration 的实现类都实现了 Lifecycle 接口,在 start 过程中完成服务注册操作。

在服务订阅侧,我们在 2.4.1 节分析过 DiscoveryClient 统一了 Spring Cloud 服务发现的操作。其中,CompositeDiscoveryClient 是一个特殊的 DiscoveryClient 实现:

```java
public class CompositeDiscoveryClient implements DiscoveryClient {

    private final List<DiscoveryClient> discoveryClients;

    public CompositeDiscoveryClient(List<DiscoveryClient> discoveryClients) {
        AnnotationAwareOrderComparator.sort(discoveryClients);
        this.discoveryClients = discoveryClients;
    }

    @Override
    public List<ServiceInstance> getInstances(String serviceId) {
        if (this.discoveryClients != null) {
            for (DiscoveryClient discoveryClient : this.discoveryClients) {
                List<ServiceInstance> instances = discoveryClient.getInstances(serviceId);
                if (instances != null && !instances.isEmpty()) {
                    return instances;
                }
            }
        }
        return Collections.emptyList();
    }

    @Override
    public List<String> getServices() {
        LinkedHashSet<String> services = new LinkedHashSet<>();
        if (this.discoveryClients != null) {
            for (DiscoveryClient discoveryClient : this.discoveryClients) {
```

```
            List<String> serviceForClient = discoveryClient.getServices();
            if (serviceForClient != null) {
                services.addAll(serviceForClient);
            }
        }
    }
    return new ArrayList<>(services);
}
```

在 getInstances 方法中，会聚合所有的 DiscoveryClient 实现类找到的服务名，也会遍历每个 DiscoveryClient 查询服务名对应的实例信息。

下面在一个应用里分别加上 Nacos（com.alibaba.cloud:spring-cloud-starter-alibaba-nacos-discovery）和 Eureka（org.springframework.cloud:spring-cloud-starter-netflix-eureka-client）依赖，用来完成双注册双订阅。

应用启动后，会出现以下报错信息：

```
***************************
APPLICATION FAILED TO START
***************************

Description:

Field registration in
org.springframework.cloud.client.serviceregistry.ServiceRegistryAutoConfiguration$S
erviceRegistryEndpointConfiguration required a single bean, but 2 were found:
    - nacosRegistration: defined by method 'nacosRegistration' in class path resource
[com/alibaba/cloud/nacos/registry/NacosServiceRegistryAutoConfiguration.class]
    - eurekaRegistration: defined in BeanDefinition defined in class path resource
[org/springframework/cloud/netflix/eureka/EurekaClientAutoConfiguration$Refreshable
EurekaClientConfiguration.class]
```

从这个报错信息可以很明显地看出，ServiceRegistryAutoConfiguration 自动化配置类的内部类 ServiceRegistryEndpointConfiguration 内部依赖一个 Registration Bean，但是在 Nacos 和 Eureka

依赖内部分别会构造 NacosRegistration 和 EurekaRegistration，这样会出现 ServiceRegistry-EndpointConfiguration 并不知道要注入哪个 Registration Bean 的问题。同理，AutoService-RegistrationAutoConfiguration 内部的 AutoServiceRegistration Bean 也会引起一样的问题。

为了解决这个问题，可以在配置文件里过滤这两个自动化配置类：

```
spring.autoconfigure.exclude=org.springframework.cloud.client.serviceregistry.ServiceRegistryAutoConfiguration,org.springframework.cloud.client.serviceregistry.AutoServiceRegistrationAutoConfiguration
```

加上该配置之后，还需要通过@EnableConfigurationProperties 注解让 AutoServiceRegistration-Properties 配置类生效。这是因为所有的 AutoServiceRegistration 实现类在构造过程中都需要这个配置类 Bean。

有了这两个条件之后，即可享受双注册双订阅模式。

2.5.2　案例：使用双注册双订阅模式将 Eureka 注册中心迁移到 Nacos 注册中心

假设某公司原先使用 Eureka 作为注册中心，Nacos 开源之后，该公司想把 Eureka 替换成 Nacos 注册中心，要求在这个过程中对客户没有任何影响，也不能造成业务损失。

对于这个场景，可以使用双注册双订阅方案来完成任务。如图 2-19 所示，这是一个 3 个阶段的过程图。

第 1 阶段：Eureka 作为注册中心，Provider 完成服务注册，Consumer 完成服务发现。

第 2 阶段：双注册双订阅的核心阶段，该阶段内部包括以下 4 个操作。

- 上线新的 Provider（拥有双注册能力），这时 Eureka 注册中心的 Provider 有两个实例。
- 下线旧的 Provider，下线之后由于新 Provider 也会注册到 Eureka 上，这时旧的 Consumer 可以找到新 Provider 的实例。
- 上线新的 Consumer（拥有双订阅能力），新 Consumer 可以订阅 Nacos 和 Eureka 集群的服务实例，这时可以订阅到 Nacos 上的服务实例。
- 下线旧的 Consumer。

第 3 阶段：Eureka 下线，使用 Nacos 替换 Eureka 作为新的注册中心，Provider 和 Consumer 的服务注册和服务发现操作只与 Nacos 交互。

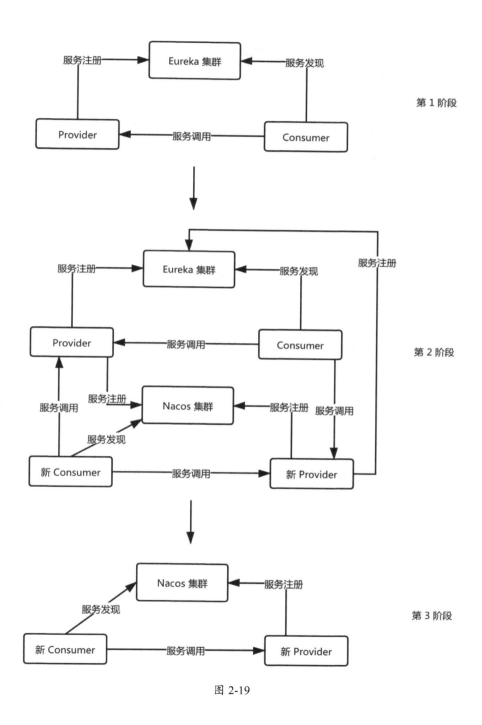

图 2-19

在本节提供的例子中，读者可以配合本书示例使用以下步骤来模拟这个场景：

① 启动 spring-cloud-netflix-eureka-provider 和 spring-cloud-netflix-eureka-consumer，模拟一开始使用 Eureka 作为注册中心的场景。

② 启动 spring-cloud-eureka-nacos-provider 应用，这个应用拥有双注册能力。

③ 下线 spring-cloud-netflix-eureka-provider 应用。

④ 启动 spring-cloud-eureka-nacos-consumer 应用。

⑤ 下线 spring-cloud-netflix-eureka-consumer 应用。

第 3 章
负载均衡与服务调用

本章将介绍客户端负载均衡，针对这个概念讲解 RPC 框架 Spring Cloud 和 Apache Dubbo 的负载均衡实现原理并进行对比。然后在服务调用层面对 Spring Cloud 客户端 RestTemplate 和 OpenFeign 的原理进行讲解，也对 Spring Cloud Alibaba 提供的 Dubbo Spring Cloud 进行分析，Dubbo Spring Cloud 可以让 RestTemplate 或 OpenFeign 调用 Dubbo 服务。

最后配合应用灰度发布案例介绍在 Spring Cloud 中，如何通过 RestTemplate、OpenFeign 及 Netflix Ribbon 实现对流量的控制。

3.1 负载均衡原理

本书第 2 章提到了注册中心存在的意义，同时也演示了使用 DiscoveryClient 或 ReactiveDiscoveryClient 获取注册中心上的服务对应的所有实例。在使用过程中，我们发现一个问题：在注册中心返回的实例列表中，我们要选择哪个实例呢？

```
// my-service 服务对应的 ServiceInstance 列表
List<ServiceInstance> serviceInstances = discoveryClient.getInstances("my-service");
// 构造负载均衡客户端组件
LoadBalancerClient client = new LoadBalancerClient(serviceInstances);
// 负载均衡组件根据特定算法选择某一个实例
Service Instance = client.chooseOne();
// 使用 RestTemplate 完成服务调用
String result = new RestTemplate().getForObject(
    "http://" + serviceInstance.getHost() + ":" + serviceInstance.getPort() + "/path",
    String.class);
```

这就需要使用负载均衡算法解决这个问题（以上代码中，LoadBalancerClient 相当于一个客户端负载均衡组件）。

常见的负载均衡算法如下：

- 随机（Random）算法：在实例列表中随机选择某个实例。
- 轮询（RoundRobin）算法：循环取下一个。比如，一共有 3 个实例，第一次取第 1 个，第二次取第 2 个，第三次取第 3 个，第四次取第 1 个，以此类推。
- 最少连接数（Least Connections）算法：每次取连接数最少的实例。
- 一致性哈希（Consistent Hashing）算法：基于一致性哈希算法总是将相同参数的请求落在同一个实例上。
- 权重随机（Weightd Random）算法：比如有 4 个实例，实例 A 权重为 10，实例 B 权重为 30，实例 C 权重为 40，实例 D 权重为 20。每次随机取 100 内的整数，若结果在 1~10 之间，则选择实例 A，在 11~40 之间选择实例 B，在 41~80 之间选择实例 C，在 81~100 之间选择实例 D。

Spring Cloud 在特性列表中声明了支持 Load Balancing（负载均衡），这里的负载均衡指的是客户端负载均衡。

 提示：维基百科对负载均衡是这样定义的，负载均衡（Load Balancing）是一种计算机技术，用来在多个计算机（计算机集群）、网络连接、CPU、磁盘驱动器或其他资源中分配负载，以达到最优化资源使用、最大化吞吐率、最小化响应时间，同时避免出现过载的目的。

客户端负载均衡表示这个操作是在客户端进行的，客户端获取了所有的实例列表，并且根据算法选择其中一个实例。

有了负载均衡特性之后，在开发过程中无须获取 ServiceInstance 列表，而是直接进行服务调用，因为 Spring Cloud 在底层屏蔽了负载均衡的逻辑。

负载均衡是一个通用的特性，所有的 RPC 框架都会有这个概念的实现。比如，Apache Dubbo 内部提供了 LoadBalance 组件来完成负载均衡。

Spring Cloud 官方也提供了两种客户端负载均衡的实现：

- Spring Cloud LoadBalancer：2019 年 7 月 3 日，在 Hoxton.M1 的发布公告上，Spring 宣布更新该项目来代替 Netflix Ribbon。
- Spring Cloud Netflix Ribbon：Netflix 开发的一款客户端负载均衡组件。

由于 Netflix Ribbon 和 Spring Cloud Netflix Ribbon 组件都已进入维护模式，目前官方推荐使用 Spring Cloud LoadBalancer（后文简称 SCL）。目前 SCL 的功能跟 Ribbon 相比还比较弱（比如，Ribbon 提供了熔断特性，当同一个实例发生 3 次错误后，就会被"熔断"一段时间，这段时间内该实例永远不会被选择）。笔者现阶段还是推荐使用 Ribbon，预计 SCL 在未来的版本中会逐渐更新更多的功能，并达到可以替换 Ribbon 的状态。

3.2　Spring Cloud LoadBalancer 负载均衡组件

SCL 作为新一代 Spring Cloud 客户端负载均衡的实现，若要使用，需要在 pom 里加上依赖：

```
<dependency>
    <groupId>org.springframework.cloud</groupId>
    <artifactId>spring-cloud-starter-loadbalancer</artifactId>
</dependency>
```

SCL 相关的代码在 spring-cloud-commons 模块中，相关类和接口的关系如图 3-1 所示。

图 3-1 中，这些类和定义的含义如下：

- ServiceInstanceChooser：服务实例选择器，根据服务名获取一个服务实例（ServiceInstance）。
- LoadBalancerClient：客户端负载均衡器，继承 ServiceInstanceChooser，会根据 ServiceInstance 和 Request 请求信息执行最终的结果。

- BlockingLoadBalancerClient：基于 Spring Cloud LoadBalancer 的 LoadBalancerClient 默认实现。
- RibbonLoadBalancerClient：基于 Netflix Ribbon 的 LoadBalancerClient 实现。

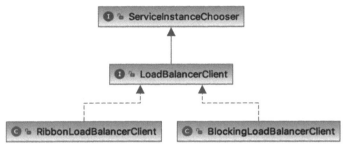

图 3-1

1. ServiceInstanceChooser

org.springframework.cloud.client.loadbalancer.ServiceInstanceChooser 服务实例选择器的定义如下：

```
public interface ServiceInstanceChooser {

    /**
     * 根据服务名得到一个服务实例
     * @param serviceId 服务名
     * @return 对应服务名下的一个服务实例
     */
    ServiceInstance choose(String serviceId);

}
```

我们自定义一个 ServiceInstanceChooser 的实现类来完成负载均衡操作（由于客户端负载均衡器 LoadBalancerClient 跟 Request 请求信息有着强耦合的关系，会涉及 RestTemplate 或 OpenFeign 的一些概念，相关内容将在后面介绍）：

```
public class RandomServiceInstanceChooser implements ServiceInstanceChooser {

    private final DiscoveryClient discoveryClient;
```

```java
    private final Random random;

    public RandomServiceInstanceChooser(DiscoveryClient discoveryClient) {
        this.discoveryClient = discoveryClient;
        random = new Random();
    }

    @Override
    public ServiceInstance choose(String serviceId) {
        List<ServiceInstance> serviceInstanceList =
            discoveryClient.getInstances(serviceId);
        return serviceInstanceList.get(random.nextInt(serviceInstanceList.size()));
    }
}
```

接下来在 Controller 里使用 RandomServiceInstanceChooser（内部使用随机算法）获取 ServiceInstance，再使用 RestTemplate 进行服务调用：

```java
@GetMapping("/customChooser")
public String customChooser() {
    ServiceInstance serviceInstance =
randomServiceInstanceChooser.choose(serviceName);
    return restTemplate.getForObject(
        "http://" + serviceInstance.getHost() + ":" + serviceInstance.getPort() +
"/", String.class);
}
```

执行 Shell 脚本，调用 Controller 的 customChooser 路径，返回结果如下：

```
30.55.192.60:8083
30.55.192.60:8081
30.55.192.60:8082
30.55.192.60:8081
30.55.192.60:8083
30.55.192.60:8080
```

```
30.55.192.60:8081
30.55.192.60:8081
...
```

我们可以看到，RandomServiceInstanceChooser 和 Controller 里的 customChooser 方法这两段代码内部先使用 DiscoveryClient 基于服务名获取这个服务对应的所有 ServiceInstance 集合，然后根据负载均衡算法从这个集合中得到一个 ServiceInstance，最后基于获取的 ServiceInstance 里的 IP 和端口使用 RestTemplate 发起 HTTP 调用。仔细想想，这两段代码还有更进一步的简化空间：直接使用 RestTemplate 根据服务名进行调用，屏蔽 ServiceInstance 的获取细节。

```
restTemplate.getForObject("http://serviceId/", String.class);
```

在实际情况下，使用 Spring Cloud 确实可以直接基于服务名进行服务调用。这是因为 Spring Cloud 扩展了 RestTemplate，只需要在定义 RestTemplate Bean 时加上@LoadBalanced 注解，就可以基于服务名进行服务调用：

```
@Bean
@LoadBalanced
public RestTemplate restTemplate() {
    return new RestTemplate();
}
```

这个神秘的@LoadBalanced 注解在底层做了什么事情呢？我们来分析一下。

2. @LoadBalanced

spring-cloud-commons 模块中的 META-INF/spring.factories 文件里存在 LoadBalancerAutoConfiguration 这个自动化配置类，根据工厂加载机制会被 ApplicationContext 加载。该自动化配置类内部的 Bean 构造代码如下：

```
...
public class LoadBalancerAutoConfiguration {

    @LoadBalanced
    @Autowired(required = false)
    private List<RestTemplate> restTemplates = Collections.emptyList();        // ①
```

```
    @Bean
    public SmartInitializingSingleton
loadBalancedRestTemplateInitializerDeprecated(
            final ObjectProvider<List<RestTemplateCustomizer>>
restTemplateCustomizers) {    //②
        return () -> restTemplateCustomizers.ifAvailable(customizers -> {
            for (RestTemplate restTemplate :
LoadBalancerAutoConfiguration.this.restTemplates) {
                for (RestTemplateCustomizer customizer : customizers) {
                    customizer.customize(restTemplate);    //③
                }
            }
        });
    }
}
```

上述代码中：

① 获取 ApplicationContext 中所有被@LoadBalanced 注解修饰的 RestTemplate。

② List<RestTemplateCustomizer>是 ApplicationContext 存在的 RestTemplateCustomizer Bean 的集合。

③ 遍历代码①处得到的 RestTemplate 集合，并使用 RestTemplateCustomizer 集合给每个 RestTemplate 定制。

这个 RestTemplateCustomizer 定制的时候做了哪些操作呢？LoadBalancerAutoConfiguration 内部的 LoadBalancerInterceptorConfig 配置类中定义了 RestTemplateCustomizer：

```
@Configuration(proxyBeanMethods = false)
@ConditionalOnMissingClass("org.springframework.retry.support.RetryTemplate") // ①
static class LoadBalancerInterceptorConfig {

    @Bean
    public LoadBalancerInterceptor ribbonInterceptor(    // ②
            LoadBalancerClient loadBalancerClient,
            LoadBalancerRequestFactory requestFactory) {
        return new LoadBalancerInterceptor(loadBalancerClient, requestFactory);
```

```
}

@Bean
@ConditionalOnMissingBean
public RestTemplateCustomizer restTemplateCustomizer(    // ③
        final LoadBalancerInterceptor loadBalancerInterceptor) {    // ④
    return restTemplate -> {
        List<ClientHttpRequestInterceptor> list = new ArrayList<>(
                restTemplate.getInterceptors());
        list.add(loadBalancerInterceptor);    // ⑤
        restTemplate.setInterceptors(list);
    };
}
```

上述代码中：

① 条件注解。LoadBalancerInterceptorConfig 配置类只有在 ClassLoader 不存在 RetryTemplate（Spring Retry 框架提供的模板类）时才会生效。

② 定义 LoadBalancerInterceptor Bean，这个拦截器继承 ClientHttpRequestInterceptor，可以被添加到 RestTemplate 的拦截器列表中。

③ 定义 RestTemplateCustomizer Bean，会在 LoadBalancerAutoConfiguration 里的 RestTemplate-Customizer 列表中存在。

④ LoadBalancerInterceptor 参数是代码②处创建的 Bean。

⑤ 使用 lambda 表达式在 RestTemplate 的拦截器列表中添加 LoadBalancerInterceptor 拦截器。

> 提示：如果 ClassLoader 存在 RetryTemplate，会触发另外一个配置类：RetryInterceptorAutoConfiguration。该配置类内部的操作与 LoadBalancerInterceptorConfig 配置类唯一的区别就是构造了 RetryLoadBalancerInterceptor 拦截器（跟 LoadBalancerInterceptor 相比，在 RestTemplate 调用失败的情况下会进行重试操作）。

看到这里，大家应该明白了。@LoadBalanced 直接修饰的 RestTemplate 会被添加一个

LoadBalancerInterceptor 拦截器。接下来进入 LoadBalancerInterceptor 拦截器，看它内部做了哪些操作，其代码如下：

```java
public class LoadBalancerInterceptor implements ClientHttpRequestInterceptor {

    private LoadBalancerClient loadBalancer;

    private LoadBalancerRequestFactory requestFactory;

    public LoadBalancerInterceptor(LoadBalancerClient loadBalancer,
            LoadBalancerRequestFactory requestFactory) { // ①
        this.loadBalancer = loadBalancer;
        this.requestFactory = requestFactory;
    }

    public LoadBalancerInterceptor(LoadBalancerClient loadBalancer) {
        this(loadBalancer, new LoadBalancerRequestFactory(loadBalancer));
    }

    @Override
    public ClientHttpResponse intercept(final HttpRequest request, final byte[] body,
            final ClientHttpRequestExecution execution) throws IOException {
        final URI originalUri = request.getURI();
        String serviceName = originalUri.getHost();  // ②
        Assert.state(serviceName != null,
                "Request URI does not contain a valid hostname: " + originalUri);
        return this.loadBalancer.execute(serviceName,   // ③
                this.requestFactory.createRequest(request, body, execution));
    }
}
```

上述代码中：

① LoadBalancerInterceptor 构造器需要 LoadBalancerClient 和 LoadBalancerRequestFactory 参数（默认会在 LoadBalancerAutoConfiguration 里被构造，开发者可以进行覆盖）。前者根据负载均衡请求（LoadBalancerRequest）和服务名做真正的服务调用，后者构造负载均衡请求

（LoadBalancerRequest），构造过程中会使用 LoadBalancerRequestTransformer 对请求做一些自定义的转换操作（默认情况下，LoadBalancerRequestTransformer 接口无任何实现类，开发者可以根据业务构造 Bean 进行 Request 的转换操作）。

② 服务名使用 URI 中的 host 信息。

③ 使用 LoadBalancerClient 客户端负载均衡器做真正的服务调用。

3. LoadBalancerClient

LoadBalancerClient（客户端负载均衡器）会根据负载均衡请求和服务名执行真正的负载均衡操作，在介绍 SCL 类和接口关系时也提到过这个接口，该接口的具体定义如下：

```java
public interface LoadBalancerClient extends ServiceInstanceChooser {

    /**
     * 使用负载均衡得到的 ServiceInstance 为指定的服务执行请求
     * @param serviceId 服务名
     * @param request 负载均衡请求
     * @param <T> response 返回的具体类型
     * @throws IOException 执行过程中可能发生的 IO 异常
     * @return 基于选中的服务实例 ServiceInstance 在 LoadBalancerRequest 回调中返回的结果
     */
    <T> T execute(String serviceId, LoadBalancerRequest<T> request) throws IOException;

    /**
     * 使用负载均衡得到的 ServiceInstance 为指定的服务执行请求
     * @param serviceId 服务名
     * @param serviceInstance 服务实例
     * @param request 负载均衡请求
     * @param <T> response 返回的具体类型
     * @throws IOException 执行过程中可能发生的 IO 异常
     * @return 基于选中的服务实例 ServiceInstance 在 LoadBalancerRequest 回调中返回的结果
     */
    <T> T execute(String serviceId, ServiceInstance serviceInstance,
            LoadBalancerRequest<T> request) throws IOException;

    /**
```

```
 * 使用服务实例 ServiceInstance 中的 host 和 port 属性构造出真正的 URI。
 * 比如 http://myservice/path/to/service 这个 URI 里的 myservice 服务对应的 host 是
 * 127.0.0.1，port 为 8080，最后会被构造成真正的 URI：
 * http://127.0.0.1:8080/path/to/service
 * @param instance 服务实例，会根据服务实例里的属性替换带有服务的 URI 里的服务名
 * @param original 带有服务名的老的 URI
 * @return 重新构造的 URI
 */
URI reconstructURI(ServiceInstance instance, URI original);
}
```

- reconstructURI 方法。这个方法用于重新构造 URI。比如，要访问 nacos-provider-lb 服务下的 "/" 路径，这个 URI 为 http://nacos-provider-lb/。nacos-provider-lb 服务在注册中心有 10 个服务实例，某个服务实例 ServiceInstance 的 IP 为 192.168.1.100，端口为 8080。那么重新构造的真正 URI 为 http://192.168.1.100:8080/。
- execute 方法。有两个重载方法，其中一个方法比另外一个方法多了 ServiceInstance 服务实例参数。没有 ServiceInstance 参数的方法内部会通过 choose 方法（父接口 ServiceInstanceChooser 提供）使用负载均衡算法得到一个 ServiceInstance，然后调用带有 ServiceInstance 参数的 execute 方法。

execute 方法的源码如下：

```
default <T> T execute(String serviceId, LoadBalancerRequest<T> request) throws IOException {
    ServiceInstance serviceInstance = choose(serviceId);
    return execute(serviceId, serviceInstance, request);
}

<T> T execute(String serviceId, ServiceInstance serviceInstance,
        LoadBalancerRequest<T> request) throws IOException;
```

LoadBalancerRequest 表示一次负载均衡请求，会被 LoadBalancerRequestFactory 构造。构造出的负载均衡请求实现类是 ServiceRequestWrapper（内部基于服务实例和请求信息构造出真正的 URI）。然后根据 LoadBalancerRequestTransformer 做二次加工。LoadBalancerRequest 接口定义如下：

```
public interface LoadBalancerRequest<T> {
    T apply(ServiceInstance instance) throws Exception;
}
```

LoadBalacerRequestFactory 构造负载均衡请求的过程如下:

```
public class LoadBalancerRequestFactory {
    ...
    public LoadBalancerRequest<ClientHttpResponse> createRequest(
    final HttpRequest request, final byte[] body,
    final ClientHttpRequestExecution execution) {
        return instance -> {
            HttpRequest serviceRequest = new ServiceRequestWrapper(request, instance,
                    this.loadBalancer);
            if (this.transformers != null) {
                for (LoadBalancerRequestTransformer transformer: this.transformers) {
                    serviceRequest = transformer.transformRequest(serviceRequest,
                            instance);
                }
            }
            return execution.execute(serviceRequest, body);
        };
    }
}
```

LoadBalancerClient 默认的实现类为基于 SCL 的 BlockingLoadBalancerClient，其定义如下：

```
public class BlockingLoadBalancerClient implements LoadBalancerClient {

    private final LoadBalancerClientFactory loadBalancerClientFactory;

    public BlockingLoadBalancerClient(
            LoadBalancerClientFactory loadBalancerClientFactory) {
        this.loadBalancerClientFactory = loadBalancerClientFactory; // ①
    }
```

```java
@Override
public <T> T execute(String serviceId, LoadBalancerRequest<T> request)
        throws IOException {
    ServiceInstance serviceInstance = choose(serviceId); // ②
    if (serviceInstance == null) {
        throw new IllegalStateException("No instances available for " + serviceId);
    }
    return execute(serviceId, serviceInstance, request);
}

@Override
public <T> T execute(String serviceId, ServiceInstance serviceInstance,
        LoadBalancerRequest<T> request) throws IOException {
    try {
        return request.apply(serviceInstance); // ③
    }
    catch (IOException iOException) {
        throw iOException;
    }
    catch (Exception exception) {
        ReflectionUtils.rethrowRuntimeException(exception);
    }
    return null;
}

@Override
public URI reconstructURI(ServiceInstance serviceInstance, URI original) {
    return LoadBalancerUriTools.reconstructURI(serviceInstance, original); // ④
}

@Override
public ServiceInstance choose(String serviceId) {
    ReactiveLoadBalancer<ServiceInstance> loadBalancer = loadBalancerClientFactory
            .getInstance(serviceId);// ⑤
    if (loadBalancer == null) {
        return null;
```

```
        }
        Response<ServiceInstance> loadBalancerResponse = Mono.from(loadBalancer.choose())
                .block();
        if (loadBalancerResponse == null) {
            return null;
        }
        return loadBalancerResponse.getServer();
    }
}
```

上述代码中：

① BlockingLoadBalancerClient 构造函数依赖 LoadBalancerClientFactory。LoadBalancer-ClientFactory 是一个用于创建 ReactiveLoadBalancer 的工厂类，LoadBalancerClientFactory 内部维护着一个 Map，该 Map 用于保存各个服务的 ApplicationContext（Map 的 key 是服务名）。每个 ApplicationContext 内部维护对应服务的一些配置和 Bean。

② 没有 ServiceInstance 参数的 execute 方法内部会调用 choose 方法获取 ServiceInstance，然后调用另外一个重载的 execute 方法。

③ 有 ServiceInstance 参数的 execute 方法把负载均衡的操作直接委托给 LoadBalancerRequest 负载均衡请求处理。

④ 根据 URI 和找到的服务实例 ServiceInstance 重新构造一个 URI，这个过程被封装在 LoadBalancerUriTools 工具类里。

⑤ 代码②处提到的 choose 方法会返回服务实例 ServiceInstance。choose 方法首先会根据服务名和 LoadBalancerClientFactory 得到该服务名所对应的 ReactiveLoadBalancer Bean，然后调用 ReactiveLoadBalancer 的 choose 方法得到服务实例 ServiceInstance。

4. Spring Cloud LoadBalancer 总结

下面对 Spring Cloud LoadBalancer 内容做个总结。

① @LoadBalanced 注解修饰 RestTemplate 后，会根据 RestTemplateCustomized 注解给 RestTemplate 做定制化操作。这个定制化操作一定含有一个添加 LoadBalancerInterceptor 负载均衡拦截器的操作。此外，我们还可以扩展添加符合业务需求的自定义定制化操作。整个过程

如图 3-2 所示。

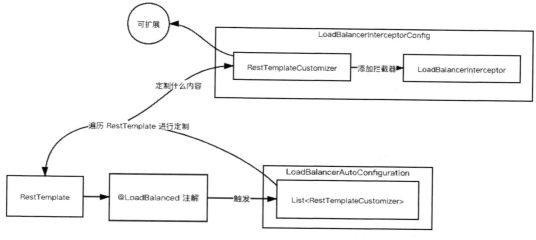

图 3-2

② LoadBalancerInterceptor 负载均衡拦截器拦截的背后会通过 LoadBalancerClient 的 execute 方法完成最终的调用。execute 需要两个参数，一个是服务名，请求信息中的 host 即表示服务名；另一个是负载均衡请求（LoadBalancerRequest），通过 LoadBalancer- RequestFactory 构造完成。整个过程如图 3-3 所示。

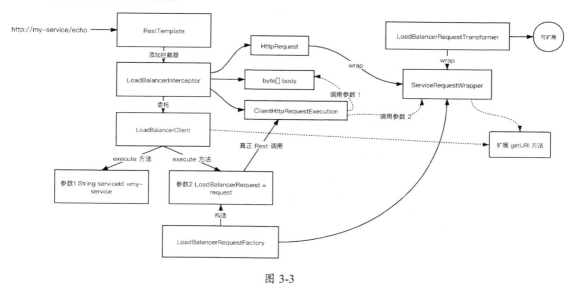

图 3-3

Spring Cloud LoadBalancer 还提供了 @LoadBalancerClient 注解用于进行自定义的配置操作，如自定义负载均衡算法（默认是轮询算法）、自定义 ServiceInstanceListSupplier（默认会缓存服务实例列表）。

```
@Configuration
@LoadBalancerClient(value = "nacos-provider-lb", configuration =
MyLoadBalancerConfiguration.class)
class LoadBalanceConfiguration { }
```

@LoadBalancerClient 注解有 3 个属性，分别是 value:String、name:String 和 configuration:Class[]。name 和 value 属性表示同一个含义，即服务名，且只能设置其中一个属性。上述代码中，nacos-provider-lb 对应的是服务名，每个服务名拥有单独的自定义配置。

 提示：@LoadBalancerClients 注解的 defaultConfiguration 属性表示默认的配置类，所有的 BlockingLoadBalancerClient 都会使用这些配置类里的配置。

configuration 属性表示配置类，配置类中返回的 Bean 会替换 LoadBalancerClientConfiguration 配置类中已经存在的 Bean（前文提到 LoadBalancerClientFactory 内部维护着一个 Map，用于保存各个服务的 ApplicationContext。每个 ApplicationContext 构造的时候都会加上 LoadBalancer-ClientConfiguration 配置类）。比如，LoadBalancerClientConfiguration 配置类中负载均衡策略为默认的轮询策略：

```
// LoadBalancerClientConfiguration.java
@Bean
@ConditionalOnMissingBean    //①
public ReactorLoadBalancer<ServiceInstance> reactorServiceInstanceLoadBalancer(
        Environment environment,
        LoadBalancerClientFactory loadBalancerClientFactory) {
    String name = environment.getProperty(LoadBalancerClientFactory.PROPERTY_NAME);
    return new RoundRobinLoadBalancer    // ②
(loadBalancerClientFactory.getLazyProvider(name,
        ServiceInstanceListSupplier.class), name);
}
```

上述代码中：

① 默认的负载均衡策略 Bean 被 ConditionalOnMissingBean 注解修饰，表示开发者配置的配置类优先级更高。

② 默认使用轮询负载均衡策略。

我们可以在 MyLoadBalancerConfiguration 配置类中定义一个新的 ReactorLoadBalancer 来覆盖默认的负载均衡策略：

```java
public class MyLoadBalancerConfiguration {

    @Bean
    public ReactorLoadBalancer<ServiceInstance> reactorServiceInstanceLoadBalancer(
        Environment environment,
        LoadBalancerClientFactory loadBalancerClientFactory) {
        String name = environment.getProperty(LoadBalancerClientFactory.PROPERTY_NAME);
        return new RandomLoadBalancer(loadBalancerClientFactory.getLazyProvider(name,
            ServiceInstanceListSupplier.class), name);
    }

}
```

RandomLoadBalancer 是一个自定义的实现随机算法的负载均衡策略：

```java
public class RandomLoadBalancer implements ReactorServiceInstanceLoadBalancer {

    private ObjectProvider<ServiceInstanceListSupplier> serviceInstanceListSupplierProvider;

    private final String serviceId;

    private final Random random;

    public RandomLoadBalancer(
        ObjectProvider<ServiceInstanceListSupplier> serviceInstanceListSupplierProvider, String serviceId) {
        this.serviceInstanceListSupplierProvider = serviceInstanceListSupplierProvider;
```

```
    this.serviceId = serviceId;
    this.random = new Random();
}

@Override
public Mono<Response<ServiceInstance>> choose(Request request) {
    ServiceInstanceListSupplier supplier = serviceInstanceListSupplierProvider
        .getIfAvailable(NoopServiceInstanceListSupplier::new);
    return supplier.get().next().map(this::getInstanceResponse);
}

private Response<ServiceInstance> getInstanceResponse(
    List<ServiceInstance> instances) {
    if (instances.isEmpty()) {
        return new EmptyResponse();
    }
    ServiceInstance instance = instances.get(random.nextInt(instances.size()));

    return new DefaultResponse(instance);
}
}
```

3.3 Netflix Ribbon 负载均衡

Netflix Ribbon 是 Netflix 开源的客户端负载均衡组件，在 Spring Cloud LoadBalancer 出现之前，它是 Spring Cloud 生态里唯一的负载均衡组件。目前市场上绝大多数的 Spring Cloud 应用还是使用 Ribbon 作为其负载均衡组件。

3.3.1 RibbonLoadBalancerClient

Netflix Ribbon 对应的 LoadBalancerClient 实现类为 RibbonLoadBalancerClient，定义如下：

```
public class RibbonLoadBalancerClient implements LoadBalancerClient {
```

```java
    private SpringClientFactory clientFactory;    // ①

    public RibbonLoadBalancerClient(SpringClientFactory clientFactory) {
        this.clientFactory = clientFactory;    // ②
    }

    @Override
    public URI reconstructURI(ServiceInstance instance, URI original) {
        Assert.notNull(instance, "instance can not be null");
        String serviceId = instance.getServiceId();
        RibbonLoadBalancerContext context = this.clientFactory
                .getLoadBalancerContext(serviceId);

        URI uri;
        Server server;    //③
        if (instance instanceof RibbonServer) {
            RibbonServer ribbonServer = (RibbonServer) instance; // ④
            server = ribbonServer.getServer();
            uri = updateToSecureConnectionIfNeeded(original, ribbonServer);
        }
        else {
            server = new Server(instance.getScheme(), instance.getHost(),
                    instance.getPort());    // ⑤
            IClientConfig clientConfig = clientFactory.getClientConfig(serviceId);
            ServerIntrospector serverIntrospector = serverIntrospector(serviceId); //⑥
            uri = updateToSecureConnectionIfNeeded(original, clientConfig,
                    serverIntrospector, server);
        }
        return context.reconstructURIWithServer(server, uri); // ⑦
    }

    @Override
    public ServiceInstance choose(String serviceId) {
        return choose(serviceId, null);
    }
```

```java
public ServiceInstance choose(String serviceId, Object hint) {
    Server server = getServer(getLoadBalancer(serviceId), hint); // ⑧
    if (server == null) {
        return null;
    }
    return new RibbonServer(serviceId, server, isSecure(server, serviceId),
            serverIntrospector(serviceId).getMetadata(server)); //⑨
}

@Override
public <T> T execute(String serviceId, LoadBalancerRequest<T> request)
        throws IOException {
    return execute(serviceId, request, null);
}

public <T> T execute(String serviceId, LoadBalancerRequest<T> request, Object hint)
        throws IOException {
    ILoadBalancer loadBalancer = getLoadBalancer(serviceId);
    Server server = getServer(loadBalancer, hint); // ⑩
    if (server == null) {
        throw new IllegalStateException("No instances available for " + serviceId);
    }
    RibbonServer ribbonServer = new RibbonServer(serviceId, server,
            isSecure(server, serviceId),
            serverIntrospector(serviceId).getMetadata(server));

    return execute(serviceId, ribbonServer, request);
}

@Override
public <T> T execute(String serviceId, ServiceInstance serviceInstance,
        LoadBalancerRequest<T> request) throws IOException {
    Server server = null;
    if (serviceInstance instanceof RibbonServer) {
        server = ((RibbonServer) serviceInstance).getServer();
    }
```

```
        if (server == null) {
            throw new IllegalStateException("No instances available for " + serviceId);
        }

        RibbonLoadBalancerContext context = this.clientFactory
                .getLoadBalancerContext(serviceId);
        RibbonStatsRecorder statsRecorder = new RibbonStatsRecorder(context, server);// ⑪

        try {
            T returnVal = request.apply(serviceInstance); // ⑫
            statsRecorder.recordStats(returnVal); //⑬
            return returnVal;
        }
        catch (IOException ex) {
            statsRecorder.recordStats(ex); // ⑭
            throw ex;
        }
        catch (Exception ex) {
            statsRecorder.recordStats(ex); // ⑮
            ReflectionUtils.rethrowRuntimeException(ex);
        }
        return null;
    }
    ...
}
```

上述代码中：

① SpringClientFactory 是一个与 LoadBalancerClientFactory 作用类似的工厂类，其内部维护着一个 Map，这个 Map 用于保存各个服务的 ApplicationContext（Map 的 key 表示服务名）。每个 ApplicationContext 内部维护对应服务的一些配置和 Bean。

② SpringClientFactory 在 RibbonAutoConfiguration 自动化配置类中被构造，可以通过构造器注入的方式注入。

③ reconstructURI 方法与 Spring Cloud LoadBalancer 中的 BlockingLoadBalancerClient 实现完全不一样。BlockingLoadBalancerClient 直接委托给 LoadBalancerUriTools#reconstructURI 方

法实现，其内部使用 ServiceInstance 进行相应的属性替换；而 RibbonLoadBalancerClient 内部基于新的类 com.netflix.loadbalancer.Server（表示一个服务器实例，内部有 host、port、schema、zone 等属性）来实现。

④ RibbonServer 是 RibbonLoadBalancerClient 的内部类，其实现了 ServiceInstance 接口，内部维护着一个 Server 属性，同时还有其他 3 个属性：serviceId:String（服务名）、secure:boolean（是否使用 HTTPS）和 metadata:Map<String, String>（服务器实例的元数据）。

⑤ 基于 ServiceInstance 构造 Server。

⑥ ServerIntrospector 接口可以根据 Server 调用 isSecure 和 getMetadata 方法获取 secure 和 metadata 信息（Server 类跟 ServiceInstance 没有直接关系，通过 ServerIntrospector 接口得到的这两个属性跟 ServiceInstance 接口中定义的 isSecure 和 getMetadata 方法返回的结果匹配）。

⑦ 使用 RibbonLoadBalancerContext#reconstructURIWithServer 方法基于 Server 和老的 URI 重新构造新的 URI。这里的 RibbonLoadBalancerContext 是服务名对应的 ApplicationContext 里存在的实例。

⑧ choose 方法返回负载均衡策略得到的最终实例，将负载均衡的操作委托给 ILoadBalancer 接口的实现类，默认的实现为 ZoneAwareLoadBalancer。这里的 ILoadBalancer 是服务名对应的 ApplicationContext 里存在的实例。

⑨ 返回的服务实例 ServiceInstance 使用 RibbonServer，secure 和 metadata 属性使用 ServerIntrospector 获取。

⑩ 服务调用的过程与 choose 方法选择服务实例的步骤一致，根据 ILoadBalancer 得到最终的实例，再调用另外一个 execute 重载方法。

⑪ 每次服务调用都会使用 RibbonStatsRecorder 内部的 ServerStats 对象进行数据统计。每个实例都有独立的 ServerStats 对象。

⑫ 真正的服务调用操作，使用 LoadBalancerRequest 完成。

⑬ 服务调用成功，进行状态记录。

⑭ 服务调用发生 IOException 异常，进行异常状态记录。

⑮ 服务调用发生 Exception 异常，进行异常状态记录。

SCL 和 Netflix Ribbon 对应功能的对比如表 3-1 所示。

表 3-1

功能/组件	Spring Cloud LoadBalancer	Netflix Ribbon
负载均衡	ReactiveLoadBalancer & ServiceInstanceListSupplier	ILoadBalancer & IRule
服务实例	ServiceInstance	ServiceInstance & RibbonServer & Server & ServerIntrospector
服务调用统计信息	-	ServerStats & LoadBalancerStats

接下来对 Netflix Ribbon 的核心类进行分析。

3.3.2 RibbonServer 和 Server

Spring Cloud LoadBalancer 对服务实例的处理封装到 ServiceInstance 中，Netflix Ribbon 在 ServiceInstance 的基础上引入了 Server 的概念。

Server 表示一个服务器实例。比如，一台阿里云 ECS，ID 为 iZ2ze1orftocdqwA12F1，IP 为 123.57.68.101，注册到注册中心的服务名是 my-service，端口为 8080，这些值都可以被设置到 Server 对应的属性中。

Server 类定义如下：

```
public class Server {

    public static final String UNKNOWN_ZONE = "UNKNOWN";
    private String host; // 域名
    private int port = 80; // 端口，默认 80
    private String scheme; // schema，如 http 或 https
    private volatile String id; // 服务器实例 ID
    private volatile boolean isAliveFlag; // 是否还存活，如 ping 不通，则可能会设置为 false
    private String zone = UNKNOWN_ZONE; // 服务实例所在的 zone
    private volatile boolean readyToServe = true; // 是否可以对外提供服务

    private MetaInfo simpleMetaInfo = new MetaInfo() { // meta 信息，云厂商可以有不同的操作。默认除 getInstanceId 方法返回服务器实例 ID 外，其他都返回 null。
        @Override
        public String getAppName() {
            return null;
```

```
        }

        @Override
        public String getServerGroup() {
            return null;
        }

        @Override
        public String getServiceIdForDiscovery() {
            return null;
        }

        @Override
        public String getInstanceId() {
            return id;
        }
    };
    ...
}
```

RibbonServer 是 RibbonLoadBalancerClient 中的内部类，其内部持有 Server 属性，实现 ServiceInstance 接口。ServiceInstance 接口中定义的方法使用 Server 对应的属性实现，getServiceId()、isSecure()和 getMetadata()这 3 个方法直接使用 serviceId、secure 和 metadata 属性返回。

```
// RibbonLoadBalancerClient.java
public static class RibbonServer implements ServiceInstance {

    private final String serviceId;

    private final Server server;

    private final boolean secure;

    private Map<String, String> metadata;

    public RibbonServer(String serviceId, Server server) {
```

```java
        this(serviceId, server, false, Collections.emptyMap());
    }

    public RibbonServer(String serviceId, Server server, boolean secure,
            Map<String, String> metadata) {
        this.serviceId = serviceId;
        this.server = server;
        this.secure = secure;
        this.metadata = metadata;
    }

    @Override
    public String getInstanceId() {
        return this.server.getId();
    }

    @Override
    public String getServiceId() {
        return this.serviceId;
    }

    @Override
    public String getHost() {
        return this.server.getHost();
    }

    @Override
    public int getPort() {
        return this.server.getPort();
    }

    @Override
    public boolean isSecure() {
        return this.secure;
    }

    @Override
```

```java
public URI getUri() {
    return DefaultServiceInstance.getUri(this);
}

@Override
public Map<String, String> getMetadata() {
    return this.metadata;
}

public Server getServer() {
    return this.server;
}

@Override
public String getScheme() {
    return this.server.getScheme();
}

@Override
public String toString() {
    final StringBuilder sb = new StringBuilder("RibbonServer{");
    sb.append("serviceId='").append(serviceId).append('\'');
    sb.append(", server=").append(server);
    sb.append(", secure=").append(secure);
    sb.append(", metadata=").append(metadata);
    sb.append('}');
    return sb.toString();
}
}
```

secure 和 metadata 属性由 ServerIntrospector 接口返回，各个注册中心实现各自的 ServerIntrospector 接口。

3.3.3　ServerIntrospector

ServerIntrospector 接口基于 Server 对象确定服务实例是否是 HTTPS 协议，以及获取服务

实例中的元数据信息,其定义如下:

```java
public interface ServerIntrospector {
    // 是否使用 HTTPS 协议
    boolean isSecure(Server server);
    // 获取元数据信息
    Map<String, String> getMetadata(Server server);

}
```

各个注册中心都有具体的实现类。比如 Nacos 注册中心提供的 NacosServerIntrospector；Eureka 注册中心提供的 EurekaServerIntrospector；Consul 注册中心提供的 ConsulServerIntrospector；Zookeeper 注册中心提供的 ZookeeperServerIntrospector。其中，NacosServerIntrospector 的实现源码如下:

```java
public class NacosServerIntrospector extends DefaultServerIntrospector {

    @Override
    public Map<String, String> getMetadata(Server server) {
        if (server instanceof NacosServer) { // ①
            return ((NacosServer) server).getMetadata();
        }
        return super.getMetadata(server); // ②
    }

    @Override
    public boolean isSecure(Server server) {
        if (server instanceof NacosServer) { // ③
            return Boolean.valueOf(((NacosServer) server).getMetadata().get("secure"));
        }

        return super.isSecure(server); // ④
    }

}
```

上述代码中：

① NacosServerIntrospector 获取元数据只针对 NacosServer 类型的 Server 做处理，直接返回 NacosServer 里的 metadata 元数据。

② 如果不是 NacosServer，调用父类 DefaultServerIntrospector 的 getMetadata 方法，返回空 Map。

③ NacosServerIntrospector 只针对 NacosServer 类型判断是否是 HTTPS 协议，可根据 getMetadata 获取元数据 map 里的 secure 属性来实现。

④ 如果不是 NacosServer，调用父类 DefaultServerIntrospector 的 isSecure 方法，根据配置文件配置的端口，默认为 443 和 8443（配置 key 为 ribbon.secure-ports）。

3.3.4 ILoadBalancer

RibbonLoadBalancerClient 内部执行负载均衡请求的时候会先根据 ILoadBalancer#chooseServer 方法得到最终的 Server 实例，然后根据 LoadBalancerRequest 执行真正的服务调用。

```java
// RibbonLoadBalancerClient.java
protected Server getServer(ILoadBalancer loadBalancer, Object hint) {
    if (loadBalancer == null) {
        return null;
    }
    return loadBalancer.chooseServer(hint != null ? hint : "default");
}
```

ILoadBalancer 的 chooseServer 方法内部将负载均衡操作交给 IRule 接口完成，定义如下：

```java
public Server chooseServer(Object key) {
    if (counter == null) {
        counter = createCounter();
    }
    counter.increment();
    if (rule == null) {
        return null;
    } else {
        try {
```

```
            return rule.choose(key);
        } catch (Exception e) {
            logger.warn("LoadBalancer [{}]: Error choosing server for key {}",
name, key, e);
            return null;
        }
    }
}
```

ILoadBalancer 接口是 Netflix Ribbon 用于实现负载均衡的核心接口，其内部有一套完善的机制用于实现负载均衡：

① 维护所有的 Server 列表，可以添加、删除 Server 或更新 Server 的状态。

② 监听机制。当 Server 列表（ServerListChangeListener）或 Server 状态（ServerStatusChangeListener）发生变化的时候，会产生相应的事件。

③ 可自定义的负载均衡策略 IRule。

④ 可自定义的服务实例健康状态检查方式 IPing（针对单个 Server 如何检查是否健康）。

⑤ 可自定义的服务实例健康状态检查策略 IPingStrategy（针对所有的 Server 列表如何检查，比如，可以过滤特定的 Server，可以并行检查）。

⑥ 可自定义的 Server 列表过滤器（ServerListFilter），可以基于 Server 列表过滤出新的 Server 列表。

⑦ 可自定义的 Server 列表获取方式（ServerList），用于获取注册中心服务对应的实例。

⑧ 可自定义的 Server 列表更新机制（ServerListUpdater），默认会使用一个调度线程池每 30s 从注册中心获取一次实例信息。

⑨ 负载均衡器中各个服务实例当前的统计信息（LoadBalancerStats）。

Netflix Ribbon 提供了@RibbonClient 注解，用于进行自定义的配置操作，上述所有可自定义的组件均可替换：

```
@Configuration
@RibbonClient(name = "nacos-provider-lb", configuration =
MyLoadBalancerConfiguration.class)
```

```
class LoadBalanceConfiguration { }
```

@RibbonClient 注解与@LoadBalancerClient 注解的作用几乎一样，同样拥有 value:String、name:String 和 configuration:Class[]这 3 个属性。name 和 value 属性表示同一个含义，即服务名，且只能设置其中一个属性。上述代码中，nacos-provider-lb 对应的是服务名，每个服务名拥有单独的自定义配置。

> **提示**：@RibbonClients 注解的 defaultConfiguration 属性表示默认的配置类，所有的 RibbonLoadBalancerClient 都会使用这些配置类里的配置。

configuration 属性表示配置类，配置类中返回的 Bean 会替换 RibbonClientConfiguration 配置类中已经存在的 Bean（前文提到 SpringClientFactory 内部维护着一个 Map，这个 Map 用于保存各个服务的 ApplicationContext。每个 ApplicationContext 构造的时候都会加上 RibbonClient-Configuration 配置类）。比如 RibbonClientConfiguration 配置类中负载均衡策略为 ZoneAvoidanceRule（一种基于可用区 zone 和过滤规则的负载均衡策略）：

```java
// RibbonClientConfiguration.java
@Bean
@ConditionalOnMissingBean
public IRule ribbonRule(IClientConfig config) {
    if (this.propertiesFactory.isSet(IRule.class, name)) {
        return this.propertiesFactory.get(IRule.class, config, name);
    }
    ZoneAvoidanceRule rule = new ZoneAvoidanceRule();
    rule.initWithNiwsConfig(config);
    return rule;
}
```

默认的 IPing 规则及 Server 列表更新机制定义如下：

```java
// RibbonClientConfiguration.java
@Bean
@ConditionalOnMissingBean
public IPing ribbonPing(IClientConfig config) {
    if (this.propertiesFactory.isSet(IPing.class, name)) {
        return this.propertiesFactory.get(IPing.class, config, name);
```

```
    }
    return new DummyPing();
}
...
@Bean
@ConditionalOnMissingBean
public ServerListUpdater ribbonServerListUpdater(IClientConfig config) {
    return new PollingServerListUpdater(config);
}
```

3.3.5 ServerList

ServerList 接口是一个获取 Server 列表的规范,其内部定义了两个方法:

```
public interface ServerList<T extends Server> {
    // 获取初始化 Server 列表
    public List<T> getInitialListOfServers();
    // 获取最新的 Server 列表
    public List<T> getUpdatedListOfServers();
}
```

各个注册中心都有 ServerList 实现类。比如,Nacos 注册中心提供的 NacosServerList;Eureka 注册中心提供的 ConfigurationBasedServerList;Consul 注册中心提供的 ConsulServerList;Zookeeper 注册中心提供的 ZookeeperServerList。其中,NacosServerList 的实现源码如下:

```
public class NacosServerList extends AbstractServerList<NacosServer> {

    private NacosDiscoveryProperties discoveryProperties;

    private String serviceId;

    public NacosServerList(NacosDiscoveryProperties discoveryProperties) {
        this.discoveryProperties = discoveryProperties;
    }

    @Override
```

```java
public List<NacosServer> getInitialListOfServers() {
    return getServers();
}

@Override
public List<NacosServer> getUpdatedListOfServers() {
    return getServers();
}

private List<NacosServer> getServers() {
    try {
        String group = discoveryProperties.getGroup();
        List<Instance> instances = discoveryProperties.namingServiceInstance()
                .selectInstances(serviceId, group, true);
        return instancesToServerList(instances);
    }
    catch (Exception e) {
        throw new IllegalStateException(
          "Can not get service instances from nacos, serviceId=" + serviceId, e);
    }
}

private List<NacosServer> instancesToServerList(List<Instance> instances) {
    List<NacosServer> result = new ArrayList<>();
    if (CollectionUtils.isEmpty(instances)) {
        return result;
    }
    for (Instance instance : instances) {
        result.add(new NacosServer(instance));
    }

    return result;
}

public String getServiceId() {
    return serviceId;
```

```
    }

    @Override
    public void initWithNiwsConfig(IClientConfig iClientConfig) {
        this.serviceId = iClientConfig.getClientName();
    }
}
```

Spring Cloud Alibaba Nacos Discovery 内部定义了 NacosRibbonClientConfiguration 自动化配置类，默认会使用 NacosServerIntrospector 和 NacosServerList 替换 RibbonClientConfiguration 中默认的 Default-ServerIntrospector 和 ConfigurationBasedServerList：

```
@Configuration(proxyBeanMethods = false)
@ConditionalOnRibbonNacos
public class NacosRibbonClientConfiguration {

    @Autowired
    private PropertiesFactory propertiesFactory;

    @Bean
    @ConditionalOnMissingBean
    public ServerList<?> ribbonServerList(IClientConfig config,
            NacosDiscoveryProperties nacosDiscoveryProperties) {
        if (this.propertiesFactory.isSet(ServerList.class, config.getClientName())) {
            ServerList serverList = this.propertiesFactory.get(ServerList.class, config,
                    config.getClientName());
            return serverList;
        }
        NacosServerList serverList = new NacosServerList(nacosDiscoveryProperties);
        serverList.initWithNiwsConfig(config);
        return serverList;
    }

    @Bean
    @ConditionalOnMissingBean
```

```
    public NacosServerIntrospector nacosServerIntrospector() {
        return new NacosServerIntrospector();
    }
}
```

3.3.6 ServerListUpdater

Server 列表更新机制 ServerListUpdater 对于服务调用非常重要,它决定着 Server 列表是否更新及时,如果 Server 列表还保留着已经下线的 Server,那么调用会直接报错。

默认的 ServerListUpdater 为 PollingServerListUpdater,内部使用调度线程池默认每隔 30s 进行一个 UpdateAction 操作:

```
// PollingServerListUpdater.java
public class PollingServerListUpdater implements ServerListUpdater {

    ...

    @Override
    public synchronized void start(final UpdateAction updateAction) {
        if (isActive.compareAndSet(false, true)) { // ①
            final Runnable wrapperRunnable = new Runnable() {
                @Override
                public void run() {
                    if (!isActive.get()) {
                        if (scheduledFuture != null) {
                            scheduledFuture.cancel(true); // ②
                        }
                        return;
                    }
                    try {
                        updateAction.doUpdate(); // ③
                        lastUpdated = System.currentTimeMillis();
                    } catch (Exception e) {
                        logger.warn("Failed one update cycle", e);
                    }
```

```
            }
        };

        scheduledFuture = getRefreshExecutor().scheduleWithFixedDelay(
                wrapperRunnable,
                initialDelayMs,
                refreshIntervalMs,
                TimeUnit.MILLISECONDS
        ); //④
    } else {
        logger.info("Already active, no-op");
    }
}

...

}

// DynamicServerListLoadBalancer.java
protected final ServerListUpdater.UpdateAction updateAction = new
ServerListUpdater.UpdateAction() {
    @Override
    public void doUpdate() {
        updateListOfServers(); // ⑤
    }
};

public void updateListOfServers() {
    List<T> servers = new ArrayList<T>();
    if (serverListImpl != null) {
        servers = serverListImpl.getUpdatedListOfServers(); //⑥
        LOGGER.debug("List of Servers for {} obtained from Discovery client: {}",
                getIdentifier(), servers);

        if (filter != null) {
            servers = filter.getFilteredListOfServers(servers); // ⑦
```

```
            LOGGER.debug("Filtered List of Servers for {} obtained from Discovery 
client: {}",
                    getIdentifier(), servers);
        }
    }
    updateAllServerList(servers); // ⑧
}
```

上述代码中：

① 确定是否已经开启调度线程池。

② 如果开关还未启动，取消调度线程池任务。

③ Runnable 内部调用 updateAction 的 doUpdate 方法，开始更新操作。具体的更新在 updateListOfServers 方法中。

④ 开启调度线程池，默认的时间间隔是 3000 ms。

⑤ DynamicServerListLoadBalancer 内部维护着一个 UpdateAction 属性，其内部会调用 updateListOfServers 方法。

⑥ 调用 ServerList 的 getUpdatedListOfServers 方法获取最新的 Server 列表。

⑦ 如果 ServerListFilter 属性存在，进行 Server 过滤。

⑧ 更新 Server 列表，同时记录统计信息。

使用 ServerListUpdater 一定要注意合理配置参数，想象有这么一个场景（IPing 默认为 DummyPing（DummyPing 的 isAlive 方法直接返回 true，不做健康检查））：用户注册一个服务且只有两个实例，客户端进行调用，其中 1 个实例撤销注册，这时最坏的情况下客户端要等待 30s 才会知道有一个实例已经撤销，在这 30s 内，如果有请求进来，就会发生调用报错。

3.3.7　ServerStats

ServerStats 内部记录了每个 Server 的统计信息和一些配置信息。

① 总请求数（totalRequests）。

② 请求错误数量（successiveConnectionFailureCount）。注意：这里的请求错误数量会一

直递增，如果某个请求成功，会重置数量为 0。

③ 请求错误阈值（connectionFailureThreshold），默认为 3 个。当请求错误数达到该阈值时，该实例会进入 blackOut（停电，不对外提供服务）状态。

④ 熔断时间因子（circuitTrippedTimeoutFactor），默认为 10。

请求错误数与请求错误阈值差的绝对值（超过 16 就取 16）左移 1 位，再乘以熔断时间因子，得到熔断时间：

```
int diff = (failureCount - threshold) > 16 ? 16 : (failureCount - threshold);
int blackOutSeconds = (1 << diff) * circuitTrippedTimeoutFactor.get();
```

⑤ 熔断最大时间（maxCircuitTrippedTimeout），默认为 30s。

当根据熔断时间因子得到的熔断时间大于熔断最大时间时，取熔断最大时间：

```
if (blackOutSeconds > maxCircuitTrippedTimeout.get()) {
    blackOutSeconds = maxCircuitTrippedTimeout.get();
}
```

⑥ 总熔断停电时间（totalCircuitBreakerBlackOutPeriod）。

⑦ 最近一次连接建立的时间戳（lastAccessedTimestamp）。

⑧ 第一次发生请求的时间戳（firstConnectionTimestamp）。

⑨ 响应时间（RT），包括每次请求最大 RT、最小 RT、平均 RT 和标准差 RT。

⑩ 最近一次连接失败时间戳（lastConnectionFailedTimestamp）。

⑪ 当前活跃请求数（activeRequestsCount）即连接刚建立到请求结束之间算的活跃请求数。

⑫ 活跃请求的计算超时时间（activeRequestsCountTimeout）。如果某个请求超过 60s 后才响应，会将活跃请求数置为 0。

⑬ 最近一次请求的时间戳（lastActiveRequestsCountChangeTimestamp），即连接刚建立会更新，请求结束会更新。

LoadBalancerStats 内部维护着每个 Server 和对应的 ServerStats。ServerStats 所有对外暴露的操作全部通过 LoadBalancerStats 完成，LoadBalancerStats 还会汇总所有的 ServerStats 里的统

计数据。

RibbonClientConfiguration 中默认的负载均衡策略是 ZoneAvoidanceRule。这个 Rule 内部有服务实例的过滤规则，其中有一个规则是 AvailabilityPredicate，这个规则内部会判断 Server 的状态来判断是否过滤掉这个 Server：

```java
// AvailabilityPredicate.java
@Override
public boolean apply(@Nullable PredicateKey input) {
    LoadBalancerStats stats = getLBStats();
    if (stats == null) {
        return true;
    }
    return !shouldSkipServer(stats.getSingleServerStat(input.getServer()));
}

private boolean shouldSkipServer(ServerStats stats) {
    if ((CIRCUIT_BREAKER_FILTERING.get() && stats.isCircuitBreakerTripped())
            || stats.getActiveRequestsCount() >= activeConnectionsLimit.get()) {
        return true;
    }
    return false;
}
```

shouldSkipServer 方法内部会根据 ServerStats 内部的熔断状态选择是否过滤掉 Server。

3.3.8 Netflix Ribbon 配置项总结

Netflix Ribbon 的配置项如表 3-2 所示，合理搭配能够避免一些问题。

表 3-2

配置 key	作　　用	默　认　值
niws.loadbalancer.<service-name>.connection-FailureCountThreshold	请求错误连接数阈值。到达阈值后会进入熔断状态，熔断状态的实例会被 ZoneAvoidanceRule 策略过滤	3 次

续表

配置 key	作用	默认值
niws.loadbalancer.\<service-name\>.circuitTrip-TimeoutFactorSeconds	熔断时间因子。请求错误数与请求错误阈值差的绝对值（超过 16 则取 16）左移 1 位，再乘以熔断时间因子，得到熔断时间	10
niws.loadbalancer.\<service-name\>.circuitTrip-MaxTimeoutSeconds	熔断最大时间。当根据熔断时间因子得到的熔断时间大于熔断最大时间时，取熔断最大时间	30s
niws.loadbalancer.serverStats.expire.minutes	LoadBalancerStats 内部 Server 和 ServerStats 的缓存时间	1800s
niws.loadbalancer.availabilityFilteringRule.filterCircuitTripped	AvailabilityPredicate 内部是否需要过滤进入熔断状态的 Server	true
niws.loadbalancer.availabilityFilteringRule.activeConnectionsLimit	AvailabilityPredicate 内部是否需要过滤出活跃连接数超过该配置值的 Server	Integer.MAX_VALUE
niws.loadbalancer.serverStats.activeRequestsCount.effectiveWindowSeconds	活跃请求的计算超时时间。如果某个请求超过该配置项后才响应，会将活跃请求数置为 0	600s
\<service-name\>.ribbon.zoneAffinity.maxLoadPerServer	平均负载阈值。使用实现类为 ZoneAffinityServerListFilter 的服务列表过滤器 ServerListFilter 的情况下，未被熔断的所有实例的平均负载与该阈值比对的条件可以作为是否需要使用相同可用区的 Server 的条件之一	0.6
\<service-name\>.ribbon.zoneAffinity.maxBlackOutServesrPercentage	熔断实例占比。使用实现类为 ZoneAffinityServerListFilter 的服务列表过滤器 ServerListFilter 的情况下，熔断实例在所有实例下的占比与该阈值比对的条件可以作为是否需要使用相同可用区的 Server 的条件之一	0.8

续表

配置 key	作用	默认值
<service-name>.ribbon.zoneAffinity.minAvailableServers	可用实例阈值（除去熔断实例后的所有实例）。使用实现类为 ZoneAffinityServerListFilter 的服务列表过滤器 ServerListFilter 的情况下，可用实例个数（除去熔断实例后的所有实例）与该阈值比对的条件可以作为是否需要使用相同可用区的 Server 的条件之一	2
<service-name>.（服务名可选）ribbon.ReadTimeout	请求读取超时时间（服务名可以不配置，没有服务的配置表示全局配置，带服务名的表示配置只针对这个服务生效）	1000ms
<service-name>.（服务名可选）ribbon.ConnectTimeout	请求连接超时时间（服务名可以不配置，没有服务的配置表示全局配置，带服务名的表示配置只针对这个服务生效）	1000ms
<service-name>.（服务名可选）ribbon.GZipPayload	是否启用 GZip 压缩（服务名可以不配置，没有服务的配置表示全局配置，带服务名的表示配置只针对这个服务生效）	true
<service-name>.（服务名可选）ribbon.OkToRetryOnAllOperations	是否重试所有的操作（服务名可以不配置，没有服务的配置表示全局配置，带服务名的表示配置只针对这个服务生效）	false
<service-name>.（服务名可选）ribbon.MaxTotalConnections	最大支持的请求数（服务名可以不配置，没有服务的配置表示全局配置，带服务名的表示配置只针对这个服务生效）	200
<service-name>.（服务名可选）ribbon.PoolKeepAliveTime	Keep-Alive 时间设置（服务名可以不配置，没有服务的配置表示全局配置，带服务名的表示配置只针对这个服务生效）	900s

续表

配置 key	作 用	默 认 值
<service-name>.（服务名可选）ribbon.PoolKeepAliveTimeUnits	最大 Keep-Alive 时间单位（服务名可以不配置，没有服务的配置表示全局配置，带服务名的表示配置只针对这个服务生效）	s

由于 Ribbon 相关的配置对于不同的 HTTP 客户端会有不同的作用，本书不再一一说明。如果读者对具体参数的内容感兴趣，可以参考这些类或接口：CommonClientConfigKey、HttpClientConfiguration、OkHttpRibbonConfiguration、HttpClientRibbonConfiguration。

3.3.9 Ribbon 缓存时间

Netflix Ribbon 提供的配置项 `ribbon.ServerListRefreshInterval` 表示客户端主动与注册中心拉取最新的服务实例数据的时间间隔。这个时间间隔的默认配置值是 30s，意味着：如果注册中心里某个服务对应的实例已经下线，但是客户端的刷新时间间隔还未达到默认值，那么客户端就不会触发主动拉取服务实例数据的逻辑，从而导致客户端订阅到已经下线的实例，此时发起服务根据负载均衡策略得到的实例是已经下线的实例，那么服务调用会发送超时异常（这个实例已经下线，发起连接会引起网络连接超时）。

 提示：单击 Nacos Dashboard 提供的服务实例"下线"按钮后并不会立即生效，这是因为 Ribbon 的刷新时间间隔还未达到所导致的结果，并不是 Nacos 的问题。

模拟 Ribbon 缓存时间使用默认值在实例下线后发生调用错误的过程如下：

（1）Ribbon 刷新时间没有配置，默认为 30s。在这个模拟调用错误的过程中，假设刷新时间是每分钟的第 0s 和第 30s 刷新。

（2）第 5s 的时候，服务对应的实例 IP 为 192.168.1.1，出现故障被迫下线。

（3）IP 为 192.168.1.1 的实例下线后，注册中心发现这个 IP 心跳检测异常，在 15s 摘除该实例。

（4）由于第 30s 才主动刷新服务实例，在 5~30s 这段时间内调用到这个 IP 的请求都会报错（在 5~15s 之间的调用错误是注册中心心跳机制的问题导致这个故障实例没下线引起的，在 15~30s 时的调用错误是 Ribbon 客户端没有主动拉取服务注册数据导致客户端获取到故障实例引起的）。

Ribbon 主动拉取注册中心数据关键的代码在 PollingServerListUpdater 类上，这是 ServerListUpdater 接口的默认实现类。我们在 3.3.6 节中已分析过这个类。

很明显，PollingServerListUpdater 这个 ServerListUpdater 接口的默认实现类实现的逻辑并不优雅。Eureka 提供了 ServerListUpdater 接口的另一个实现类 EurekaNotificationServerListUpdater，其内部基于事件通知的方式主动去拉取数据。

EurekaNotificationServerListUpdater 内部的核心代码（接收到 CacheRefreshedEvent 事件后通过线程池提交主动拉取配置数据的任务）如下：

```java
if (event instanceof CacheRefreshedEvent) {
    ...
    refreshExecutor.submit(new Runnable() {
        @Override
        public void run() {
            try {
                updateAction.doUpdate();
                lastUpdated.set(System.currentTimeMillis());
            } catch (Exception e) {
                logger.warn("Failed to update serverList", e);
            } finally {
                updateQueued.set(false);
            }
        }
    });
}
```

3.4　Dubbo LoadBalance 负载均衡

Apache Dubbo 是一款高性能 Java RPC 框架，其内部也拥有负载均衡的功能，定义如下：

```java
@SPI(RandomLoadBalance.NAME)
public interface LoadBalance {
```

```
@Adaptive("loadbalance")
<T> Invoker<T> select(List<Invoker<T>> invokers, URL url, Invocation invocation)
throws RpcException;
}
```

LoadBalance 接口只有一个 select 方法，会从一堆 Invoker 列表中根据负载均衡算法得到唯一的 Invoker。Dubbo 负载均衡的上一阶段路由会得到这个 Invoker 列表。Spring Cloud 内部则是通过 ILoadBalancer 获取实例列表。Dubbo Router 接口的定义如下：

```
public interface Router extends Comparable<Router> {

    ...

    <T> List<Invoker<T>> route(List<Invoker<T>> invokers, URL url, Invocation
invocation) throws RpcException;

    ...

}
```

Spring Cloud 与 Apache Dubbo 在路由和负载均衡侧的功能对比如表 3-3 所示。

表 3-3

功能/框架	Spring Cloud	Apache Dubbo
负载均衡	ReactiveLoadBalancer(SCL) 或 IRule(Ribbon)	LoadBalance
路由	ServiceInstanceListSupplier(SCL) 或 ILoadBalancer(Ribbon)	Router
容错机制	ServerStats(Ribbon) & ILoadBalancer	Cluster
服务实例刷新机制	DiscoveryClient(SCL) 或 ServerListUpdater(Ribbon)	NotifyListener
健康检查	IPing(Ribbon)	不处理（注册中心实现），可以依靠 fault tolerant 机制过滤不健康的实例

3.5 OpenFeign：声明式 Rest 客户端

在实际的开发工作中，我们会接触更多的上层服务调用过程（RestTemplate 或即将提到的 OpenFeign），负载均衡为服务调用提供底层能力。

回顾一下前面介绍过 RestTemplate 的构造可以通过加上@LoadBalanced 注解，使其拥有基于服务名进行服务调用的能力。

想象一下这个场景：一个复杂的微服务系统涉及上百个服务名，如果使用 RestTemplate 发起服务，服务信息的维护会非常麻烦。

Spring Cloud 提供了另外一个 Rest 客户端 OpenFeign，用于解决这个问题（每个服务对应一个声明式接口）。OpenFeign 是一种声明式接口方式的 Rest 客户端，只需要定义接口和对应的方法及注解，底层就会自动生成动态代理，发起 Rest 请求并获得最终结果。

3.5.1 OpenFeign 概述

若要使用 OpenFeign，需要在 pom 里加入对应的依赖：

```xml
<dependency>
    <groupId>org.springframework.cloud</groupId>
    <artifactId>spring-cloud-starter-openfeign</artifactId>
</dependency>
```

引入依赖之后，再添加 @EnableFeignClients 注解到启动类：

```java
@EnableFeignClients
@SpringBootApplication
public class NacosConsumer4OpenFeign {
    ...
}
```

然后定义接口并使用@FeignClient 注解修饰。@FeignClient 注解需要指定 name 属性指明调用哪个服务，接口的方法声明映射服务对外暴露的方法：

```java
@FeignClient(name = "nacos-provider-lb")
interface EchoService {
```

```
@GetMapping("/")
String echo();
}
```

最后通过@Autowired 注解直接注入定义的接口,发起服务调用即可。

@EnableFeignClients 注解的作用是扫描被@FeignClient 注解修饰的类,内部提供了一些属性,比如 basePackages:String[]、basePackageClasses:Class<?>[]和 clients:Class<?>[]。这些属性的目的只有一个,就是找出需要扫描的包路径,然后根据包名在扫描的类里找出被@FeignClient 注解修饰的类。

如果@EnableFeignClients 注解设置了 clients 属性,那么直接获取这些类对应的包名即可,否则获取 value、basePackages,以及 basePackageClasses 属性对应的包名(字符串类型属性直接当作包名,Class 类型获取 class 的 package),如果找到的包名为空,那么会使用@EnableFeignClients 注解修饰的类的包名(上述例子中没有设置任何属性,会使用入口类的包名)。

@EnableFeignClients 注解对外还暴露了一个 defaultConfiguration:Class<?>[]属性,这个属性表示默认的配置类。

@FeignClient 注解对外也暴露了一个 configuration:Class<?>[]属性。读者看到这个场景时,有没有似曾相识的感觉?没错,这套加载机制跟@LoadBalancerClient 和@LoadBalancerClients 注解,以及@RibbonClient 和@RibbonClients 注解的作用是一样的,解决的都是配置类的优先级问题。

配置类中返回的 Bean 会替换 FeignClientsConfiguration 配置类中已经存在的 Bean(FeignContext 内部维护着一个 Map,用于保存各个服务的 ApplicationContext。每个 ApplicationContext 构造的时候都会加上 FeignClientsConfiguration 配置类)。比如,FeignClientsConfiguration 配置类中默认的 Contract 实现类为 SpringMvcContract,把接口中 SpringMvc 的注解解析成 MethodMetadata:

```
@Bean
@ConditionalOnMissingBean
public Contract feignContract(ConversionService feignConversionService) {
    return new SpringMvcContract(this.parameterProcessors, feignConversionService);
}
```

@FeignClient 注解的其他属性如下：

- value:String：服务名。比如 service-provider、http://service-provider。如果 EchoService 中配置了 value=service-provider，那么调用 echo 方法的 URL 为 http://service-provider/echo；如果配置了 value=https://service-provider，那么调用 echo 方法的 URL 为 https://service-provider/divide。
- serviceId:String：该属性已过期，但还能用，其作用跟 value 一致。
- name:String：跟 value 属性的作用一致。
- qualifier:String：给@FeignClient 注解修饰的 Bean 设置的 Spring Bean 名称。
- url:String：绝对路径，用于替换服务名，优先级比服务名高。比如，EchoService 中如果配置了 url=aaa，则调用 echo 方法的 URL 为 http://aaa/echo；如果配置了 url=https://aaa，则调用 echo 方法的 URL 为 https://aaa/divide。
- decode404:boolean：默认是 false，表示对于一个 HTTP 状态码为 404 的请求是否需要进行解码。默认为 false，表示不进行解码，当成一个异常处理。设置为 true 之后，遇到 HTTP 状态码为 404 的 Response 还是会解析这个请求的 body。
- fallback:Class<?>：默认值是 void.class，表示 fallback 类，需要实现 FeignClient 对应的接口，当调用方法发生异常时，会调用这个 Fallback 类对应的 FeignClient 接口方法。如果配置了 fallback 属性，就会把 Fallback 类包装在一个默认的 FallbackFactory 实现类 FallbackFactory.Default 上，而不使用 fallbackFactory 属性对应的 FallbackFactory 实现类。
- fallbackFactory:Class<?>：默认值是 void.class，表示生产 fallback 类的 Factory，可以实现 feign.hystrix.FallbackFactory 接口，FallbackFactory 内部会针对一个 Throwable 异常返回一个 Fallback 类进行 fallback 操作。
- path:String：请求路径。在服务名或 URL 与 requestPath 之间。
- primary:boolean：默认是 true，表示当前这个 FeignClient 生成的 Bean 是否是 primary。如果在 ApplicationContext 中存在一个实现 EchoService 接口的 Bean，但注入时并不会使用该 Bean，因为 FeignClient 生成的 Bean 是 primary。
- contextId:String：FeignContext 里维护着 map 的 key，如果没设置，会使用 name 属性。

3.5.2 OpenFeign 对 JAX-RS 的支持

在 3.5.1 节中，@FeignClient 注解接口内部的方法都是被 SpringMVC 相关参数所修饰的。OpenFeign 接口声明的定义还支持 JAX-RS 注解，若要使用 JAX-RS 注解修饰 OpenFeign 接口，需要在 pom 中加上依赖：

```xml
<dependency>
    <groupId>io.github.openfeign</groupId>
    <artifactId>feign-jaxrs</artifactId>
</dependency>
<dependency>
    <groupId>javax.ws.rs</groupId>
    <artifactId>jsr311-api</artifactId>
</dependency>
```

@FeignClient 修饰的接口中的方法使用 JAX-RS 注解：

```java
@FeignClient(name = "nacos-provider-lb", configuration = MyOpenFeignConfiguration.class)
interface EchoService {

    @GET
    @Path("/")
    String echo();

}
```

这里 MyOpenFeignConfiguration 配置类内部的 Contract 实现类为 JAXRSContract：

```java
@Bean
public Contract feignContract() {
    return new JAXRSContract();
}
```

3.5.3 OpenFeign 底层执行原理

OpenFeign 声明的接口不论是 SpringMVC 还是 JAX-RS，其底层都会把接口解析成方法元数据（MethodMetadata），再通过动态代理生成接口的代理，并基于 MethodMetadata 进行 Rest 调用。

@EnableFeignClients 注解提供的包名与 @FeignClient 注解修饰的接口找到所有的接口，

并基于这些接口构造 FeignClientFactoryBean 这个 FactoryBean。

FactoryBean 内部真正构造的对象是一个 Proxy，这个 Proxy 是通过 Targeter#target 构造出来的。Targeter 内部构造通过 Feign.Builder#build 方法完成，build 方法返回的是一个 Feign 对象。默认情况下返回的是 ReflectiveFeign 这个 Feign 对象的子类：

```
@Override
public <T> T newInstance(Target<T> target) {
  Map<String, MethodHandler> nameToHandler = targetToHandlersByName.apply(target);
  Map<Method, MethodHandler> methodToHandler = new LinkedHashMap<Method, MethodHandler>();
  List<DefaultMethodHandler> defaultMethodHandlers = new LinkedList<DefaultMethodHandler>();

  for (Method method : target.type().getMethods()) {
    if (method.getDeclaringClass() == Object.class) {
      continue;
    } else if (Util.isDefault(method)) {
      DefaultMethodHandler handler = new DefaultMethodHandler(method);
      defaultMethodHandlers.add(handler);
      methodToHandler.put(method, handler);
    } else {
      methodToHandler.put(method, nameToHandler.get(Feign.configKey(target.type(), method)));
    }
  }
  InvocationHandler handler = factory.create(target, methodToHandler);
  T proxy = (T) Proxy.newProxyInstance(target.type().getClassLoader(),
      new Class<?>[] {target.type()}, handler);

  for (DefaultMethodHandler defaultMethodHandler : defaultMethodHandlers) {
    defaultMethodHandler.bindTo(proxy);
  }
  return proxy;
}
```

从这段代码可以看到，InvocationHandler、Proxy 这些 JDK 内置的动态代理类完成了这个操作。

OpenFeign 整体的构造时序如图 3-4 所示。OpenFeign 目前已经集成了 Sentinel、Hystrix、Resilience4j 框架，其相关内容会在 5.4.3 节和 5.5.2 节中介绍。

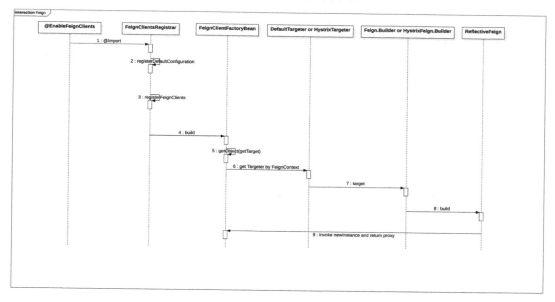

图 3-4

3.6　Dubbo Spring Cloud：服务调用的新选择

Dubbo Spring Cloud 是 Spring Cloud Alibaba 项目内部提供的一个可以使用 Spring Cloud 客户端 RestTemplate 或 OpenFeign 调用 Dubbo 服务的模块。

Apache Dubbo 和 Spring Cloud 是两套架构完全不同的开发框架。在讲解 Dubbo Spring Cloud 之前，我们先来看这个问题：Apache Dubbo 暴露的服务都是接口级别的，而 Spring Cloud 暴露的服务是应用级别的。RestTemplate 或 OpenFeign 发起调用服务都会有对应的 URL Path、Query Parameter、Header 等内容（这是 HTTP 协议调用），如何让这些内容关联 Dubbo 服务呢？

针对上述问题，Dubbo Spring Cloud 实现了以应用为粒度的注册机制，每个 Dubbo 应用注册到注册中心后有且仅有一个服务。那么原先以接口为维度的那些接口信息去哪里了？

Dubbo Spring Cloud 定义了 DubboMetadataService 元数据服务的概念。这是一个专门用于存储 Dubbo 服务的元数据接口。DubboMetadataService 接口定义如下：

```java
public interface DubboMetadataService {

    // 最核心的接口，用于获取 Dubbo 服务的 Rest 元数据
    String getServiceRestMetadata();

    // 返回所有 Dubbo 服务的 ServiceKey 集合
    Set<String> getAllServiceKeys();

    // 返回所有对外暴露的 Dubbo 服务。key 为 ServiceKey，value 为 URL 的 json 格式
    Map<String, String> getAllExportedURLs();

    // 基于接口名分组及版本获取到 URL 的 json 格式
    String getExportedURLs(String serviceInterface, String group, String version);

}
```

核心方法 getServiceRestMetadata 获取 Dubbo 服务的 Rest 元数据是指：当一个 Dubbo 服务同时也被 SpringMVC 相关注解修饰时，SpringMVC 相关注解修饰的内容就是这些 Rest 元数据。这些 Rest 元数据由 RestMethodMetadata 类修饰，比如，这个 Dubbo 服务 RestService 接口，其定义如下：

```java
@Service
@RestController
public class SpringRestService implements RestService {

    @Override
    @GetMapping("/param")
    public String param(@RequestParam String param) {
        return param;
    }

}
```

RestService 服务对应的 Rest 元数据内容如下：

```
RestMethodMetadata{method=MethodMetadata{name='param',
returnType='java.lang.String', params=[MethodParameterMetadata{index=0,
name='param', type='java.lang.String'}], method=public java.lang.String
com.alibaba.cloud.dubbo.service.SpringRestService.param(java.lang.String)},
request=RequestMetadata{method='GET', path='/param', params={param=[{param}]},
headers=[], consumes=[], produces=[]}, urlIndex=null, bodyIndex=null,
headerMapIndex=null, queryMapIndex=null, queryMapEncoded=false,
returnType='java.lang.String', bodyType='null', indexToName={0=[param]},
formParams=[], indexToEncoded={}}
```

除了 SpringMVC 相关注解，当 Dubbo 服务自身也暴露 Rest 协议的时候，这些 JAX-RS 相关注解修饰的内容也会被解析成 Rest 元数据。比如，这个 Dubbo 服务 RestService 接口的代码如下：

```
@Service(protocol = { "dubbo", "rest" })
@Path("/")
public class StandardRestService implements RestService {

    @Override
    @Path("param")
    @GET
    public String param(@QueryParam("param") String param) {
        return param;
    }

}
```

RestService 服务对应的 Rest 元数据内容如下：

```
RestMethodMetadata{method=MethodMetadata{name='param',
returnType='java.lang.String', params=[MethodParameterMetadata{index=0,
name='param', type='java.lang.String'}], method=public java.lang.String
com.alibaba.cloud.dubbo.service.SpringRestService.param(java.lang.String)},
request=RequestMetadata{method='GET', path='/param', params={param=[{param}]},
headers=[], consumes=[], produces=[]}, urlIndex=null, bodyIndex=null,
```

```
headerMapIndex=null, queryMapIndex=null, queryMapEncoded=false,
returnType='java.lang.String', bodyType='null', indexToName={0=[param]},
formParams=[], indexToEncoded={}}
```

从这两个例子的 RestMethodMetadata 内容可看出，SpringMVC 和 JAX-RS 的 Rest 元数据是一致的。

Rest 元数据出现的意义是为了匹配 RestTemplate 或 OpenFeign 的 HTTP 协议内容，匹配 HTTP 协议内容的目的是为了解决本节一开始提到的让 HTTP 协议内容关联上 Dubbo 服务。

使用 RestTemplate 或 OpenFeign 调用 Dubbo 服务会经历以下过程：

（1）根据服务名得到注册中心的 Dubbo 服务 DubboMetadataService。

（2）使用 DubboMetadataService 里提供的 getServiceRestMetadata 方法获取要使用的 Dubbo 服务和对应的 Rest 元数据。

（3）基于 Dubbo 服务和 Rest 元数据构造 GenericService。

（4）服务调用过程中使用 GenericService 发起泛化调用。

下面是使用 Dubbo Spring Cloud 调用 Dubbo 服务的开发步骤。

（1）引入 spring-cloud-starter-dubbo 依赖。

```
<dependency>
    <groupId>com.alibaba.cloud</groupId>
    <artifactId>spring-cloud-starter-dubbo</artifactId>
</dependency>
```

加上依赖后，注册中心改成 spring-cloud 注册中心：

```
dubbo.registry.address=spring-cloud://localhost
```

（2）Provider 端接口加上 SpringMVC 相关注解或使用 JAX-RS 暴露 Rest 协议。

① 加上 SpringMVC 相关注解。

```
@Service(version = "1.0.0")
@RestController
```

```java
public class SpringRestService implements RestService {

    @Override
    @GetMapping("/param")
    public String param(@RequestParam String param) {
        log("/param", param);
        return param;
    }

    @Override
    @PostMapping("/params")
    public String params(@RequestParam int a, @RequestParam String b) {
        log("/params", a + b);
        return a + b;
    }
}
```

② 使用 JAX-RS 暴露 Rest 协议。

配置文件暴露 rest 协议：

```
dubbo.protocols.rest.name=rest
dubbo.protocols.rest.port=9090
dubbo.protocols.rest.server=netty
```

接口使用 JAX-RS 注解修饰：

```java
@Service(version = "1.0.0", protocol = { "dubbo", "rest" })
@Path("/")
public class StandardRestService implements RestService {

    @Override
    @Path("param")
    @GET
    public String param(@QueryParam("param") String param) {
        log("/param", param);
        return param;
```

```
    }

    @Override
    @Path("params")
    @POST
    public String params(@QueryParam("a") int a, @QueryParam("b") String b) {
        log("/params", a + b);
        return a + b;
    }
}
```

（3）Consumer 客户端加上 @DubboTransported 注解。

RestTemplate 和 OpenFeign 客户端都支持 @DubboTransported 注解。

RestTemplate 的使用方式如下：

```
@Bean
@LoadBalanced
@DubboTransported
public RestTemplate restTemplate() {
    return new RestTemplate();
}
```

OpenFeign 的使用方式如下：

```
@FeignClient("nacos-provider-lb")
@DubboTransported(protocol = "dubbo")
public interface DubboFeignRestService {

    @GetMapping("/param")
    String param(@RequestParam("param") String param);

    @PostMapping("/params")
    String params(@RequestParam("b") String paramB, @RequestParam("a") int paramA);

}
```

（4）使用 RestTemplate 或 OpenFeign 调用 Dubbo 服务。

使用 RestTemplate 调用的方式如下：

```
restTemplate.getForEntity("http://dubbo-provider-service/param?param=deepinspringcloud",
String.class);
```

使用 OpenFeign 调用的方式如下：

```
dubboFeignRestService.param("deepinspringcloud");
```

3.7 再谈路由和负载均衡

基于 Netflix Ribbon 的 Spring Cloud 负载均衡设计了以下两个核心接口：

- 路由对应的 ILoadBalancer 接口，获取服务的 Server 实例列表。
- 负载均衡对应的 IRule 接口，从服务的 Server 实例列表中根据负载均衡算法获取一个实例。

Spring Cloud 应用的流量控制本质上就是对 Server 列表的控制：

- 自定义 ILoadBalancer 接口，重写获取 Server 列表的逻辑（找出与当前请求匹配的 Server 列表）。
- 自定义 IRule 接口，从所有的 Server 列表里找出与当前请求匹配的 Server。

很明显，第一种基于 ILoadBalancer 的方式更加合理。我们来看 Dubbo 路由 Router 的实现。Route 方法会从 Invoker 列表中过滤一批 Invoker，得到另一批 Invoker 列表：

```
public interface Router extends Comparable<Router> {

    <T> List<Invoker<T>> route(List<Invoker<T>> invokers, URL url, Invocation
invocation) throws RpcException;

}
```

然后 Router 在 RouterChain 里被使用：

```
public class RouterChain<T> {

    public List<Invoker<T>> route(URL url, Invocation invocation) {
        List<Invoker<T>> finalInvokers = invokers;
        for (Router router : routers) {
            finalInvokers = router.route(finalInvokers, url, invocation);
        }
        return finalInvokers;
    }
}
```

RouterChain 中的 Router 列表可以随意被添加，开发者可以基于 SPI 添加各式各样的 Router。

笔者认为 Duboo 在路由侧的实现更加优雅。在 Spring Cloud 的设计中，Ribbon 的路由设计与 Request（流量）请求信息是解耦的，而 Dubbo 的 Router 与 Invocation（流量）是绑定的，这意味着路由过程可以直接基于流量特征进行动态操作，无须引入类似 ThreadLocal 的方式来传递流量特征。

另外，我们在 3.3.8 节介绍 ServerStats 时提到过 Spring Cloud 默认提供的 ZoneAvoidanceRule 这个 IRule 负载均衡策略，它内部会依赖 ServerStats 去根据 Server 状态摘除异常节点。

Spring Cloud 提供的其他 IRule 负载均衡策略并没有这个能力，如果想在自定义的 IRule 负载均衡也拥有摘除异常节点的能力，需要在代码里配合 ServerStats 使用。

3.8 案例：应用流量控制

流量控制指的是根据一些流量特征，控制其流向下游的动作。如图 3-5 所示，consumer 调用 provider 的场景中，满足 Query Parameter 中 name=jim 的流量只会路由到 192.168.1.1 这个 IP 的 provider 实例；满足 HEADER 中 Test=1 的流量只会路由到 192.168.1.2 这个 IP 的 provider 实例；满足 cookie 中 lang=zh-cn 的流量只会路由到 192.168.1.3 这个 IP 的 provider 实例。当然，也可以按照流量的比例路由到特定的实例上。

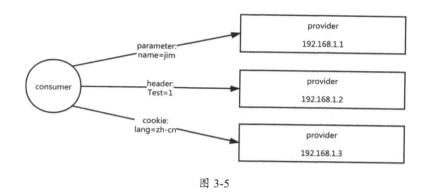

图 3-5

3.8.1 流量控制应用的业务场景

流量控制可以应用在很多业务场景中,比如金丝雀发布、同机房优先路由、标签路由、全链路灰度等。

金丝雀发布场景如图 3-6 所示,其应用只有新和老两个版本,根据流量特征(图 3-6 按照流量比例)将部分流量流入新版本验证,确定新版本稳定后再全量发布。

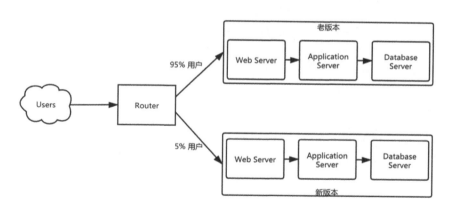

图 3-6

同机房优先路由场景如图 3-7 所示,当公司规模扩大之后,应用会跨机房部署来达到高可用的目的。由于异地跨机房调用出现的网络延迟问题,需要确保服务消费方能优先调用相同机房的服务消费方,这就需要同机房优先路由场景。

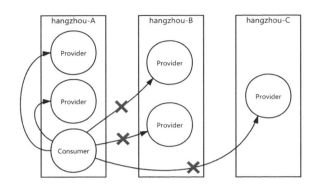

图 3-7

全链路灰度场景如图 3-8 所示,当公司规模扩大之后,微服务数量会增多。微服务数量众多的情况下进行灰度发布会发现整个链路非常长。全链路灰度解决的问题是保证特定流量能够路由到所有的特殊灰度版本。

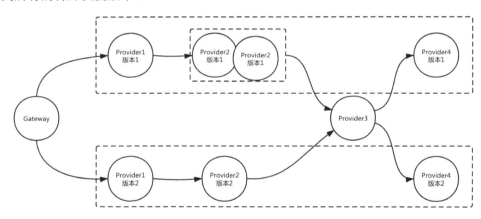

图 3-8

根据这些场景总结出流量控制的前提是需要以下两个能力。

- 流量识别能力:Spring Cloud 默认的服务调用使用 HTTP 协议,在服务调用的过程中需要识别 HTTP 协议的内容后再决定路由。
- 实例打标能力:每个实例都需要被标记,比如,金丝雀场景下的实例需要被标记成老版本和新版本;同机房优先路由场景下的实例需要被标记成机房信息;全链路灰度场景下的实例需要被标记成灰度。

3.8.2 使用 Netflix Ribbon 完成应用灰度发布

应用流量控制得出的流量识别能力和实例打标能力分析如下：

- 流量识别能力：在 3.3 节分析 Netflix Ribbon 架构时，无论是 ILoadBalancer 作为路由，还是 IRule 作为负载均衡，这些组件的定义跟 HTTP 请求信息解耦，在 Ribbon 内部无法解析 HTTP 请求信息。这时需要通过 ThreadLocal 来完成 HTTP 请求解析结果的透传。
- 实例打标能力：在实例的 metadata（元数据）中加上标签信息。通过 IRule 获取 Server 列表并根据这些 Server 中元数据的标签信息决定路由情况。

定义 RibbonRequestContext 请求上下文，内部维护着一个 Map，该 Map 用于透传解析的请求信息内容如下：

```java
public class RibbonRequestContext {

    private final Map<String, String> attr = new HashMap<>();

    public String put(String key, String value) {
        return attr.put(key, value);
    }

    public String remove(String key) {
        return attr.remove(key);
    }

    public String get(String key) {
        return attr.get(key);
    }

}
```

自定义 ThreadLocal 持有者内部维护着 RibbonRequestContext 信息，具体如下：

```java
public class RibbonRequestContextHolder {

    private static ThreadLocal<RibbonRequestContext> holder = new ThreadLocal<RibbonRequestContext>() {
```

```java
    @Override
    protected RibbonRequestContext initialValue() {
        return new RibbonRequestContext();
    }
};

public static RibbonRequestContext getCurrentContext() {
    return holder.get();
}

public static void setCurrentContext(RibbonRequestContext context) {
    holder.set(context);
}

public static void clearContext() {
    holder.remove();
}
}
```

有了透传能力之后，我们模拟一个 Consumer 调用两个 Provider 实例的过程。如图 3-9 所示，Provider 实例是灰度实例，另一个是正常实例，满足 Header 里 Gray=true 的请求会路由到灰度实例，否则路由到正常实例。

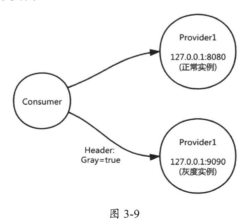

图 3-9

接下来编写流量识别能力的代码。定义 GrayInterceptor，用于拦截请求，并基于请求的特征判断是否是一次灰度调用，代码如下：

```java
public class GrayInterceptor implements ClientHttpRequestInterceptor {

    @Override
    public ClientHttpResponse intercept(HttpRequest request, byte[] body, ClientHttpRequestExecution execution)
        throws IOException {
        if (request.getHeaders().containsKey("Gray")) {
            String value = request.getHeaders().getFirst("Gray");
            if (value.equals("true")) {
                RibbonRequestContextHolder.getCurrentContext().put("Gray", Boolean.TRUE.toString());
            }
        }
        return execution.execute(request, body);
    }
}
```

GrayInterceptor 需要被加到 RestTemplate 的拦截器列表里：

```java
@Bean
@LoadBalanced
public RestTemplate restTemplate() {
    RestTemplate restTemplate = new RestTemplate();
    restTemplate.getInterceptors().add(new GrayInterceptor());
    return restTemplate;
}
```

然后编写实例打标能力的代码。直接通过 spring.cloud.nacos.discovery.metadata.* 配置项决定元数据内容：

```
# 正常实例配置内容
spring.cloud.nacos.discovery.metadata.gray=false
# 灰度实例配置项内容
spring.cloud.nacos.discovery.metadata.gray=true
```

最后发起服务调用的时候传递 Header 中 Gray 的信息，如果为 true，只会路由到灰度实例，否则路由到正常实例。

上述示例针对的是 RestTemplate 客户端的流量识别能力，OpenFeign 客户端同理，可以自定义 RequestInterceptor 拦截器进行流量识别：

```java
public class GrayRequestInterceptor implements RequestInterceptor {
    @Override
    public void apply(RequestTemplate template) {
        if (template.headers().containsKey("Gray")) {
            String value = template.headers().get("Gray").iterator().next();
            if (value.equals("true")) {
                RibbonRequestContextHolder.getCurrentContext().put("Gray", Boolean.TRUE.toString());
            }
        }
    }
}
```

OpenFeign 的底层也是使用 Ribbon，所以在路由层不需要做任何修改。

接下来使用 curl 命令查看效果。下面两个命令分别使用 RestTemplate 和 OpenFeign 发起服务调用，结果都会路由到灰度实例上（如果没有带上 Gray:true 这个 Header，结果都会路由到正常实例上）：

```
curl -s -H "Gray:true" http://localhost:8888/echoFeign
curl -s -H "Gray:true" http://localhost:8888/echo
```

大家有没有发现？这里的拦截器针对流量识别的代码都是写"死"的，只针对 Header 识别。很明显，这个需要优化。能否动态修改这些识别规则呢？答案肯定是可以的，第 4 章介绍的内容可以解决这个问题。

第 4 章

配置管理

本章将介绍 Spring、Spring Boot、Spring Cloud 在配置管理上的编程模型抽象及其使用方式，在此基础上从"如何从配置中心获取配置"和"动态刷新配置后如何生效"两个层面讲解 Spring Cloud 配置管理的原理。然后对 Spring Cloud 自身提供的配置组件 Spring Cloud Config Server 和 Spring Cloud Config Client 进行深入讲解。最后通过灰度发布案例结合配置中心完成应用灰度发布规则动态更新的案例，以加深读者对配置中心作用的理解。

4.1 配置中心背景概述

在第 1 章已经介绍过"统一所有的配置格式"是 Spring Boot 产生的背景之一。每个 Spring Boot 应用的基本配置都会在 resources 目录下的 application.properties 或 application.yml 等文件中。

Spring Boot 所有的配置都是跟应用绑定的，配置文件会被打包到 JAR 中跟随应用一起部署到服务器或容器内。这种方式会存在如下问题：

- 配置数据存在安全泄露问题，只要进入服务器或容器内，就可查看所有的配置。比如，配置中存在数据库连接信息，会被直接泄露。
- 配置在应用启动后，无法动态更新，只能手动更新后再重启应用。
- 一个应用部署多份实例的时候，需要维护多套配置，容易造成数据不一致的问题。
- 配置没有管理方式。如更新配置后，线上发生问题需要回滚，必须手动将新的配置改回老的配置。

要解决上述问题，需要一个配置中心，其方案如下：

- 所有的配置数据在配置中心内，进入服务器或容器内也无法知道配置信息。
- 配置数据从配置中心获取，并动态修改后，配置中心会推送新的配置数据到各个应用上。
- 所有的实例配置都从配置中心获取，不存在数据不一致的问题。
- 配置中心只需要提供版本管理的功能即可，在配置数据下发的过程中出现问题，可以立即回滚到上一个版本。

目前 Alibaba Nacos、Apache ZooKeeper、HashiCorp Consul、HashiCorp Vault、Baidu Disconf 等中间件都是业界比较知名的配置中心。Spring Cloud 内部的 Spring Cloud Config 组件提供了服务器端和客户端对于配置的支持。

接下来将深入介绍 Spring 在客户端和服务器端对配置的抽象原理。

4.2　Spring/Spring Boot 与配置

Spring 在配置管理上有一个非常重要的接口：Environment。Environment 接口表示当前应用运行时的环境信息，每一个环境信息内部都有两个概念：Profile 和 Properties。Profile 表示一个逻辑隔离的分组，一个 Bean 或者一个配置只有在它们所在的 Profile 被激活的情况下才会生效；Properties 表示真正的配置内容。

Profile 的一个最佳应用就是用于区分开发环境，比如，一个 UserService 接口不同的实现类属于不同的 Profile，custom.welcome 配置项在不同的 Profile 下，其配置值不相同。下面这段代码是 Profile 最佳应用的案例：

```
public interface UserService {
```

```java
    String findAll();
}

@Profile(value = "prod")
@Service
public class UserServiceImpl implements UserService {
    @Override
    public String findAll() {
        return "User=prod";
    }
}

@Profile(value = "dev")
@Service
public class MockService implements UserService {
    @Override
    public String findAll() {
        return "User=mock";
    }
}
```

配置文件的定义如下：

```
application-dev.properties:
custom.welcome=Hello Mock Data
application-prod.properties:
custom.welcome=Hello Spring Profile
```

在 application.properties 中激活 dev Profile，代码如下：

```
spring.profiles.active=dev
```

应用启动成功后，直接打印 dev 这个 Profile 提供的信息：

```
User=mock
Hello Mock Data
```

这里需要注意两点：

- 可以有多个 Profile 被激活，比如，同时激活 profile1 和 profile2 这两个 Profile（spring.profiles.active 配置项是一个 List 集合）。可以通过 Environment 接口的 getActiveProfiles 获取激活的 Profile 列表。
- 在不激活任何 Profile 的情况下，default 会作为默认的 Profile 被激活。可以通过 getDefaultProfiles 获取默认的 Profile 列表。

Properties 表示真正的配置内容。配置内容可能来自 properties 配置文件、JVM 系统参数、操作系统环境变量、servlet 上下文参数等数据源。这些数据源被抽象成抽象类 PropertySource，每个 PropertySource 都有自己的名称（name）以及真正的配置数据（source）：

```
public abstract class PropertySource<T> {

    protected final String name;

    protected final T source;

    ...
}
```

抽象类 PropertySource 有非常多的子类，比如 PropertiesPropertySource 子类表示 properties 配置文件数据源；SystemEnvironmentPropertySource 子类表示环境变量数据源；ServletContextPropertySource 子类表示 servlet 上下文初始化参数数据源等。

所有这些数据源都会被 Environment 托管，Environment 对外提供了诸如 getProperty 方法用于获取配置，获取过程中会遍历数据源列表，直到找到对应的配置项。

这里需要注意：既然存在多个 ProperySource，那么肯定存在不同的 PropertySource 有相同的配置 key 这个场景。这个场景涉及配置的优先级问题。

1.2.4 节已经提到过 ConfigFileApplicationListener 监听器会在应用启动的过程中把所有的数据源添加到 Environment 里，这就是配置在 Spring Boot 上的使用。

4.3　Spring Cloud 与配置

Spring Cloud 与配置的关系沿用了 Spring/Spring Boot 的编程模型，并进行了一些扩展，主要新增了两部分：如何从配置中心获取配置；动态刷新配置后如何生效。

4.3.1 使用 Alibaba Nacos 体验配置的获取及动态刷新

我们先使用 Alibaba Nacos 配置中心来体验这两个部分。关于 Nacos Server 的下载和启动，请参考第 2 章相关的内容。

Nacos 启动成功后，进入配置列表。如图 4-1 所示，新建一个 Data ID 为 nacos-configuration-sample.properties、Group 为 DEFAULT_GROUP 的配置。

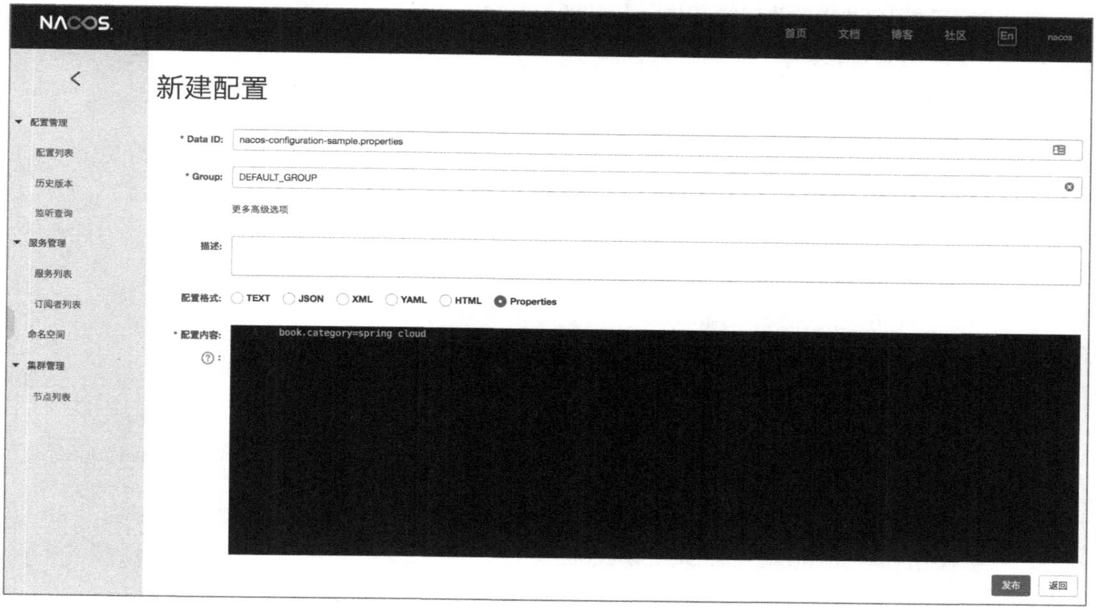

图 4-1

该配置格式属于 Properties 格式，内容如下：

```
book.category=spring cloud
```

然后在 Java 工程脚手架上选择 Nacos Configuration 模块和 Spring Web 模块，创建 spring-cloud-alibaba-nacos-configuration 项目。创建完毕后在 resources 目录下创建 bootstrap.properties 文件：

```
spring.application.name=nacos-configuration-sample
```

```
server.port=8080
# Nacos 配置中心地址
spring.cloud.nacos.config.server-addr=localhost:8848
```

创建启动类和对外暴露服务的 ConfigurationController，内容如下：

```
@SpringBootApplication
public class NacosConfiguration {

    public static void main(String[] args) {
        SpringApplication.run(NacosConfiguration.class, args);
    }

    @RestController
    class ConfigurationController {

        @Autowired
        ApplicationContext applicationContext;

        @GetMapping("/config")
        public String config() {
            StringBuilder sb = new StringBuilder();
            sb.append("env.get('book.category')=" + applicationContext.getEnvironment()
                .getProperty("book.category", "unknown"));
            return sb.toString();
        }

    }

}
```

访问 http://localhost:8080/config，得到结果如下：

```
env.get('book.category')=spring cloud
```

配置值是从 Nacos 配置中心获取的，符合预期。

接下来体验配置动态更新的过程。我们在 Nacos Console 上修改刚才创建的配置信息的配置值，把 book.category 配置项对应的值改成 spring cloud alibaba，然后访问 http://localhost:8080/config，得到结果：spring cloud alibaba。配置动态更新后生效。

体验完 Alibaba Nacos 配置中心后，下面深入介绍这两个过程的背后 Spring Cloud 到底做了哪些事情。

4.3.2 从 Spring Cloud 配置中心获取配置的原理

从 Alibaba Nacos 配置中心的体验代码来看，有一个 bootstrap.properties 文件，该文件配置了 Nacos 配置中心的地址，这个文件会让 Spring Cloud 启动 Bootstrap 阶段在 Nacos 配置中心获取配置。

Spring Cloud Bootstrap 阶段的概念：在 Bootstrap 阶段会构造 ApplicationContext，这个 ApplicationContext 加载配置的过程会基于 bootstrap.properties 或 bootstrap.yml 文件（spring.config.name 为 bootstrap）去加载文件，在加载文件的过程中，Spring Cloud 有一套机制（PropertySourceLocator 接口的定义）来构造数据源 PropertySource；其余过程跟 Spring Boot 是一致的。Bootstrap 阶段构造的 ApplicationContext 会作为正常阶段的 ApplicationContext 的父类（parent），有了这一层父子关系之后，如果不能从子 ApplicationContext 获取配置，就会从父 ApplicationContext 获取。

不少开发者在使用如 Spring Cloud Alibaba Nacos Config、Consul Config 或 ZooKeeper Config 读取配置中心的配置时都会遇到读取不到配置的问题，绝大部分原因是 bootstrap.properties 或 bootstrap.yml 配置文件中没有配置配置中心导致的。

其实这个原因很简单：我们使用配置中心前必须先对配置中心的地址进行配置，这个地址需要在配置文件里配置。Spring Cloud Bootstrap 阶段优先级高，会先读取配置中心的配置，这些配置在下一次正常的 ApplicationContext 启动时使用。Bootstrap 阶段从配置中心获取配置的时候肯定需要知道配置中心的地址。因此，在 Bootstrap 配置文件中需要对配置中心的地址进行配置。

spring-cloud-context 模块内部的 META-INF/spring.factories 添加了一个 BootstrapApplicationListener（实现了 ApplicationListener 接口），用于监听 ApplicationEnvironmentPreparedEvent 事件（Environment 刚创建，ApplicationContext 未创建时会触发该事件），收到该事件后进入 Bootstrap 阶段，从配置中心加载配置。BootstrapApplicationListener 的源码如下：

```
public class BootstrapApplicationListener
```

```java
    implements ApplicationListener<ApplicationEnvironmentPreparedEvent>, Ordered {

...

    @Override
    public void onApplicationEvent(ApplicationEnvironmentPreparedEvent event) {
        ConfigurableEnvironment environment = event.getEnvironment();
        if (!environment.getProperty("spring.cloud.bootstrap.enabled", Boolean.class,
                true)) {    //①
            return;
        }
        if (environment.getPropertySources().contains(BOOTSTRAP_PROPERTY_SOURCE_NAME)) {  //②
            return;
        }
        ConfigurableApplicationContext context = null;
        String configName = environment
                .resolvePlaceholders("${spring.cloud.bootstrap.name:bootstrap}"); // ③
        for (ApplicationContextInitializer<?> initializer : event.getSpringApplication()
                .getInitializers()) {
            if (initializer instanceof ParentContextApplicationContextInitializer) {
                context = findBootstrapContext(
                        (ParentContextApplicationContextInitializer) initializer,
                        configName); // ④
            }
        }
        if (context == null) {
            context = bootstrapServiceContext(environment, event.getSpringApplication(),
                    configName); // ⑤
            event.getSpringApplication()
                    .addListeners(new CloseContextOnFailureApplicationListener(context)); // ⑥
        }

        apply(context, event.getSpringApplication(), environment); //⑦
    }
```

```
...

private ConfigurableApplicationContext bootstrapServiceContext(
        ConfigurableEnvironment environment, final SpringApplication application,
        String configName) {
    StandardEnvironment bootstrapEnvironment = new StandardEnvironment();
    MutablePropertySources bootstrapProperties = bootstrapEnvironment
            .getPropertySources();
    for (PropertySource<?> source : bootstrapProperties) {
        bootstrapProperties.remove(source.getName()); // ⑧
    }
    String configLocation = environment
            .resolvePlaceholders("${spring.cloud.bootstrap.location:}");
    Map<String, Object> bootstrapMap = new HashMap<>();
    bootstrapMap.put("spring.config.name", configName);
    bootstrapMap.put("spring.main.web-application-type", "none");
    if (StringUtils.hasText(configLocation)) {
        bootstrapMap.put("spring.config.location", configLocation);
    }
    bootstrapProperties.addFirst(
            new MapPropertySource(BOOTSTRAP_PROPERTY_SOURCE_NAME, bootstrapMap)); // ⑨
    for (PropertySource<?> source : environment.getPropertySources()) {
        if (source instanceof StubPropertySource) {
            continue;
        }
        bootstrapProperties.addLast(source);
    }
    SpringApplicationBuilder builder = new SpringApplicationBuilder()
            .profiles(environment.getActiveProfiles()).bannerMode(Mode.OFF)
            .environment(bootstrapEnvironment)
            .registerShutdownHook(false).logStartupInfo(false)
            .web(WebApplicationType.NONE); // ⑩
    final SpringApplication builderApplication = builder.application();
    if (builderApplication.getMainApplicationClass() == null) {
        builder.main(application.getMainApplicationClass());
```

```
        }
        if (environment.getPropertySources().contains("refreshArgs")) {
            builderApplication
                    .setListeners(filterListeners(builderApplication.getListeners()));
        }
        builder.sources(BootstrapImportSelectorConfiguration.class); // ⑪
        final ConfigurableApplicationContext context = builder.run(); //⑫
        context.setId("bootstrap"); // ⑬
        addAncestorInitializer(application, context); // ⑭
        bootstrapProperties.remove(BOOTSTRAP_PROPERTY_SOURCE_NAME);
        mergeDefaultProperties(environment.getPropertySources(), bootstrapProperties);
        return context;
    }
    ...
}
```

上述代码中：

① 在 Bootstrap 阶段有个开关控制。配置项为 spring.cloud.bootstrap.enabled，默认是打开的。这个配置项在 bootstrap.properties 或 application.properties 配置文件内配置无效，因为 BootstrapApplicationListener 的优先级比 ConfigFileApplicationListener 高。所以在系统环境变量或启动参数中（比如-Dspring.cloud.bootstrap.enabled=false）添加该配置项会有效。

② 防止经历两遍 Bootstrap 阶段，Bootstrap 阶段执行的过程中会添加一个名称为 bootstrap 的 PropertySource（参考代码⑨处，当子 ApplicationContext 也执行 BootstrapApplicationListener 监听器时，不会再执行下面的逻辑。所以后面讲解的所有部署都在 Bootstrap 阶段执行。

③ 获取 spring.cloud.bootstrap.name 配置项的配置值，如果获取不到，则默认使用 bootstrap。这个配置表示读取的配置文件名称（同理，需要在系统环境变量或启动参数中添加该配置项才会有效），所以默认会读取 bootstrap.properties 或 bootstrap.yml 文件里的配置。后续会作为名称为 bootstrap 的 PropertySource 里的一个配置留存。

④ 遍历 ApplicationContextInitializer 初始化器，只执行 ParentContextApplicationContext-Initializer。有开发者可能会用到单层甚至多层父子关系，并且手动设置 boostrap ApplicationContext 作为 parent（SpringApplicationBuilder#parent(ConfigurableApplicationContext parent)方法），其目的就是找到已经构造的 bootstrap ApplicationContext。

⑤ 没有找到 bootstrap ApplicationContext，则调用 bootstrapServiceContext 方法构造 bootstrap ApplicationContext。

⑥ 给子 ApplicationContext 添加 CloseContextOnFailureApplicationListener 监听器，其目的是让子 ApplicationContext 启动时关闭父 ApplicationContext。

⑦ 给子 ApplicationContext 添加一些额外的 ApplicationContextInitializer 初始化器，比如 EnvironmentDecryptApplicationInitializer，用于加密或解密配置中心里的配置。

⑧ bootstrapServiceContext 方法的内容是构造 bootstrap ApplicationContext。这里先构造一个 StandardEnvironment（默认加上系统参数和启动参数的 PropertySource 数据源），然后删除内部所有的数据源（子 ApplicationContext 也有系统参数和启动参数的数据源，所以这里删除重复的数据源），后续再添加新的数据源。这个 Environment 会作为 boostrap ApplicationContext 的 Environment。

⑨ 添加一个名称为 bootstrap 的 PropertySource 到 boostrap ApplicationContext 中，其中包括三个配置项，分别为 spring.main.web-application-type=none，表示父 ApplicationContext 不是一个 Web 应用；代码③处的 spring.cloud.bootstrap.name 配置项；spring.config.location=${spring.cloud.bootstrap.location:}，添加配置信息所在的目录（默认识别 4 种目录：classpath:/、classpath:/config/、file:./和 file:./config/）。

⑩ 构造 SpringApplicationBuilder，后续构建出的 ApplicationContext 就是 bootstrap ApplicationContext。bootstrap ApplicationContext 关闭了 banner 打印的开关，内部的 Environment 使用代码⑧处构建的 StandardEnvironment，同时关闭启动日志打印的开关，最后不注册 ShutdownHook（这些操作在子 ApplicationContext 里完成，不是 bootstrap ApplicationContext 的目的）。bootstrap ApplicationContext 是一个非 Web ApplicationContext。

⑪ 注册 BootstrapImportSelectorConfiguration 配置类。这个配置类非常关键，它会使用工厂加载机制找出 key 为 org.springframework.cloud.bootstrap.BootstrapConfiguration 的配置类，并进行加载。其中，PropertySourceBootstrapConfiguration 配置会被加载，内部会使用 PropertySourceLocator 加载配置中心里的配置。

⑫ 使用 SpringApplicationBuilder.run 方法得到 boostrap ApplicationContext。

⑬ boostrap ApplicationContext 的 ID 为 bootstrap。

⑭ 为子 ApplicationContext 添加 AncestorInitializer（如果子 ApplicationContext 已经存在 AncestorInitializer，则建立关系），内部会建立父子关系。

从 BootstrapApplicationListener 代码的分析过程看，在配置中心加载配置的过程是在 BootstrapImportSelectorConfiguration 配置类中完成的。这个配置类内部会引入 BootstrapImportSelector。下面是 BootstrapImportSelector 的源码：

```java
public class BootstrapImportSelector implements EnvironmentAware,
DeferredImportSelector {

    ...

    @Override
    public String[] selectImports(AnnotationMetadata annotationMetadata) {
        ClassLoader classLoader = Thread.currentThread().getContextClassLoader();

        List<String> names = new ArrayList<>(SpringFactoriesLoader
                .loadFactoryNames(BootstrapConfiguration.class, classLoader)); // ①
        names.addAll(Arrays.asList(StringUtils.commaDelimitedListToStringArray(
                this.environment.getProperty("spring.cloud.bootstrap.sources",
"")))); //②

        List<OrderedAnnotatedElement> elements = new ArrayList<>();
        for (String name : names) {
            try {
                elements.add(
                        new OrderedAnnotatedElement(this.metadataReaderFactory, name));
            }
            catch (IOException e) {
                continue;
            }
        }
        AnnotationAwareOrderComparator.sort(elements);    // ③

        String[] classNames = elements.stream().map(e -> e.name).toArray(String[]::new);
```

```
    return classNames;
}
...
}
```

上述代码中:

① 使用工厂加载机制从 META-INF/spring.factories 中找出 key 为 `org.springframework.cloud.bootstrap.BootstrapConfiguration` 的配置类。

② 可以通过配置项为 spring.cloud.bootstrap.sources 新增配置类。

③ 对所有找出的配置类进行排序。

下面是 spring-cloud-context 模块里 META-INF/spring.factories 文件中 key 为 `org.springframework.cloud.bootstrap.BootstrapConfiguration` 对应的配置类:

```
org.springframework.cloud.bootstrap.BootstrapConfiguration=\
org.springframework.cloud.bootstrap.config.PropertySourceBootstrapConfiguration,\
org.springframework.cloud.bootstrap.encrypt.EncryptionBootstrapConfiguration,\
org.springframework.cloud.autoconfigure.ConfigurationPropertiesRebinderAutoConfiguration,\
org.springframework.boot.autoconfigure.context.PropertyPlaceholderAutoConfiguration
```

其中,PropertySourceBootstrapConfiguration 内部会获取 PropertySourceLocator 列表,用于加载配置中心的配置,并封装到 PropertySource 列表,之后将该列表添加到 Environment 里的 PropertySource 列表中:

```
@Configuration(proxyBeanMethods = false)
@EnableConfigurationProperties(PropertySourceBootstrapProperties.class)
public class PropertySourceBootstrapConfiguration implements
        ApplicationContextInitializer<ConfigurableApplicationContext>, Ordered {

    ...

    @Autowired(required = false)
```

```java
    private List<PropertySourceLocator> propertySourceLocators = new ArrayList<>(); // ①

    @Override
    public void initialize(ConfigurableApplicationContext applicationContext) {
        CompositePropertySource composite = new OriginTrackedCompositePropertySource(
                BOOTSTRAP_PROPERTY_SOURCE_NAME); //②
        AnnotationAwareOrderComparator.sort(this.propertySourceLocators);
        boolean empty = true;
        ConfigurableEnvironment environment = applicationContext.getEnvironment(); //③
        for (PropertySourceLocator locator : this.propertySourceLocators) {
            PropertySource<?> source = null;
            source = locator.locate(environment); //④
            if (source == null) {
                continue;
            }
            logger.info("Located property source: " + source);
            composite.addPropertySource(source); // ⑤
            empty = false;
        }
        if (!empty) {
            MutablePropertySources propertySources = environment.getPropertySources();
            String logConfig = environment.resolvePlaceholders("${logging.config:}");
            LogFile logFile = LogFile.get(environment);
            if (propertySources.contains(BOOTSTRAP_PROPERTY_SOURCE_NAME)) {
                propertySources.remove(BOOTSTRAP_PROPERTY_SOURCE_NAME); // ⑥
            }
            insertPropertySources(propertySources, composite); // ⑦
            reinitializeLoggingSystem(environment, logConfig, logFile);
            setLogLevels(applicationContext, environment);
            handleIncludedProfiles(environment);
        }
    }

    ...

}
```

上述代码中：

① 获取 ApplicationContext 中所有的 PropertySourceLocator Bean。

② 新建一个 name 为 bootstrapProperties 的 CompositePropertySource 数据源。CompositePropertySource 内部维护着 PropertySource 数据源集合。新建这个 CompositePropertySource 的目的是添加所有从 PropertySourceLocator 加载的数据源。

③ 得到 bootstrap ApplicationContext 里的 Environment。

④ 遍历所有的 PropertySourceLocator 列表，并调用它的 locate 方法获取 PropertySource。

⑤ 遍历加载得到的 PropertySource 添加到代码②处新建的 CompositePropertySource 集合内。

⑥ 先从数据源列表中移除 key 为 bootstrapProperties 的数据源，防止后续重复添加。

⑦ 将代码②处从 PropertySourceLocator 加载到的 CompositePropertySource 数据源添加到 bootstrap ApplicationContext 的 Environment 里。

若使用 Alibaba Nacos 配置中心，那么 spring-cloud-alibaba-nacos-config 模块中对应的 spring.factories 会加载 key 为 org.springframework.cloud.bootstrap.BootstrapConfiguration 的 NacosConfigBootstrapConfiguration 配置类。该配置类内部会构造 NacosPropertySourceLocator 这个 PropertySourceLocator Bean。

NacosPropertySourceLocator 的作用就是从 Nacos 配置中心获取配置，定义如下：

```
@Order(0)
public class NacosPropertySourceLocator implements PropertySourceLocator {

    ...

    @Override
    public PropertySource<?> locate(Environment env) {
        nacosConfigProperties.setEnvironment(env);
        ConfigService configService = nacosConfigManager.getConfigService(); // ①

        if (null == configService) {
            log.warn("no instance of config service found, can't load config from nacos");
            return null;
```

```
        }
        long timeout = nacosConfigProperties.getTimeout();
        nacosPropertySourceBuilder = new NacosPropertySourceBuilder(configService,
                timeout);
        String name = nacosConfigProperties.getName();

        String dataIdPrefix = nacosConfigProperties.getPrefix();
        if (StringUtils.isEmpty(dataIdPrefix)) {
            dataIdPrefix = name;
        }

        if (StringUtils.isEmpty(dataIdPrefix)) {
            dataIdPrefix = env.getProperty("spring.application.name");
        }

        CompositePropertySource composite = new CompositePropertySource(
                NACOS_PROPERTY_SOURCE_NAME); // ②

        loadSharedConfiguration(composite); // ③
        loadExtConfiguration(composite); //④
        loadApplicationConfiguration(composite, dataIdPrefix, nacosConfigProperties,
env); // ⑤

        return composite;
    }
    ...
}
```

上述代码中：

① 得到 ConfigService，用于从 Nacos 配置中心获取配置。

② 构造 CompositePropertySource，后续 Nacos 配置中心加载的配置都会构造成 PropertySource，并添加到该 CompositePropertySource 中。

③ 加载共享配置。

④ 加载扩展配置。

⑤ 加载应用配置。

4.3.3 Spring Cloud 配置动态刷新

Spring Cloud 配置动态刷新的知识点较多，我们先抛出 Spring Cloud 配置相关的一些概念，在此基础上体验配置动态刷新，最后对这个过程的原理进行分析。

1. RefreshEvent 事件和 EnvironmentChangeEvent 事件

Spring Cloud 配置动态刷新机制基于事件监听机制，涉及以下两个事件。

- RefreshEvent 事件：配置刷新事件。接收到此事件后应用会构造一个临时的 ApplicationContext（会加上 BootstrapApplicationListener 和 ConfigFileApplicationListener，这意味着从配置中心和配置文件重新获取配置数据）。构造完毕后，新的 Environment 里的 PropertySource 会跟原先的 Environment 里的 PropertySource 进行比对并覆盖。
- EnvironmentChangeEvent 事件：环境变化事件。接收到此事件表示应用里的配置数据已经发生改变。EnvironmentChangeEvent 事件里维护着一个配置项 keys 集合，当配置动态修改后，配置值发生变化后的 key 会设置到事件的 keys 集合中。

下面介绍一下 RefreshEvent 事件，在 bootstrap.properties 配置文件中添加一行新的配置：

```
book.author=jim
```

ConfigurationController 中的 config 方法内部新增获取 book.author 配置项的代码如下：

```java
@GetMapping("/config")
public String config() {
    StringBuilder sb = new StringBuilder();
    sb.append("env.get('book.category')=" + applicationContext.getEnvironment()
        .getProperty("book.category", "unknown"))
        .append("<br/>env.get('book.author')=" + applicationContext.getEnvironment()
            .getProperty("book.author", "unknown"));
    return sb.toString();
}
```

ConfigurationController 新增 event 方法，用于发送 RefreshEvent 事件：

```
@GetMapping("/event")
public String event() {
    applicationContext.publishEvent(new RefreshEvent(this, null, "just for test"));
    return "send RefreshEvent";
}
```

最后新增一个事件接收器用于接收 EnvironmentChangeEvent 事件：

```
@Component
class EventReceiver implements ApplicationListener<EnvironmentChangeEvent> {

    @Override
    public void onApplicationEvent(EnvironmentChangeEvent event) {
        System.out.println(event.getKeys());
    }
}
```

访问 http://localhost:8080/config，得到如下结果：

```
env.get('book.category')=spring cloud
env.get('book.author')=jim
```

进入 target/classes 目录下，修改 bootstrap.properties 内容，将 book.author=jim 配置改成 book.author=jim fang，然后访问 http://localhost:8080/event，发送 RefreshEvent 事件。

事件发送完毕后，Console 打印 [book.author]，表示只有 book.author 配置项里的配置值发生了改变。

访问 http://localhost:8080/config，得到如下结果：

```
env.get('book.category')=spring cloud
env.get('book.author')=jim fang
```

从这个例子可以看到，本地文件修改并触发了 RefreshEvent 事件之后，Environment 获取的配置是更新后的配置，配置动态刷新后生效。

2. @RefreshScope

在前面的代码中，配置的获取都是通过 Environment 完成的，类中的属性动态修改是否会生效呢？

我们给 ConfigurationController 加上一个 bookAuthor 属性，通过@Value 属性读取配置。然后在 config 方法的返回值里添加 bookCategory：

```java
@RestController
class ConfigurationController {

    @Autowired
    ApplicationContext applicationContext;

    @Value("${book.author:unknown}")
    String bookAuthor;

    @GetMapping("/config")
    public String config() {
        StringBuilder sb = new StringBuilder();
        sb.append("env.get('book.category')=" + applicationContext.getEnvironment()
            .getProperty("book.category", "unknown"))
            .append("<br/>env.get('book.author')=" + applicationContext.getEnvironment()
                .getProperty("book.author", "unknown"))
            .append("<br/>bookAuthor=" + bookAuthor);
        return sb.toString();
    }

    @GetMapping("/event")
    public String event() {
        applicationContext.publishEvent(new RefreshEvent(this, null, "just for test"));
        return "send RefreshEvent";
    }
}
```

执行相同的操作（修改 target/classes 目录下 bootstrap.properties 配置文件里的配置值，调用/event 接口，最后调用/config 接口查看配置），通过 Environment 可以动态获取 book.author 配置项里的配置值，但是 bookAuthor 属性的值还是旧配置，无法支持动态刷新。这时需要添加一个@RefreshScope 注解去修饰 ConfigurationController：

```
@RestController
@RefreshScope
class ConfigurationController {
    ...
}
```

@RefreshScope 注解的作用就是使其修饰的类在收到 RefreshEvent 事件的时候被销毁，再次获取这个类的时候会重新构造，重新构造意味着重新解析表达式，这也代表着获取最新的配置。

@RefreshScope 注解定义如下：

```
@Target({ ElementType.TYPE, ElementType.METHOD })
@Retention(RetentionPolicy.RUNTIME)
@Scope("refresh")
@Documented
public @interface RefreshScope {

    ScopedProxyMode proxyMode() default ScopedProxyMode.TARGET_CLASS;

}
```

@RefreshScope 注解对应的 scope 值为 refresh。Spring Cloud 扩展了 Spring 对于 scope 的定义。默认情况下，Spring 的 scope 值如表 4-1 所示。

表 4-1

scope 值	作　　用
singleton	全局只有一个实例。每次获取的相同类型的 Bean 都是同一个实例
prototype	每次获取 Bean 都创建一个新的实例
request	同一个 Request 作用域内会返回之前保留的 Bean，否则重新创建 Bean
session	同一个 Session 作用域内会返回之前保留的 Bean，否则重新创建 Bean
global session	Portlet 环境下多个 Portlet 共享的 Session 作用域内会返回之前保留的 Bean，否则重新创建 Bean

Spring Cloud 新增了 scope 值为 refresh 类型的定义，表示 Bean 支持配置动态刷新。

Spring Cloud 对这个 scope 值为 refresh 的 Bean 做了哪些操作能够使其支持配置动态刷新呢？

答案就在 org.springframework.cloud.context.scope.refresh.RefreshScope 类中。RefreshScope 类继承了 GenericScope 类，默认的构造方法中，将父类 GenericScope 的 name 属性设置为 refresh。RefreshScope 类实现了 BeanDefinitionRegistryPostProcessor 接口，对满足 scope 为 refresh 条件的 BeanDefinition 做了一些修改：把这个 Bean 的类型修改成 LockedScopedProxyFactoryBean。LockedScopedProxyFactoryBean 类继承 ScopedProxyFactoryBean 类，相比父类，它多了一层锁的操作，以确保每个方法的执行都加锁，不存在并发问题。

每次 RefreshEvent 事件发送完毕之后，都会触发 RefreshScope 的 refreshAll 方法，该方法内部会在 Spring 上下文里销毁所有 scope 为 refresh 的 Bean（被 @RefreshScope 注解修饰），销毁之后，下次获取这些 Bean 的时候会重新构造一遍（意味着会重新解析表达式，这也代表着会获取最新的配置）。destroy 方法调用完毕之后会发送一个 RefreshScopeRefreshedEvent 事件（笔者认为这个事件可以解释为 scope 值为 refresh 的 Bean 发生刷新事件）。

由于 LockedScopedProxyFactoryBean 内部的每个操作都会加锁，因此，调用 ConfigurationController 的 event 方法时会获取锁，event 方法内部发送的 RefreshEvent 事件会触发 RefreshScope#destroy 方法，destroy 方法内部也会获取同一个锁，这就会出现死锁现象。所以，需要将发送 RefreshEvent 事件的方法移到另外一个没有 @RefreshScope 注解的 EventController 中：

```java
@RestController
class EventController {

    @Autowired
    ApplicationContext applicationContext;

    @GetMapping("/event")
    public String event() {
        applicationContext.publishEvent(new RefreshEvent(this, null, "just for test"));
        return "send RefreshEvent";
    }

}
```

修改 target/classes 目录下 bootstrap.properties 配置文件中 book.author 配置项的配置值，调用 /event 接口发送 RefreshEvent 事件，最后调用 /config 接口查看配置，我们发现，不论是通过 Environment 获取的配置，还是类中根据 @Value 注解得到的配置，全部变成了修改后的值：

```
env.get('book.category')=spring cloud
env.get('book.author')=jim fang
bookAuthor=jim fang
```

3. @ConfigurationProperties

Spring Cloud 对 @ConfigurationProperties 注解修饰的配置类做了特殊处理。当触发 Environment ChangeEvent 事件的时候，这些配置类会进行重绑定（rebind）操作以获取最新的配置。

定义被 @ConfigurationProperties 注解修饰的 BookProperties 配置类：

```
@ConfigurationProperties(prefix = "book")
public class BookProperties {

    private String category;

    private String author;

    ...
}
```

ConfigurationController 内部注入 BookProperties 并在 config 内打印：

```
@GetMapping("/config")
public String config() {
    StringBuilder sb = new StringBuilder();
    sb.append("env.get('book.category')=" + applicationContext.getEnvironment()
        .getProperty("book.category", "unknown"))
        .append("<br/>env.get('book.author')=" + applicationContext.getEnvironment()
            .getProperty("book.author", "unknown"))
        .append("<br/>bookAuthor=" + bookAuthor)
        .append("<br/>bookProperties=" + bookProperties);
    return sb.toString();
}
```

修改 target/classes 目录下 bootstrap.properties 配置文件中 book.author 配置项的配置值，调用 /event 接口，最后调用 /config 接口查看配置，BookProperties 配置类也发生了变化：

```properties
env.get('book.category')=spring cloud
env.get('book.author')=jim fang
bookAuthor=jim fang
bookProperties=BookProperties{category='spring cloud', author='jim fang'}
```

4. 配置动态刷新原理

Spring Cloud 配置动态刷新的三种使用方式如下：

- 通过 Environment 获取的配置，比如 env.getProperty("book.category")。
- 被 @ConfigurationProperties 注解修饰的配置类，这些配置类生效的原因是触发了 EnvironmentChangeEvent 事件。
- 被 @RefreshScope 注解修饰的类，这些 scope 值为 refresh 的类初始化时经过了特殊处理，当触发 RefreshEvent 事件后会重新构造。

RefreshEvent 配置刷新事件是配置动态刷新的核心类，配置中心在适配的过程中会涉及这个类。

我们看一下 spring-cloud-starter-alibaba-nacos-config 模块动态刷新相关的代码：

```
public class NacosContextRefresher
        implements ApplicationListener<ApplicationReadyEvent>,
ApplicationContextAware {

    ...

    @Override
    public void onApplicationEvent(ApplicationReadyEvent event) {
        if (this.ready.compareAndSet(false, true)) { // ①
            this.registerNacosListenersForApplications();
        }
    }

    private void registerNacosListenersForApplications() {
        if (isRefreshEnabled()) {
            for (NacosPropertySource propertySource : NacosPropertySourceRepository
```

```java
                .getAll()) { // ②
            if (!propertySource.isRefreshable()) {
                continue;
            }
            String dataId = propertySource.getDataId();
            registerNacosListener(propertySource.getGroup(), dataId);
        }
    }
}

private void registerNacosListener(final String groupKey, final String dataKey) {
    String key = NacosPropertySourceRepository.getMapKey(dataKey, groupKey);
    Listener listener = listenerMap.computeIfAbsent(key, // ③
            lst -> new AbstractSharedListener() {
                @Override
                public void innerReceive(String dataId, String group,
                        String configInfo) {
                    refreshCountIncrement();
                    nacosRefreshHistory.addRefreshRecord(dataId, group, configInfo);
                    applicationContext.publishEvent(
                            new RefreshEvent(this, null, "Refresh Nacos config")); // ④
                    if (log.isDebugEnabled()) {
                        log.debug(String.format(
                                "Refresh Nacos config group=%s,dataId=%s, configInfo=%s",
                                group, dataId, configInfo));
                    }
                }
            });
    try {
        configService.addListener(dataKey, groupKey, listener); // ⑤
    }
    catch (NacosException e) {
        log.warn(String.format(
                "register fail for nacos listener ,dataId=[%s],group=[%s]", dataKey,
```

```
                    groupKey), e);
            }
        }
        ...
}
```

上述代码中：

① 状态判断，防止注册两次。

② 遍历所有的 Nacos 配置项（dataId 和 group 组成一个配置项）。

③ 构造 Nacos 配置监听器。

④ 监听器每次收到 Nacos 配置更新的推送事件后会发送 RefreshEvent 配置刷新事件。

⑤ 为 Nacos 配置项添加监听器。

上述代码说明 RefreshEvent 配置刷新事件在收到 Nacos 配置推送的时候会被触发，这也验证了之前分析的配置动态刷新内容。

RefreshEvent 配置刷新事件是配置动态刷新的核心类，下面从这个类入手分析整个过程。

RefreshEventListener 监听器监听 RefreshEvent 事件，代码如下：

```java
public class RefreshEventListener implements SmartApplicationListener {

    private ContextRefresher refresh; //①

    private AtomicBoolean ready = new AtomicBoolean(false);

    public RefreshEventListener(ContextRefresher refresh) {
        this.refresh = refresh;
    }

    @Override
    public boolean supportsEventType(Class<? extends ApplicationEvent> eventType) {
        return ApplicationReadyEvent.class.isAssignableFrom(eventType)
```

```java
                || RefreshEvent.class.isAssignableFrom(eventType);// ②
    }

    @Override
    public void onApplicationEvent(ApplicationEvent event) {
        if (event instanceof ApplicationReadyEvent) {
            handle((ApplicationReadyEvent) event); //③
        }
        else if (event instanceof RefreshEvent) {
            handle((RefreshEvent) event); //④
        }
    }

    public void handle(ApplicationReadyEvent event) {
        this.ready.compareAndSet(false, true);
    }

    public void handle(RefreshEvent event) {
        if (this.ready.get()) {
            log.debug("Event received " + event.getEventDesc());
            Set<String> keys = this.refresh.refresh(); //⑤
            log.info("Refresh keys changed: " + keys);
        }
    }
}
```

上述代码中：

① ContextRefresher 为配置动态刷新的另外一个核心类，内部执行真正的刷新操作。在 RefreshAutoConfiguration 自动化配置类中构造。

② 支持 ApplicationReadyEvent 事件和 RefreshEvent 事件。ApplicationReadyEvent 事件意味着应用已经准备好对外提供服务（ApplicationContext 和 Environment 都已经全部初始化完毕）。

③ 处理 ApplicationReadyEvent 事件，将内部的一个 ready 属性从 false 更改为 true。

④ 处理 RefreshEvent 事件。

⑤ ApplicationReadyEvent 事件触发之后才可继续进行，调用 ContextRefresher 的 refresh 方法进行配置刷新。

ContextRefresher 的代码如下：

```java
public class ContextRefresher {

    ...

    private ConfigurableApplicationContext context;

    private RefreshScope scope;

    public synchronized Set<String> refresh() {
        Set<String> keys = refreshEnvironment(); // ①
        this.scope.refreshAll();// ②
        return keys;
    }

    public synchronized Set<String> refreshEnvironment() {
        Map<String, Object> before = extract(
                this.context.getEnvironment().getPropertySources());// ③
        addConfigFilesToEnvironment(); // ④
        Set<String> keys = changes(before,
                extract(this.context.getEnvironment().getPropertySources())).keySet(); //⑤
        this.context.publishEvent(new EnvironmentChangeEvent(this.context, keys)); // ⑥
        return keys;
    }

    ConfigurableApplicationContext addConfigFilesToEnvironment() {
        ConfigurableApplicationContext capture = null;
        try {
            StandardEnvironment environment = copyEnvironment(
                    this.context.getEnvironment()); // ⑦
            SpringApplicationBuilder builder = new SpringApplicationBuilder(Empty.class)
                    .bannerMode(Mode.OFF).web(WebApplicationType.NONE)
```

```java
                .environment(environment);
        builder.application()
                .setListeners(Arrays.asList(new BootstrapApplicationListener(),
                        new ConfigFileApplicationListener()));
        capture = builder.run(); // ⑧
        if (environment.getPropertySources().contains(REFRESH_ARGS_PROPERTY_SOURCE)) {
            environment.getPropertySources().remove(REFRESH_ARGS_PROPERTY_SOURCE);
        }
        MutablePropertySources target = this.context.getEnvironment()
                .getPropertySources();
        String targetName = null;
        for (PropertySource<?> source : environment.getPropertySources()) {
            String name = source.getName();
            if (target.contains(name)) {
                targetName = name;
            }
            if (!this.standardSources.contains(name)) {
                if (target.contains(name)) {
                    target.replace(name, source); // ⑨
                }
                else {
                    if (targetName != null) {
                        target.addAfter(targetName, source);
                    }
                    else {
                        target.addFirst(source);
                        targetName = name;
                    }
                }
            }
        }
    }
    finally {
        ConfigurableApplicationContext closeable = capture;
        while (closeable != null) {// ⑩
```

```
            try {
                closeable.close();
            }
            catch (Exception e) {
            }
            if (closeable.getParent() instanceof ConfigurableApplicationContext) {
                closeable = (ConfigurableApplicationContext) closeable.getParent();
            }
            else {
                break;
            }
        }
    }
    return capture;
}
...
}
```

上述代码中：

① refresh 首先调用内部的 refreshEnvironment 方法，返回值是一个 String 集合 keys，keys 内部存储着动态刷新过的配置项的 key。

② 调用 RefreshScope 的 refreshAll 方法，该方法内部会销毁@RefreshScope 注解修饰的类（这些类下次使用时会重新构造，即重新获取配置信息），并触发 RefreshScope-RefreshedEvent 事件。

③ 动态刷新配置之前，获取旧的 PropertySource 列表中所有的配置，并存储到 Map 中。

④ 在真正的动态配置刷新过程中，内部会构造一个临时的 ApplicationContext，这意味着要重新加载配置，配置重新加载后会把旧的配置替换成新的配置。详细内容可以看⑦~⑩处的代码。

⑤ 动态刷新配置之后与刷新之前发生变化的对比结果（只存储配置信息的 key），得到的集合则是这些发生变化的配置信息的 key 集合。这也是代码①处得到的内容。

⑥ 触发 EnvironmentChangeEvent 事件，以及应用的 ApplicationContext 及配置发生变化的 key 集合。

⑦ 在动态刷新配置的过程中，首先基于应用 ApplicationContext 复制一份 Environment（复制过程中，一些配置会基于原先的 Env 里的内容，比如，activeProfiles 和 defaultProfiles），这个 Environment 会作为后续临时的 ApplicationContext 里的 Environment。

⑧ 构造临时的 ApplicationContext，ApplicationContext 内部有 BootstrapApplicationListener 和 ConfigFileApplicationListener。这两个 Listener 会遵循 Spring Boot 里的配置文件加载规范，以及 Spring Cloud 里的 Bootstrap 阶段。

⑨ 新旧 PropertySource 对比判断是否进行覆盖（系统参数、启动参数、JNDI、Servlet 参数不会覆盖），其判断条件是数据源的名称是否相同。

⑩ 递归逐一关闭临时构建的 ApplicationContext。

从这段代码可以看出，ContextRefresh 内部核心的操作就是构造一个临时的 ApplicationContext，从而使应用重新加载配置（本地和配置中心都要重新加载），加载配置后，再与旧的配置进行比对并替换。

在代码②处中，RefreshScope 的 refreshAll 方法内部销毁的过程如下：

```java
public class RefreshScope extends GenericScope implements ApplicationContextAware,
        ApplicationListener<ContextRefreshedEvent>, Ordered {

    ...

    public RefreshScope() {
        super.setName("refresh"); // ①
    }

    ...

    @ManagedOperation(description = "Dispose of the current instance of bean name "
            + "provided and force a refresh on next method execution.")
    public boolean refresh(String name) {
        if (!name.startsWith(SCOPED_TARGET_PREFIX)) {
            name = SCOPED_TARGET_PREFIX + name;
```

```
        }
        if (super.destroy(name)) {// ②
            this.context.publishEvent(new RefreshScopeRefreshedEvent(name)); // ③
            return true;
        }
        return false;
    }

    @ManagedOperation(description = "Dispose of the current instance of all beans "
            + "in this scope and force a refresh on next method execution.")
    public void refreshAll() { // ④
        super.destroy();
        this.context.publishEvent(new RefreshScopeRefreshedEvent());
    }
}
```

上述代码中：

① 设置父类 GenericScope 的 name 属性为 refresh，满足 scope 值为 refresh（被@RefreshScope 修饰）的 Bean 在初始化过程中的类型被修改为 LockedScopedProxyFactoryBean。

② 调用父类 GenericScope 的 destroy 方法（父类还有一个未带 name 参数的 destroy 方法，带 name 参数的方法表示只销毁满足这个 name 的 Bean）去销毁@RefreshScope 注解修饰的 Bean。

③ destroy 方法调用成功，则触发 RefreshScopeRefreshedEvent 事件。

④ 与代码②、代码③处相同，唯一区别是销毁所有@RefreshScope 注解修饰的 Bean。

读者在这里是不是发现@ConfigurationProperties 注解修饰的配置类的自动刷新还没有讲解。这与 EnvironmentChangeEvent 事件有关，EnvironmentChangeEvent 事件在刚才分析 ContextRefresher 的过程中提到过，它在触发的时候会带上配置动态修改的具体的 key 集合。我们来看一下 ConfigurationPropertiesRebinder（由 ConfigurationPropertiesRebinderAutoConfiguration 自动化配置类构造）内部的逻辑：

```
public class ConfigurationPropertiesRebinder
        implements ApplicationContextAware,
ApplicationListener<EnvironmentChangeEvent> {
```

```java
    private ConfigurationPropertiesBeans beans;   // ①

    private ApplicationContext applicationContext;

    ...

    @ManagedOperation
    public void rebind() {
        this.errors.clear();
        for (String name : this.beans.getBeanNames()) {
            rebind(name);   // ②
        }
    }

    @Override
    public void onApplicationEvent(EnvironmentChangeEvent event) {
        if (this.applicationContext.equals(event.getSource())
                || event.getKeys().equals(event.getSource())) {     // ③
            rebind();
        }
    }
}
```

上述代码中：

① ConfigurationPropertiesBeans 内部维护着所有被@ConfigurationProperties 注解修饰的 Bean。

② 所有被 @ConfigurationProperties 注解修饰的 Bean 进行重绑定操作（如果 Environment 已经发生了变化，解析出的属性必然也是最新的）。

③ EnvironmentChangeEvent 事件内部 source 代表的 ApplicationContext 是应用本身的 ApplicationContext（ContextRefresh 在刷新过程中传递的 source 是应用本身的 ApplicationContext）。

至此，Spring Cloud 配置动态刷新源码解析完毕。下面来看一个 Eureka 的小知识，相关代码如下：

```java
// EurekaDiscoveryClientConfiguration.java
@Configuration(proxyBeanMethods = false)
@ConditionalOnClass(RefreshScopeRefreshedEvent.class)
protected static class EurekaClientConfigurationRefresher
        implements ApplicationListener<RefreshScopeRefreshedEvent> {

    @Autowired(required = false)
    private EurekaClient eurekaClient;

    @Autowired(required = false)
    private EurekaAutoServiceRegistration autoRegistration;

    public void onApplicationEvent(RefreshScopeRefreshedEvent event) {
        if (eurekaClient != null) {
            eurekaClient.getApplications();
        }
        if (autoRegistration != null) {
            this.autoRegistration.stop();
            this.autoRegistration.start();
        }
    }
}
```

EurekaDiscoveryClientConfiguration 自动化配置类里的一个内部类监听了 RefreshScope-RefreshedEvent 事件，当配置动态刷新的时候会下线该应用对应的实例，然后重新上线（这里认为 Eureka 的配置可能也进行了修改，所以重新注册）。

4.4 Spring Cloud Config Server/Client

Spring Cloud 提供了配置相关的规范，其自身也拥有一套 Spring Cloud Config Server 和 Spring Cloud Config Client 组件，用于配置的读取和动态刷新。

4.4.1 Spring Cloud Config Server

Spring Cloud Config Server 相当于存储配置的一个配置中心，对外也提供了 HTTP API 用于配置的获取。若想使用 Config Server，可以在 Spring Boot 的启动类上使用@EnableConfigServer 注解，代码如下：

```
@SpringBootApplication
@EnableConfigServer
public class ConfigServerApplication {

    public static void main(String[] args) {
        SpringApplication.run(ConfigServerApplication.class);
    }

}
```

Spring Cloud Config Server 抽象了 EnvironmentRepository 接口用于获取配置，代码如下：

```
public interface EnvironmentRepository {

    Environment findOne(String application, String profile, String label);

    default Environment findOne(String application, String profile, String label,
            boolean includeOrigin) {
        return findOne(application, profile, label);
    }

}
```

从接口可以看到，获取 Environment 有 3 个维度：application、profile 和 label。

这里得到的 Environment 并不是 Spring Core 里的 Environment 接口，而是 Spring Cloud 封装的一个类，包括如下属性：

- name：String 配置文件名，对应 application 这个维度。
- profiles：String[] 生效的 active profile，对应 profile 这个维度。
- label：String 标签，对应 label 这个维度。
- propertySources：List<PropertySource> 加载的数据源集合。
- version：String 版本号。

- state：String 状态。

目前 EnvironmentRepository 具体的实现有 Git（JGitEnvironmentRepository）、SVN（SvnKit-EnvironmentRepository）、File System 文件系统（NativeEnvironmentRepository）、JDBC（JdbcEnvironmentRepository）、Redis（RedisEnvironmentRepository）、AWS S3 云存储（AwsS3-Environment-Repository）等。

Spring Cloud Config Server 内部带有一个对外暴露 HTTP API 的 EnvironmentController。这个 controller 提供对外暴露的各个方法会获取配置信息，在获取过程中调用 EnvironmentRepository 的 findOne 方法，代码如下：

```java
@RestController
@RequestMapping(method = RequestMethod.GET,
        path = "${spring.cloud.config.server.prefix:}")
public class EnvironmentController {

    @RequestMapping("/{name}/{profiles:.*[^-].*}")
    public Environment defaultLabel(@PathVariable String name,
            @PathVariable String profiles) {
        return getEnvironment(name, profiles, null, false);
    }

    @RequestMapping(path = "/{name}/{profiles:.*[^-].*}",
            produces = EnvironmentMediaType.V2_JSON)
    public Environment defaultLabelIncludeOrigin(@PathVariable String name,
            @PathVariable String profiles) {
        return getEnvironment(name, profiles, null, true);
    }

    @RequestMapping("/{name}/{profiles}/{label:.*}")
    public Environment labelled(@PathVariable String name, @PathVariable String profiles,
            @PathVariable String label) {
        return getEnvironment(name, profiles, label, false);
    }

    @RequestMapping(path = "/{name}/{profiles}/{label:.*}",
```

```java
            produces = EnvironmentMediaType.V2_JSON)
    public Environment labelledIncludeOrigin(@PathVariable String name,
            @PathVariable String profiles, @PathVariable String label) {
        return getEnvironment(name, profiles, label, true);
    }

    public Environment getEnvironment(String name, String profiles, String label,
            boolean includeOrigin) {
        if (name != null && name.contains("(_)")) {
            name = name.replace("(_)", "/");
        }
        if (label != null && label.contains("(_)")) {
            label = label.replace("(_)", "/");
        }
        Environment environment = this.repository.findOne(name, profiles, label,
                includeOrigin);
        if (!this.acceptEmpty
                && (environment == null || environment.getPropertySources().isEmpty())) {
            throw new EnvironmentNotFoundException("Profile Not found");
        }
        return environment;
    }
    ...
}
```

1. NativeEnvironmentRepository

下面分析文件系统 NativeEnvironmentRepository 的具体实现。

NativeEnvironmentRepository 生效的前提是配置文件里的 active profile 需要一个 native 的 profile（NativeRepositoryConfiguration 自动化配置类生效的前提是拥有 native profile）：

```java
@Configuration(proxyBeanMethods = false)
@Profile("native")
class NativeRepositoryConfiguration {
```

```
    @Bean
    public NativeEnvironmentRepository nativeEnvironmentRepository(
            NativeEnvironmentRepositoryFactory factory,
            NativeEnvironmentProperties environmentProperties) {
        return factory.build(environmentProperties);
    }
}
```

NativeEnvironmentRepository 获取配置过程的分析如下：

```
public class NativeEnvironmentRepository
        implements EnvironmentRepository, SearchPathLocator, Ordered {

    ...

    @Override
    public Environment findOne(String config, String profile, String label,
            boolean includeOrigin) {
        SpringApplicationBuilder builder = new SpringApplicationBuilder(
                PropertyPlaceholderAutoConfiguration.class); // ①
        ConfigurableEnvironment environment = getEnvironment(profile); // ②
        builder.environment(environment);
        builder.web(WebApplicationType.NONE).bannerMode(Mode.OFF);
        if (!logger.isDebugEnabled()) {
            builder.logStartupInfo(false);
        }
        String[] args = getArgs(config, profile, label);// ③
        builder.application()
                .setListeners(Arrays.asList(new ConfigFileApplicationListener())); // ④

        try (ConfigurableApplicationContext context = builder.run(args)) {
            environment.getPropertySources().remove("profiles"); // ⑤
            return clean(new PassthruEnvironmentRepository(environment).findOne(config,
                    profile, label, includeOrigin)); // ⑥
        }
    }
```

```
        catch (Exception e) {
            String msg = String.format(
                    "Could not construct context for config=%s profile=%s label=%s includeOrigin=%b",
                    config, profile, label, includeOrigin);
            String completeMessage = NestedExceptionUtils.buildMessage(msg,
                    NestedExceptionUtils.getMostSpecificCause(e));
            throw new FailedToConstructEnvironmentException(completeMessage, e);
        }
    }

...

@Override
public Locations getLocations(String application, String profile, String label) {
    String[] locations = this.searchLocations;  // ⑦
    if (this.searchLocations == null || this.searchLocations.length == 0) {
        locations = DEFAULT_LOCATIONS;
    }
    Collection<String> output = new LinkedHashSet<String>();

    if (label == null) {   // ⑧
        label = this.defaultLabel;
    }
    for (String location : locations) {    // ⑨
        String[] profiles = new String[] { profile };    // ⑩
        if (profile != null) {
            profiles = StringUtils.commaDelimitedListToStringArray(profile);
        }
        String[] apps = new String[] { application };    // ⑪
        if (application != null) {
            apps = StringUtils.commaDelimitedListToStringArray(application);
        }
        for (String prof : profiles) {
            for (String app : apps) {    // ⑫
                String value = location;
```

```
            if (application != null) {
                value = value.replace("{application}", app);
            }
            if (prof != null) {
                value = value.replace("{profile}", prof);
            }
            if (label != null) {
                value = value.replace("{label}", label);
            }
            if (!value.endsWith("/")) {
                value = value + "/";
            }
            if (isDirectory(value)) {
                output.add(value);
            }
        }
    }
    if (this.addLabelLocations) {    //⑩
        for (String location : locations) {
            if (StringUtils.hasText(label)) {
                String labelled = location + label.trim() + "/";
                if (isDirectory(labelled)) {
                    output.add(labelled);
                }
            }
        }
    }
    return new Locations(application, profile, label, this.version,
            output.toArray(new String[0]));
}

...

}
```

上述代码中：

① 构造一个 SpringApplicationBuilder，后续用于 ApplicationContext 的创建。

② 构造一个 Environment，并往其中添加一个 name 为 profiles 的数据源 PropertySource，该数据源内部有两个配置：一个是 spring.profiles.active 参数为外部传入的 profile；另一个是 spring.main.web-application-type 为 none（非 Web 应用）。

③ 构造如下启动参数：

- spring.config.name=配置文件名。如果外部传入的 application 参数不以 application 开头，则使用 application 和外部传入的 application 参数，否则使用外部传入的 application 参数。
- spring.cloud.bootstrap.enabled=false。禁止 Bootstrap 过程。
- encrypt.failOnError=false。配置解密失败的情况下是否报错，默认为不报错，可以通过 spring.cloud.config.server.native.failOnError 配置获取。
- spring.config.location=配置文件所在目录。具体逻辑查看代码⑦~⑬。

④ Environment 中删除代码②处创建的数据源，这个数据源的数据无须对外暴露。

⑤ 返回 Environment，这里会通过 PassthruEnvironmentRepository 内部过滤掉系统参数、启动参数、JNDI、Servlet 参数等标准的数据源。

⑥ 获取 spring.cloud.config.server.native.searchLocations 配置项里配置的目录。如果没有配置，则取默认配置："classpath:/"、"classpath:/config/"、"file:./"、"file:./config/"。

⑦ 获取 spring.cloud.config.server.native.defaultLabel 标签配置。如果没有配置，则取默认配置 master。

⑧ 遍历代码⑦处得到的目录集合。

⑨ 构造 profile 集合，使用外部传入的 profile 参数，并使用","分隔。

⑩ 构造 application 集合，使用外部传入的 application 参数，并使用","分隔。

⑪ 在遍历目录集合的过程中，再遍历 profile 和 application。如果目录配置中带有 {application}、{profile} 和 {label} 通配符，则会进行替换。

⑫ 再新增一些配置文件目录：遍历代码⑦处得到的目录和外部传入的 label 字段。可通过 spring.cloud.config.server.native.addLabelLocations=false 配置取消。

举个例子，客户端传递的 application、profile 分别为 book 和 prod，并且服务器端存在以下配置：

```
spring.profiles.active=native
spring.cloud.config.server.native.searchLocations=classpath:/
```

那么最终加载的配置文件名为 application 和 book，配置目录为 classpath:/和 classpath:/master/，生效的 profile 为 prod。

如果 spring.cloud.config.server.native.searchLocations 配置为 classpath:/{label}，那么最终加载的配置文件名为 application 和 book，配置目录为 classpath:/master/（classpath:/master/master/ 没有成为目录，是因为代码⑦处的 location 包含通配符，在代码⑬处中与 label 结合后还是含有通配符，不属于目录），生效的 profile 为 prod。

2. JdbcEnvironmentRepository

下面分析数据库 JdbcEnvironmentRepository 的具体实现。

JdbcEnvironmentRepository 生效的前提是配置文件里的 spring.profiles.active 配置内容需要有 jdbc 这个 profile（JdbcRepositoryConfiguration 自动化配置类生效的前提是拥有 jdbc profile）：

```
@Configuration(proxyBeanMethods = false)
@Profile("jdbc")
@ConditionalOnClass(JdbcTemplate.class)
class JdbcRepositoryConfiguration {

    @Bean
    @ConditionalOnBean(JdbcTemplate.class)
    public JdbcEnvironmentRepository jdbcEnvironmentRepository(
            JdbcEnvironmentRepositoryFactory factory,
            JdbcEnvironmentProperties environmentProperties) {
        return factory.build(environmentProperties);
    }

}
```

JdbcEnvironmentRepository 获取配置的过程分析如下：

```java
public class JdbcEnvironmentRepository implements EnvironmentRepository, Ordered {

    private final JdbcTemplate jdbc; // ①

    ...

    @Override
    public Environment findOne(String application, String profile, String label) {
        String config = application;
        if (StringUtils.isEmpty(label)) { // ②
            label = "master";
        }
        if (StringUtils.isEmpty(profile)) { // ③
            profile = "default";
        }
        if (!profile.startsWith("default")) { // ④
            profile = "default," + profile;
        }
        String[] profiles = StringUtils.commaDelimitedListToStringArray(profile);
        Environment environment = new Environment(application, profiles, label, null,
                null);
        if (!config.startsWith("application")) { // ⑤
            config = "application," + config;
        }
        List<String> applications = new ArrayList<String>(new LinkedHashSet<>(
                Arrays.asList(StringUtils.commaDelimitedListToStringArray(config))));
        List<String> envs = new ArrayList<String>(
                new LinkedHashSet<>(Arrays.asList(profiles)));
        Collections.reverse(applications);
        Collections.reverse(envs);
        for (String app : applications) {
            for (String env : envs) { // ⑥
                Map<String, String> next = (Map<String, String>) this.jdbc.query
                        (this.sql, new Object[] { app, env, label }, this.extractor); // ⑦
                if (!next.isEmpty()) {
                    environment.add(new PropertySource(app + "-" + env, next));
```

```
                }
            }
        }
        return environment;
    }
    ...
}
```

上述代码中：

① JdbcEnvironmentRepository 依赖 JdbcTemplate，构造 JdbcEnvironmentRepository 的 `JdbcRepositoryConfiguration` 配置类生效的前提是 `ClassLoader` 里有 `JdbcTemplate` 类存在。

② label 参数不传，默认使用 master。

③ profile 参数不传，默认使用 default。

④ profile 参数如果有传递且不以 default 开头，则加上"default"前缀，后续以","分隔。这里相当于在原有 profile 参数基础上再添加一个 default profile。

⑤ application 参数如果不以 application 开头，则加上"application"前缀，后续以","分隔。这里相当于在原有 application 参数基础上再添加一个 application。

⑥ 遍历 application 和 profile 集合。

⑦ 查询数据库。默认的 SQL 语句是"SELECT KEY, VALUE from PROPERTIES where APPLICATION=? and PROFILE=? and LABEL=?"。可以通过 spring.cloud.config.server.jdbc.sql 配置项进行修改。

Config Server 启动成功后访问 http://localhost:8080/book/prod，返回如下内容：

```
{
    "name": "book",
    "profiles": ["prod"],
    "label": null,
    "version": null,
    "state": null,
```

```
    "propertySources": [{
        "name": "book-prod",
        "source": {
            "book.name": "deep in spring cloud",
            "book.author": "jim",
            "book.category": "spring cloud"
        }
    }]
}
```

3. Spring Cloud Config Server 体验

下面部署一个使用文件系统方式（NativeEnvironmentRepository）的应用来体验 Spring Cloud Config Server。

在 Java 工程脚手架上选择 Config Server、Spring Web 模块和 Spring Boot Actuator，创建 spring-cloud-config-server-file 项目。

修改 application.properties 配置文件：

```
spring.application.name=sc-config-server-file
server.port=8080

spring.profiles.active=native

spring.cloud.config.server.native.searchLocations=classpath:/
```

创建 4 个配置文件，内容分别如下：

```
# resources/book.properties
book.category=spring cloud

# resources/book-prod.properties
book.name=deep in spring cloud

# resources/master/book.properties
book.author=jim
```

```
# resources/master/master/book.properties
book.publishYear=2020
```

编写启动类:

```
@SpringBootApplication
@EnableConfigServer
public class ConfigServerFileApplication {

    public static void main(String[] args) {
        SpringApplication.run(ConfigServerFileApplication.class);
    }

}
```

启动后,访问 EnvironmentController 对外暴露的 API: http://localhost:8080/book/prod (参照之前的例子中客户端传递的 application、profile 分别为 book 和 prod):

```
{
    "name": "book",
    "profiles": ["prod"],
    "label": null,
    "version": null,
    "state": null,
    "propertySources": [{
        "name": "classpath:/book-prod.properties",
        "source": {
            "book.name": "deep in spring cloud"
        }
    }, {
        "name": "classpath:/master/book.properties",
        "source": {
            "book.author": "jim"
        }
    }, {
        "name": "classpath:/book.properties",
```

```
        "source": {
            "book.category": "spring cloud"
        }
    }]
}
```

这个 json 验证了我们之前的分析：

- book 和 name 通过参数传入。
- 被加载的配置目录为 classpath:/ 和 classpath:/master/。这两个目录下的配置文件都被加载。

如果 spring.cloud.config.server.native.searchLocations 配置为 classpath:/{label}，那么返回的 json 如下：

```
{
    "name": "book",
    "profiles": ["prod"],
    "label": null,
    "version": null,
    "state": null,
    "propertySources": [{
        "name": "classpath:/master/book.properties",
        "source": {
            "book.author": "jim"
        }
    }]
}
```

同样，这个 json 验证了另一个例子的分析：classpath:/master/master/没有成为配置目录。这里仅使用了 searchLocations 配置项。如表 4-2 所示，列出了 NativeEnvironmentRepository 其他的配置项（通过 NativeEnvironmentProperties 配置类）。

表 4-2

配 置 项	作 用	默 认 值
spring.cloud.config.server.native.failOnError	解密失败的情况下是否抛出异常	false

续表

配 置 项	作 用	默 认 值
spring.cloud.config.server.native.addLabelLocations	是否添加配置目录 searchLocation 与 label 结合的目录（例中 classpath:/master/目录就是结合的目录）	true
spring.cloud.config.server.native.defaultLabel	默认为 master，如果用户配置，则使用配置的 label	master
spring.cloud.config.server.native.searchLocations	配置文件目录	classpath:/、classpath:/config/、file:./和 file:./config/
spring.cloud.config.server.native.version	Spring Cloud Environment 对应的版本号	

Spring Cloud Config Server 对应的客户端 Spring Cloud Config Client 内部会使用 EnvironmentController 获取配置信息。

下面来看 Spring Cloud Config Client 内部的实现原理。

4.4.2　Spring Cloud Config Client

Spring Cloud Config Client 是一个读取 Config Server 配置的客户端。如果想要获取 Spring Cloud Config Server 上的配置，可以引入 Spring Cloud Config Client 组件。

Spring Cloud Config Client 内部的 ConfigServicePropertySourceLocator 在 Bootstrap 阶段会通过 RestTemplate 请求 Config Server 内部 EnvironmentController 对外暴露的接口：

```
@Order(0)
public class ConfigServicePropertySourceLocator implements PropertySourceLocator {

    private RestTemplate restTemplate;

    private ConfigClientProperties defaultProperties;

    public ConfigServicePropertySourceLocator(ConfigClientProperties defaultProperties) {
        this.defaultProperties = defaultProperties;
    }
```

```java
@Override
@Retryable(interceptor = "configServerRetryInterceptor")
public org.springframework.core.env.PropertySource<?> locate(
        org.springframework.core.env.Environment environment) {
    ConfigClientProperties properties = this.defaultProperties.override(environment); // ①
    CompositePropertySource composite = new OriginTrackedCompositePropertySource(
            "configService");
    RestTemplate restTemplate = this.restTemplate == null
            ? getSecureRestTemplate(properties) : this.restTemplate; // ②
    Exception error = null;
    String errorBody = null;
    try {
        String[] labels = new String[] { "" };
        if (StringUtils.hasText(properties.getLabel())) {
            labels = StringUtils
                    .commaDelimitedListToStringArray(properties.getLabel()); // ③
        }
        String state = ConfigClientStateHolder.getState();
        for (String label : labels) { // ④
            Environment result = getRemoteEnvironment(restTemplate, properties,
                    label.trim(), state); // ⑤
            if (result != null) {
                log(result);

                if (result.getPropertySources() != null) {
                    for (PropertySource source : result.getPropertySources()) { // ⑥
                        @SuppressWarnings("unchecked")
                        Map<String, Object> map = translateOrigins(source.getName(),
                                (Map<String, Object>) source.getSource());
                        composite.addPropertySource(
                                new OriginTrackedMapPropertySource
                                        (source.getName(), map));
                    }
                }
```

```java
                }

                if (StringUtils.hasText(result.getState())
                        || StringUtils.hasText(result.getVersion())) {
                    HashMap<String, Object> map = new HashMap<>();
                    putValue(map, "config.client.state", result.getState());
                    putValue(map, "config.client.version", result.getVersion());
                    composite.addFirstPropertySource(
                            new MapPropertySource("configClient", map)); // ⑦
                }
                return composite; // ⑧
            }
        }
        errorBody = String.format("None of labels %s found", Arrays.toString(labels));
    }
    catch (HttpServerErrorException e) {
        error = e;
        if (MediaType.APPLICATION_JSON
                .includes(e.getResponseHeaders().getContentType())) {
            errorBody = e.getResponseBodyAsString();
        }
    }
    catch (Exception e) {
        error = e;
    }
    if (properties.isFailFast()) {
        throw new IllegalStateException(
                "Could not locate PropertySource and the fail fast property is
                    set, failing" + (errorBody == null ? "" : ": " + errorBody),
                error);
    }
    logger.warn("Could not locate PropertySource: "
            + (error != null ? error.getMessage() : errorBody));
    return null;
}
```

```
    ...
}
```

上述代码中：

① 覆盖 ConfigClientProperties 内部的 3 个维度配置：application、profile 和 label。具体的覆盖方式请参考 Spring Cloud Config Client 配置表。

② 构造 RestTemplate，会根据 ConfigClientProperties 内部的配置设置一些属性。比如连接超时时间、读取超时时间、ClientHttpRequestInterceptor 拦截器等内容。

③ 读取 spring.cloud.config.label 配置，这是 label 维度的查找条件。多个 label 以 "," 分隔。

④ 遍历找到的 label。

⑤ 根据 Rest API 获取 Config Server 上的配置，得到 Spring Cloud Environment 实例。请求的 URI 由 spring.cloud.config.uri 配置得到，path 由 /{name}/{profile}/{label} 组成，分别表示 3 个查找维度。

⑥ 添加 Spring Cloud Environment 内部的数据源 PropertySource 集合到方法一开始构造的 OriginTrackedCompositePropertySource 数据源中。

⑦ 添加一个 MapPropertySource 到一开始构造的 OriginTrackedCompositePropertySource 数据源中，里面包含了 Spring Cloud Environment 内部的 version 和 state。

⑧ 返回这个 OriginTrackedCompositePropertySource，内部的数据源 PropertySource 集合将会被加载到应用的 Environment 中。

Spring Cloud Config Client 的配置项如表 4-3 所示。

表 4-3

配 置 项	作 用	默 认 值
spring.cloud.config.name	application 维度的查找条件，也就是配置文件名。如果未配置，取 spring.application.name 配置，如还未配置，则使用 application	application
spring.cloud.config.enabled	是否启用 Spring Cloud Config Client	true

续表

配 置 项	作 用	默 认 值
spring.cloud.config.profile	profile 维度的查找条件，多个 profile 以 "," 分隔	default
spring.cloud.config.label	label 维度的查找条件，多个 label 以 "," 分隔	
spring.cloud.config.username	基于 Basic Authentication 认证的 username	user
spring.cloud.config.password	基于 Basic Authentication 认证的 password	
spring.cloud.config.uri	Config Server 对外提供服务的地址。该属性是一个 String 数组，支持多个 Config Server 获取配置	http://localhost:8888
spring.cloud.config.failFast	当从 Config Server 获取配置发生错误时是否抛出异常	false
spring.cloud.config.requestReadTimeout	Config Server 读取超时时间	185s
spring.cloud.config.requestConnectTimeout	Config Server 链接超时时间	10s
spring.cloud.config.sendState	是否在 HEADER 中添加 key 为 X-Config-State 的状态。Config Client 内部有一个调度器专门维护这个状态，每次跟 Config Server 返回的 Spring Cloud Environemtn 内部的 state 字段进行比较	true
spring.cloud.config.headers	自定义 HTTP Header、Map 类型	

在 Java 工程脚手架上选择 Config Client，创建 spring-cloud-config-client 项目。

添加 bootstrap.properties 配置文件如下：

```
spring.application.name=sc-config-client
spring.cloud.config.name=book
spring.cloud.config.profile=prod
spring.cloud.config.uri=http://localhost:8080/
```

从这个配置文件可以看到，我们使用了 application 为 book、profile 为 prod，从 Config Server 地址 http://localhost:8080/ 中获取 Spring Cloud Environment 配置。

编写启动类：

```java
@SpringBootApplication
public class ConfigClientApplication {

    public static void main(String[] args) {
        new SpringApplicationBuilder().web(WebApplicationType.NONE)
            .sources(ConfigClientApplication.class).run(args);
    }

    @Autowired
    Environment env;

    @Bean
    public CommandLineRunner runner() {
        return (args) -> {
            System.out.println(
                "book.categoty=" + env.getProperty("book.category", "unknown")
            );
            System.out.println(
                "book.author=" + env.getProperty("book.author", "unknown")
            );
            System.out.println(
                "book.name=" + env.getProperty("book.name", "unknown")
            );
        };
    }
}
```

应用启动后打印：

```
book.categoty=spring cloud
book.author=jim
book.name=deep in spring cloud
```

4.4.3　Spring Cloud Config Client 与 Service Discovery 整合

Spring Cloud Config Client 提供的 spring.cloud.config.uri 配置虽然支持多个 Config Server 地址，但地址增多后，每个地址就需要手动配置，这个过程并不优雅。

Spring Cloud Config Client 将 Config Server 地址的获取跟注册中心配合（需要配置 spring.cloud.config.discovery.enabled 为 true），从注册中心获取服务名对应的所有实例，并使用这些实例对应的地址作为 Config Server 地址。默认的服务名是 configserver，可以通过 spring.cloud.config.discovery.serviceId 配置进行修改。这个过程在 DiscoveryClientConfigService-BootstrapConfiguration 类中完成：

```java
@ConditionalOnProperty(value = "spring.cloud.config.discovery.enabled",
        matchIfMissing = false) // ①
@Configuration(proxyBeanMethods = false)
@Import({ UtilAutoConfiguration.class })
@EnableDiscoveryClient  // ②
public class DiscoveryClientConfigServiceBootstrapConfiguration
        implements SmartApplicationListener {

    @Autowired
    private ConfigClientProperties config;

    ...

    @Override
    public void onApplicationEvent(ApplicationEvent event) {
        if (event instanceof ContextRefreshedEvent) {  // ③
            startup((ContextRefreshedEvent) event);
        }
        else if (event instanceof HeartbeatEvent) {  // ④
            heartbeat((HeartbeatEvent) event);
        }
    }

    public void startup(ContextRefreshedEvent event) {
        refresh();
```

```java
    }

    public void heartbeat(HeartbeatEvent event) {
        if (this.monitor.update(event.getValue())) {
            refresh();
        }
    }

    private void refresh() {
        try {
            String serviceId = this.config.getDiscovery().getServiceId();    // ⑤
            List<String> listOfUrls = new ArrayList<>();
            List<ServiceInstance> serviceInstances = this.instanceProvider
                    .getConfigServerInstances(serviceId);    // ⑥

            for (int i = 0; i < serviceInstances.size(); i++) {

                ServiceInstance server = serviceInstances.get(i);
                String url = getHomePage(server);    // ⑦

                if (server.getMetadata().containsKey("password")) {    // ⑧
                    String user = server.getMetadata().get("user");
                    user = user == null ? "user" : user;
                    this.config.setUsername(user);
                    String password = server.getMetadata().get("password");
                    this.config.setPassword(password);
                }

                if (server.getMetadata().containsKey("configPath")) {    // ⑨
                    String path = server.getMetadata().get("configPath");
                    if (url.endsWith("/") && path.startsWith("/")) {
                        url = url.substring(0, url.length() - 1);
                    }
                    url = url + path;
                }
```

```
            listOfUrls.add(url);
        }

        String[] uri = new String[listOfUrls.size()];
        uri = listOfUrls.toArray(uri);
        this.config.setUri(uri);   // ⑩

    }
    catch (Exception ex) {
        if (this.config.isFailFast()) {
            throw ex;
        }
        else {
            logger.warn("Could not locate configserver via discovery", ex);
        }
    }
}
...
}
```

上述代码中：

① spring.cloud.config.discovery.enabled 配置为 true 时，配置中心地址从注册中心获取的这个过程才会生效。

② 默认会把客户端注册到注册中心。

③ 发生 ContextRefreshedEvent 事件（ApplicationContext 初始化并刷新后的事件）后，从注册中心获取配置中心地址。

④ HeartbeatEvent 事件（注册中心实例变化后会发送心跳事件）每次发送时，都会从注册中心获取配置中心地址。

⑤ 获取 spring.cloud.config.discovery.serviceId 配置项对应的服务名，默认为 configserver。

⑥ 根据服务名得到服务实例 ServiceInstance 集合。

⑦ 遍历服务实例，获取每个实例对应的注册地址。

⑧ 如果服务实例的 metadata 中存在 password，会从 metadata 中获取 username 和 password 作为 Basic Authentication 认证的 username 和 password。

⑨ 如果 metadata 存在 configpath，会作为配置中心的 context path。

⑩ 设置注册中心地址。

4.4.4 Spring Cloud Config 配置动态刷新

下面以 NativeEnvironmentRepository 为例，在 Config Server 中将 resources/master/book.properties 配置文件中的内容改为：

book.author=jim fang

根据之前的分析，我们需要触发 RefreshEvent 事件才能动态刷新配置。RefreshEvent 事件触发后会调用 ContextRefresher 的 refresh 方法进行动态刷新。

Spring Cloud 对外提供了 RefreshEndpoint，代码如下：

```
@Endpoint(id = "refresh")
public class RefreshEndpoint {

    private ContextRefresher contextRefresher;

    public RefreshEndpoint(ContextRefresher contextRefresher) {
        this.contextRefresher = contextRefresher;
    }

    @WriteOperation
    public Collection<String> refresh() {
        Set<String> keys = this.contextRefresher.refresh();
        return keys;
    }
}
```

RefreshEndpoint 内部也调用了 ContextRefresher 的 refresh 方法。我们在 Config Client 客户端调用 RefreshEndpoint，也可以实现动态刷新（配置文件加上 management.endpoints.web.exposure.include=* 配置启用 RefreshEndpoint）。

RefreshEndpoint 对应的 ID 为 refresh，使用 curl 调用 curl -XPOST http://localhost:8081/actuator/refresh，返回["book.author"]，表示这个配置项对应的配置已经发生了改变。

试想一下：在分布式系统中，如果一个服务对应 100 个应用实例，这 100 个实例使用 Spring Cloud Config Client 加载配置，如果需要动态刷新配置，是否要在每一个实例中调用 RefreshEndpoint？这显然是不合理的。Spring Cloud Bus（消息总线）能解决这个问题，其具体内容将在第 7 章介绍。

4.5 再谈配置动态刷新

我们在 4.3.3 节已介绍过 Spring Cloud 配置动态刷新的原理。

Spring Cloud 配置动态刷新本质上是事件的触发让客户端再一次从配置中心拉取最新的配置，然后配合@RefreshScope 重新刷新 Bean（重新刷新 Spring Bean 会重新读取一遍配置），以达到配置动态刷新的目的。这种方式会带来以下问题：

- Spring Cloud 提供的 Spring Cloud Config Server/Client 组件在分布式环境下需要配合 Spring Cloud Bus（消息总线），来达到多节点配置动态刷新的目的。因此，使用 Spring Cloud Config Server/Client 组件需要额外加上消息中间件组件。
- @RefreshScope 引起的 Spring Bean 刷新可能会与其他组件冲突。比如，spring-context 模块的 scheduling 调度功能在 Spring Bean 被刷新后会失去作用。

笔者认为配置管理更应该专注在配置上，而不应该有其他操作，比如需要引入消息中间件，或者需要重新刷新 Spring Bean 等操作。

Nacos 内部的配置动态刷新原理是通过客户端维护长轮询的任务，定时拉取发生变更的配置信息，然后将最新的数据推送给客户端 Listener 的持有者。下面来看一下这个长轮询的实现原理。

Nacos 配置所有的操作通过 NacosConfigService 完成，NacosConfigService 的构造函数内部会构造一个 ClientWorker，这个 ClientWorker 维护长轮询任务：

```
public NacosConfigService(Properties properties) throws NacosException {
```

```
    String encodeTmp = properties.getProperty(PropertyKeyConst.ENCODE);
    if (StringUtils.isBlank(encodeTmp)) {
        encode = Constants.ENCODE;
    } else {
        encode = encodeTmp.trim();
    }
    initNamespace(properties);
    agent = new MetricsHttpAgent(new ServerHttpAgent(properties));
    agent.start();
    worker = new ClientWorker(agent, configFilterChainManager, properties);
}
```

ClientWorker 的构造函数会构造线程池，并且每隔 10ms 会发起 LongPollingRunnable 任务：

```
public ClientWorker(final HttpAgent agent, final ConfigFilterChainManager
configFilterChainManager, final Properties properties) {
    ...
    executor = Executors.newScheduledThreadPool(1,

    executor.scheduleWithFixedDelay(new Runnable() {
        @Override
        public void run() {
            try {
                checkConfigInfo();
            } catch (Throwable e) {
                LOGGER.error("[" + agent.getName() + "] [sub-check] rotate check
error", e);
            }
        }
    }, 1L, 10L, TimeUnit.MILLISECONDS);
}

public void checkConfigInfo() {
    int listenerSize = cacheMap.get().size();
    int longingTaskCount = (int) Math.ceil(listenerSize / ParamUtil.getPerTaskConfigSize());
    if (longingTaskCount > currentLongingTaskCount) {
```

```
        for (int i = (int) currentLongingTaskCount; i < longingTaskCount; i++) {
            executorService.execute(new LongPollingRunnable(i));
        }
        currentLongingTaskCount = longingTaskCount;
    }
}
```

LongPollingRunnable 任务内部的逻辑定义如下：

```
class LongPollingRunnable implements Runnable {
    private int taskId;

    public LongPollingRunnable(int taskId) {
        this.taskId = taskId;
    }

    @Override
    public void run() {

        List<CacheData> cacheDatas = new ArrayList<CacheData>();
        List<String> inInitializingCacheList = new ArrayList<String>();
        try {
            for (CacheData cacheData : cacheMap.get().values()) {  // ①
                if (cacheData.getTaskId() == taskId) {
                    cacheDatas.add(cacheData);
                    try {
                        checkLocalConfig(cacheData);  // ②
                        if (cacheData.isUseLocalConfigInfo()) {
                            cacheData.checkListenerMd5();
                        }
                    } catch (Exception e) {
                        LOGGER.error("get local config info error", e);
                    }
                }
```

```java
                List<String> changedGroupKeys = checkUpdateDataIds(cacheDatas,
inInitializingCacheList);    // ③

            for (String groupKey : changedGroupKeys) {
                String[] key = GroupKey.parseKey(groupKey);
                String dataId = key[0];
                String group = key[1];
                String tenant = null;
                if (key.length == 3) {
                    tenant = key[2];
                }
                try {
                    String content = getServerConfig(dataId, group, tenant, 3000L); // ④
                    CacheData cache = cacheMap.get().get(GroupKey.getKeyTenant
(dataId, group, tenant));
                    cache.setContent(content);
                    LOGGER.info("[{}] [data-received] dataId={}, group={}, 
tenant={}, md5={}, content={}",
                        agent.getName(), dataId, group, tenant, cache.getMd5(),
                        ContentUtils.truncateContent(content));
                } catch (NacosException ioe) {
                    String message = String.format(
                        "[%s] [get-update] get changed config exception. dataId=%s, 
group=%s, tenant=%s",
                        agent.getName(), dataId, group, tenant);
                    LOGGER.error(message, ioe);
                }
            }
            
            ...
            
            executorService.execute(this);
            
        } catch (Throwable e) {
        
        LOGGER.error("longPolling error : ", e);
```

```
            executorService.schedule(this, taskPenaltyTime, TimeUnit.MILLISECONDS);
        }
    }
}
```

上述代码中：

① Nacos 所有的配置对应的 dataId、group 和 tenant 租户都会缓存在 cacheMap 属性内。

② 每个 dataId/group 对应的配置项更新配置的时候都先从本地读取，本地文件目录默认是当前用户的 nacos/config 目录。这是一个容灾措施，客户端可以在 Nacos 配置中心无法工作的时候从本地获取配置。

③ checkUpdateDataIds 方法内部从服务器端获取发生了变化的 dataId 列表。

④ getServerConfig 方法内部根据 dataId 从服务器端获取最新的配置信息，然后将最新的配置信息保存到 CacheData 中。

从 Nacos 配置的源码可看出，Nacos 配置是一个纯粹的配置操作，不掺杂其他处理事项。

Spring Cloud Alibaba 体系内提供的 Nacos Config Spring Boot Starter 是一个使用 Nacos 完成配置动态刷新的 Starter。该 Starter 提供如下注解：

- @NacosValue。类似 Spring 体系的@Value 注解，并拥有动态刷新的能力，比如，@NacosValue(value = "${book.name}", autoRefreshed = true)。
- @NacosConfigListener。一个全新的基于 dataId 和 group 的注解，作用到回调方法上，比如，@NacosConfigListener(dataId = "deep.in.springcloud.method")。
- @NacosConfigurationProperties。类似 Spring 体系的 @ConfigurationProperties 体系，可以加载基于 dataId 和 group 的配置，比如，@NacosConfigurationProperties(dataId = "deep.in.springcloud.cp")。

这些注解可以理解为是 Spring 原生注解的加强版本。

4.6 案例：Spring Cloud 应用流量控制策略动态生效

我们在 3.8.2 节使用 Netflix Ribbon 完成了应用灰度发布，但是流量识别代码是写"死"的，我们只认为 Header 中 Gray=true 的流量才是灰度流量。流量识别代码如下：

```
if (request.getHeaders().containsKey("Gray")) {
    String value = request.getHeaders().getFirst("Gray");
    if (value.equals("true")) {
        RibbonRequestContextHolder.getCurrentContext().put("Gray", Boolean.TRUE.toString());
    }
}
```

接下来把这个流量识别的规则配置放到配置中心里,并使其动态生效,整个过程如图 4-2 所示。

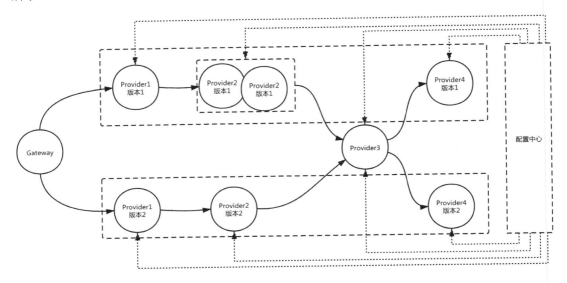

图 4-2

首先,定义流量识别规则 TrafficRule:

```
public class TrafficRule {

    private String type;

    private String name;

    private String value;
```

```
    ...
}
```

代码中，type 表示请求参数的类型，只有 param、header 这两个选项。name 是这些类型的 key，value 是这个 key 对应的值。比如 type=header、name=Gray、value=true，表示 Header 中 Gray=true 的流量才是灰度流量；type=param、name=user、value=test，表示 Query Parameter 中 user=test 的请求才是灰度流量。

其次，修改拦截器里的流量识别逻辑。

修改 OpenFeign 拦截器的代码如下：

```
public class GrayRequestInterceptor implements RequestInterceptor {

    private TrafficRule rule;

    public GrayRequestInterceptor(TrafficRule rule) {
        this.rule = rule;
    }

    @Override
    public void apply(RequestTemplate template) {
        if (rule.getType().equalsIgnoreCase("header")) {
            if (template.headers().containsKey(rule.getName())) {
                String value = template.headers().get(rule.getName()).iterator().next();
                if (value.equals(rule.getValue())) {
                    RibbonRequestContextHolder.getCurrentContext().put("Gray", Boolean.TRUE.toString());
                }
            }
        } else if (rule.getType().equalsIgnoreCase("param")) {
            if (template.queries().containsKey(rule.getName())) {
                String value = template.queries().get(rule.getName()).iterator().next();
                if (value.equals(rule.getValue())) {
                    RibbonRequestContextHolder.getCurrentContext().put("Gray", Boolean.TRUE.toString());
```

```
                }
            }
        }
    }
}
```

修改 RestTemplate 拦截器的代码如下：

```
public class GrayInterceptor implements ClientHttpRequestInterceptor {

    private TrafficRule rule;

    public GrayInterceptor(TrafficRule rule) {
        this.rule = rule;
    }

    @Override
    public ClientHttpResponse intercept(HttpRequest request, byte[] body,
ClientHttpRequestExecution execution)
        throws IOException {
        if (rule.getType().equalsIgnoreCase("header")) {
            if (request.getHeaders().containsKey(rule.getName())) {
                String value = request.getHeaders().get(rule.getName()).iterator().
next();
                if (value.equals(rule.getValue())) {
                    RibbonRequestContextHolder.getCurrentContext().put("Gray",
Boolean.TRUE.toString());
                }
            }
        } else if (rule.getType().equalsIgnoreCase("param")) {
            String query = request.getURI().getQuery();
            String[] queryKV = query.split("&");
            for(String queryItem : queryKV) {
                String[] queryInfo = queryItem.split("=");
                if (queryInfo[0].equalsIgnoreCase(rule.getName()) &&
queryInfo[1].equals(rule.getValue())) {
                    RibbonRequestContextHolder.getCurrentContext().put("Gray",
Boolean.TRUE.toString());
                }
```

```
            }
        }
        return execution.execute(request, body);
    }
}
```

最后，验证修改 Nacos 配置里的规则是否可以让灰度动态生效。

先在 Nacos 配置中心配置规则内容：

```
traffic.rule.type=header
traffic.rule.name=Gray
traffic.rule.value=true
```

访问地址查看灰度是否生效：

```
curl -s -H "Gray:true" http://localhost:8888/echoFeign
curl -s -H "Gray:true" http://localhost:8888/echo
```

然后把规则内容的 header 改成新的值：

```
traffic.rule.type=header
traffic.rule.name=Gray
traffic.rule.value=test
```

这时需要修改 Header 内容才会生效：

```
curl -s -H "Gray:test" http://localhost:8888/echoFeign
curl -s -H "Gray:test" http://localhost:8888/echo
```

第 5 章

熔断器

本章将介绍熔断器模式的知识并手动实现一个熔断器。然后介绍目前业界最流行的 3 个开源熔断器 Alibaba Sentinel、Netflix Hystrix 和 Resilience4j，以及它们与 Spring Cloud 生态的融合和熔断器之间的对比。最后介绍使用 Sentinel 保护应用以防止服务雪崩的案例，来加深读者对熔断器的理解。

5.1 熔断器模式概述

熔断器（也叫断路器）不但在软件设计中会遇到，在现实生活中也会经常遇到。例如，有装修经验的朋友一定知道强电箱，强电箱上会安装断路器，用于安全用电。当发生一些特殊情况，比如水电工师傅在厨房水槽下面安装开关面板时，在安装过程中，如果不小心碰到了打开的水龙头，当水和电接触时，强电箱上的总开关就会自动跳闸，断掉所有的电来保护水电工师傅。

再如，群租房的朋友可能遇到过这种场景，一套房里隔断的房间太多，每个房间都安装了空调。夏天到来的时候，每个房间都开着空调，导致用电负载过大，也会自动跳闸。

同样，在软件系统中也会遇到类似的情况。比如，应用依赖的外部 API 不可用，该应用一直在超时并重试，从而引发应用整体无法提供服务，如果这时能有一个类似断路器的组件（模式）让客户端在发生多次调用失败的情况下不再重试，继续调用，则可以让服务器能够健康地运行。这个组件（模式）就是熔断器模式（Circuit Breaker Pattern）。如图 5-1 所示，微软的云计算设计模式文档对断路器模式的实现进行了说明。

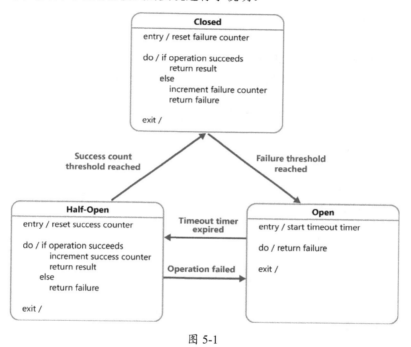

图 5-1

从图 5-1 中可以看到，熔断器有 3 种状态：Closed（关闭）、Half-Open（半开启）和 Open（开启）。

- Closed 状态：为默认状态。Circuit Breaker 内部维护着最近失败的次数（failure count）。操作每失败一次，就会增加一次失败次数。当错误数达到一定阈值时，就会从 Closed 状态进入 Open 状态。
- Open 状态：操作不会执行，而是立即失败，Circuit Breaker 内部维护着一个计时器，当时间达到一定阈值时，就会从 Open 状态进入 Half-Open 状态。
- Half-Open 状态：只要操作失败一次，立即进入 Open 状态；否则增加操作成功的次数，当成功次数达到一定阈值时，进入 Closed 状态并重置失败次数。

5.2　手动实现一个断路器

下面按照微软云计算设计模式中的说明（见图 5-1），实现一个简单的断路器。

要实现自定义的断路器，最重要的就是状态之间的转换。3 种状态之间的转换逻辑如下：

- Closed→Open：指定的时间窗口内（比如 2s 内），Closed 状态进入 Open 状态需要满足的错误个数阈值。
- Open→Half-Open：Open 状态进入 Half-Open 状态的超时时间。
- Half-Open→Closed：Half-Open 状态进入 Open 状态的成功个数阈值。

5.2.1　定义 State 枚举和 Counter 计数器类

State 枚举类型的定义如下：

```
public enum State {

    CLOSED,
    HALF_OPEN,
    OPEN

}
```

从状态转换的条件看，我们需要 4 个参数来维护状态的转换。

```
public class Config {
    // Closed 状态进入 Open 状态的错误个数阈值
    private int failureCount = 5;

    // failureCount 统计时间窗口
    private long failureTimeInterval = 2 * 1000;

    // Open 状态进入 Half-Open 状态的超时时间
    private int halfOpenTimeout = 5 * 1000;

    // Half-Open 状态进入 Open 状态的成功个数阈值
```

```
    private int halfOpenSuccessCount = 2;
}
```

时间窗口的统计需要临时保存上一次调用失败的时间戳（lastTime），该时间戳需要与当前时间进行比较，确认是否超过时间窗口；当前失败次数也需要记录，该次数会与错误个数阈值进行比较，确认是否进入 Open 状态。我们把这些统计相关的逻辑放到 Counter 类里：

```java
public class Counter {

    // Closed 状态进入 Open 状态的错误个数阈值
    private final int failureCount;

    // failureCount 统计时间窗口
    private final long failureTimeInterval;

    // 当前错误次数
    private final AtomicInteger currentCount;

    // 上一次调用失败的时间戳
    private long lastTime;

    // Half-Open 状态下成功次数
    private final AtomicInteger halfOpenSuccessCount;

    public Counter(int failureCount, long failureTimeInterval) {
        this.failureCount = failureCount;
        this.failureTimeInterval = failureTimeInterval;
        this.currentCount = new AtomicInteger(0);
        this.halfOpenSuccessCount = new AtomicInteger(0);
        this.lastTime = System.currentTimeMillis();
    }

    public synchronized int incrFailureCount() {
        long current = System.currentTimeMillis();
        if (current - lastTime > failureTimeInterval) { // 超过时间窗口，当前失败次数
重置为 0
```

```
            lastTime = current;
            currentCount.set(0);
        }
        return currentCount.getAndIncrement();
    }

    public int incrSuccessHalfOpenCount() {
        return this.halfOpenSuccessCount.incrementAndGet();
    }

    public boolean failureThresholdReached() {
        return getCurCount() >= failureCount;
    }

    public int getCurCount() {
        return currentCount.get();
    }

    public synchronized void reset() {
        halfOpenSuccessCount.set(0);
        currentCount.set(0);
    }

}
```

5.2.2 定义 CircuitBreaker 类

CircuitBreaker 类用于状态的转换和代码的执行：

```
public class CircuitBreaker {

    private State state;

    private Config config;

    private Counter counter;
```

```java
    private long lastOpenedTime;

    public CircuitBreaker(Config config) {
        this.counter = new Counter(config.getFailureCount(),
config.getFailureTimeInterval());
        this.state = CLOSED;
        this.config = config;
    }

    public <T> T run(Supplier<T> toRun, Function<Throwable, T> fallback) {
        try {
            if (state == OPEN) {
                // 判断 Half-Open 是否超时
                if (halfOpenTimeout()) {
                    return halfOpenHandle(toRun, fallback);
                }
                return fallback.apply(new DegradeException("degrade by circuit breaker"));
            } else if (state == CLOSED) {
                T result = toRun.get();
                closed();
                return result;
            } else {
                return halfOpenHandle(toRun, fallback);
            }
        } catch (Exception e) {
            counter.incrFailureCount();
            if (counter.failureThresholdReached()) { // 错误次数达到阈值，进入 Open 状态
                open();
            }
            return fallback.apply(e);
        }
    }

    private <T> T halfOpenHandle(Supplier<T> toRun, Function<Throwable, T> fallback) {
```

```java
        try {
            halfOpen(); // Closed 状态超时进入 Half-Open 状态
            T result = toRun.get();
            int halfOpenSuccCount = counter.incrSuccessHalfOpenCount();
            if (halfOpenSuccCount >= this.config.getHalfOpenSuccessCount()) {
// Half-Open 状态成功次数到达阈值，进入 Closed 状态
                closed();
            }
            return result;
        } catch (Exception e) {
            // Half-Open 状态发生一次错误进入 Open 状态
            open();
            return fallback.apply(new DegradeException("degrade by circuit breaker"));
        }
    }

    private boolean halfOpenTimeout() {
        return System.currentTimeMillis() - lastOpenedTime > config.getHalfOpenTimeout();
    }

    private void closed() {
        counter.reset();
        state = CLOSED;
    }

    private void open() {
        state = OPEN;
        lastOpenedTime = System.currentTimeMillis();
    }

    private void halfOpen() {
        state = HALF_OPEN;
    }
}
```

CircuitBreaker 对外暴露了 run 方法，run 方法对外暴露了以下两个参数：

- toRun: Supplier<T>：逻辑执行体，比如 () -> { doSomeThing(); return result; }。
- fallback：Function<Throwable, T>：熔断或发生错误之后的 fallback 处理，比如 throwable -> { System.err.println(t); }。

5.2.3 使用 CircuitBreaker 进行场景测试

下面使用 CircuitBreaker 来完成 3 个场景的测试。

（1）正常逻辑下的 run 方法

```
CircuitBreaker cb = new CircuitBreaker(new Config());
String bookName = cb.run(() -> {
    return "deep in spring cloud";
}, t -> {
    return "boom";
});
System.out.println(bookName);
```

该场景最终在控制台上打印结果"deep in spring cloud"。因为这里的 CircuitBreaker 一直处于 Closed 状态，我们在逻辑体内仅仅返回了 deep in spring cloud 字符串。

（2）发生异常逻辑下的 run 方法

```
CircuitBreaker cb = new CircuitBreaker(new Config());
RestTemplate restTemplate = new RestTemplate();
String result = cb.run(() -> {
    return restTemplate.getForObject("https://httpbin.org/status/500", String.class);
}, t -> {
    return "boom";
});
System.out.println(result);
```

该场景最终在控制台上打印结果"boom"。因为这个 HTTP 请求返回了 Response Code 为 500 的响应（https://httpbin.org/是一个对外提供 HTTP 各种操作测试的网站，这里模拟 Response

Code 为 500 的请求），RestTemplate 调用发生了 HttpServerErrorException$InternalServerError 异常。

（3）触发 Open 状态的 run 方法

```java
CircuitBreaker cb = new CircuitBreaker(new Config());
RestTemplate restTemplate = new RestTemplate();

int degradeCount = 0;

for(int index = 0; index < 10; index ++) {
    String result = cb.run(() -> {
        return restTemplate.getForObject("https://httpbin.org/status/500", String.class);
    }, t -> {
        if(t instanceof DegradeException) {
            return "degrade";
        }
        return "boom";
    });
    if(result.equals("degrade")) {
        degradeCount ++;
    }
}

System.out.println(degradeCount);
```

该场景最终在控制台上打印结果 5。因为默认情况下的失败阈值是 5 次，这 10 次请求中，前面 5 次都发生异常，第 6~10 次触发了降级，抛出的异常是 DegradeException。

在这个例子中，如果将失败阈值改成 2 次，那么最终会打印结果 8。

```java
Config config = new Config();
config.setFailureCount(2);
CircuitBreaker cb = new CircuitBreaker(config);
...
```

5.3 Spring Cloud Circuit Breaker 的技术演进

Spring Cloud 的早期版本对 Circuit Breaker 的支持仅仅提供了一个注解：@EnableCircuitBreaker。该注解的作用也仅仅是导入 EnableCircuitBreakerImportSelector 这个 Selector。Selector 继承了 SpringFactoryImportSelector，在 selectImports 方法内使用工厂加载机制 key 为 org.springframework.cloud.client.circuitbreaker.EnableCircuitBreaker 的配置类，并加载这些配置类。

Circuit Breaker 在 Spring Cloud 中没有任何接口或任何类，仅仅只是通过@EnableCircuitBreaker 注解加载熔断器组件（如 Alibaba Sentinel、Netflix Hystrix、Resilience4J）对应的自动化配置类去完成组件的加载。

Spring Cloud 在 Hoxton.M3 中新创建了 spring-cloud-circuitbreaker 模块，该模块内部的代码对 Circuit Breaker 相关的技术进行了抽象。后续在 Hoxton.RELEASE 版本发布的时候，spring-cloud-circuitbreaker 模块被移除，其相关的抽象代码被移到了 spring-cloud-commons 模块中，具体的实现代码由相应的 Spring Cloud 实现托管，比如，Sentinel 由 Spring Cloud Alibaba 托管，Hystrix 由 Spring Cloud Netflix 托管。目前 Resilience4J、Sentinel 和 Hystrix 是 Spring Cloud Circuit Breaker 官方推荐的选择。

spring-cloud-circuitbreaker 定义了 CircuitBreakerFactory，用于创建 CircuitBreaker，CircuitBreaker 对外提供了 `<T> T run(Supplier<T> toRun, Function<Throwable, T> fallback);` 方法，用于业务的执行（这两个参数与自定义的 CircuitBreaker 实现是一样的，自定义 CircuitBreaker 的代码参考了这里的实现）。如在 toRun 参数中发生异常或熔断时，就会调用 fallback 参数进行 fallback 操作。

比如，一个 Service 对外暴露的 slow 方法执行速度很慢，需要进入熔断状态，该过程就可以被 CircuitBreaker 处理，代码如下：

```java
@Service
public static class DemoControllerService {
    private RestTemplate rest;
    private CircuitBreakerFactory cbFactory;

    public DemoControllerService(RestTemplate rest,
                CircuitBreakerFactory cbFactory) {
        this.rest = rest;
```

```java
        this.cbFactory = cbFactory;
    }

    public String slow() {
        return cbFactory.create("slow").run(() ->
                rest.getForObject("/slow", String.class),
                throwable -> "fallback");
    }
}
```

spring-cloud-circuitbreaker 支持 Reactive 编程，相同的操作可以由这段 reactive 代码完成：

```java
@Service
public static class DemoControllerService {
    private ReactiveCircuitBreakerFactory cbFactory;
    private WebClient webClient;

    public DemoControllerService(WebClient webClient,
            ReactiveCircuitBreakerFactory cbFactory) {
        this.webClient = webClient;
        this.cbFactory = cbFactory;
    }

    public Mono<String> slow() {
        return webClient.get().uri("/slow").retrieve()
                .bodyToMono(String.class).transform(it -> {
            CircuitBreaker cb = cbFactory.create("slow");
            return cb.run(it, throwable ->
                            Mono.just("fallback"));
        });
    }
}
```

spring-cloud-circuitbreaker 对 Circuit Breaker 的定义抽象了以下 4 个重要的接口：

- CircuitBreakerFactory(ReactiveCircuitBreakerFactory)：用于创建 CircuitBreaker(ReactiveCircuitBreaker)熔断器。

- CircuitBreaker(ReactiveCircuitBreaker)：开发者接触最多的接口，会使用 run 方法进行业务逻辑的处理，熔断或失败后执行 fallback 逻辑。
- Customizer：自定义操作，给 CircuitBreakerFactory 配置默认的配置数据。
- ConfigBuilder：构建配置相关的内容，比如，配置时间窗口、错误阈值。

ConfigBuilder 定义如下：

```java
public interface ConfigBuilder<CONF> {

    CONF build();

}
```

ConfigBuilder 用于构建一个配置，这里的 CONF 表示 Config 配置。比如 Alibaba Sentinel 对应的实现类为 SentinelConfigBuilder：

```java
public class SentinelConfigBuilder implements
        ConfigBuilder<SentinelConfigBuilder.SentinelCircuitBreakerConfiguration> {

    private String resourceName;

    private EntryType entryType;

    private List<DegradeRule> rules;

    ...

    @Override
    public SentinelCircuitBreakerConfiguration build() {
        Assert.hasText(resourceName, "resourceName cannot be empty");
        List<DegradeRule> rules = Optional.ofNullable(this.rules)
                .orElse(new ArrayList<>());

        EntryType entryType = Optional.ofNullable(this.entryType).orElse
(EntryType.OUT);
        return new SentinelCircuitBreakerConfiguration()
```

```
            .setResourceName(this.resourceName).setEntryType(entryType)
            .setRules(rules);
    }

    public static class SentinelCircuitBreakerConfiguration {

        private String resourceName;

        private EntryType entryType;

        private List<DegradeRule> rules;

        ...

    }
}
```

SentinelCircuitBreakerConfiguration 内部类保存 Sentinel 熔断需要的配置，SentinelConfigBuilder 构造的配置也就是 SentinelCircuitBreakerConfiguration。

CircuitBreakerFactory 定义如下：

```
public abstract class CircuitBreakerFactory<CONF, CONFB extends ConfigBuilder<CONF>>
        extends AbstractCircuitBreakerFactory<CONF, CONFB> {

    public abstract CircuitBreaker create(String id);

}
```

这里的 CONF 表示 Config 配置，CONFB 表示 ConfigBuilder 接口。父类 AbstractCircuitBreakerFactory 内部定义了一些方法，具体如下：

```
public abstract class AbstractCircuitBreakerFactory<CONF, CONFB extends ConfigBuilder<CONF>> {
```

```
    private final ConcurrentHashMap<String, CONF> configurations = new
ConcurrentHashMap<>(); //①

    public void configure(Consumer<CONFB> consumer, String... ids) {
        for (String id : ids) { //②
            CONFB builder = configBuilder(id);
            consumer.accept(builder);
            CONF conf = builder.build();
            getConfigurations().put(id, conf);
        }
    }

    protected ConcurrentHashMap<String, CONF> getConfigurations() {
        return configurations;
    }

    protected abstract CONFB configBuilder(String id); //③

    public abstract void configureDefault(Function<String, CONF>
defaultConfiguration); //④

}
```

上述代码中：

① 内部使用 ConcurrentHashMap 保存配置信息。

② 提供 configure 方法，先通过 configBuilder 抽象方法得到 ConfigBuilder，再调用 Consumer 接口对这个 ConfigBuilder 做一些额外处理，最终构造出 Config 配置，并设置到代码①处提到的 ConcurrentHashMap 中。

③ 抽象方法。通过一个 ID 得到 ConfigBuilder。

④ 抽象方法，进行默认的配置。一般 CircuitBreakerFactory 实现类内部都有默认的配置，如果开发者不进行配置构造，那么就会使用默认的配置。默认的配置可以通过该方法进行修改。

CircuitBreaker 定义如下：

```java
public interface CircuitBreaker {

    default <T> T run(Supplier<T> toRun) {
        return run(toRun, throwable -> {
            throw new NoFallbackAvailableException("No fallback available.", throwable);
        });
    };

    <T> T run(Supplier<T> toRun, Function<Throwable, T> fallback);
}
```

CircuitBreaker 的定义与自定义的 CircuitBreaker 对外提供的方法一致，并且多了一个带有默认 fallback Function 的重载方法。

Customizer 定义如下：

```java
public interface Customizer<TOCUSTOMIZE> {

    void customize(TOCUSTOMIZE tocustomize);

}
```

该接口非常简单，用于自定义操作。这里的 TOCUSTOMIZE 表示自定义行为。比如，通过自定义操作可以给 CircuitBreakerFactory 进行默认的配置：

```java
@Bean
public Customizer<SentinelCircuitBreakerFactory> defaultConfig() {
    return factory -> {
        factory.configureDefault(
                id -> new SentinelConfigBuilder().resourceName(id)
                        .rules(Collections.singletonList(new DegradeRule(id)
                                .setGrade(RuleConstant.DEGRADE_GRADE_RT).setCount(100)
                                .setTimeWindow(10)))
                        .build());
    };
}
```

在后面的内容中，我们会分别介绍 Spring Cloud 官方推荐的熔断器，并使用它们进行体验。

5.4 Alibaba Sentinel

Sentinel 是阿里巴巴在 2018 年对外开源的面向分布式服务架构的轻量级高可用流量控制组件，主要以流量为切入点，从流量控制、熔断降级、系统负载保护等多个维度来帮助用户保护服务的稳定性。

目前 Sentinel 不仅对 Spring Cloud 体系有了比较好的支持，在其他领域（如 Service Mesh、MQ 等）也都有比较好的支持。如图 5-2 所示，展示了 Sentinel 目前的开源生态。

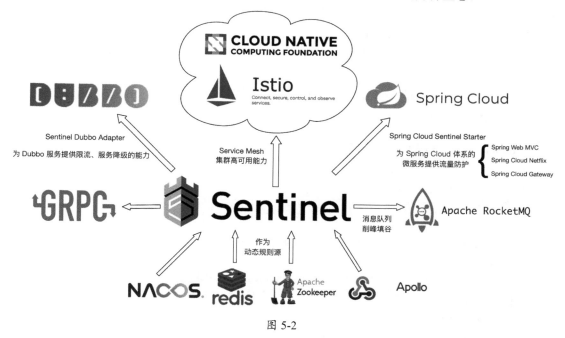

图 5-2

5.4.1 Sentinel 核心概述

Sentinel 提供了如下 3 个核心功能。

- 流量控制：任何时候到来的请求往往都是随机不可控的，而系统的处理能力是有限的。Sentinel 可以根据资源的配置对流量进行控制。

- 熔断降级：当检测到调用链路中某个资源出现不稳定的表现，例如，请求响应时间长或异常比例升高的时候，则对这个资源的调用进行限制，让请求快速失败，避免影响到其他资源，进而导致级联故障。
- 系统负载保护：提供系统维度的自适应保护能力。防止雪崩是系统防护中重要的一环。当系统负载较高的时候，如果还持续让请求进入，可能会导致系统崩溃，无法响应。在集群环境下，网络负载均衡会把本应是这台机器承载的流量转发到其他机器上。如果这个时候其他的机器也处在一个边缘状态，这个增加的流量就会导致这台机器崩溃，最后导致整个集群不可用。

笔者总结了 Sentinel 的 3 个核心概念。

- 资源：Sentinel 的核心概念。限流或熔断针对的是某个资源，这个资源可以是一个 HTTP URI、一个方法名或一个接口名。
- 规则：针对资源对应的策略。比如，限流 QPS，限流并发线程数，根据响应时间熔断等。
- 数据源：规则的数据都从数据源中加载，这个数据源可以是本地的一个文件，或者是配置中心里的一个配置。

Sentinel 核心 API 的使用方式如下：

```
Entry entry = null;
// 务必保证 finally 会被执行
try {
    // 资源名可使用任意有业务语义的字符串，注意数目不能太多（不要超过 1000 个），超出几千请作为参数传入，而不要直接作为资源名
    // EntryType 代表流量类型（inbound/outbound），其中系统规则只对 IN 类型的埋点生效
    entry = SphU.entry("/path");
    // 被保护的业务逻辑
    System.out.println("do some business");
} catch (BlockException ex) {
    // 资源访问阻止，被限流或被降级
    // 阻止后进行相应的处理操作
    System.out.println("block by sentinel: " + ex.getClass().getSimpleName());
} catch (Exception ex) {
    // 若需要配置降级规则，应通过这种方式记录业务异常
```

```
        //Tracer.traceEntry(ex, entry);
        System.out.println("business exception");
} finally {
    // 保证每个 SphU.entry 得到的 entry 与 exit 配对出现
    if (entry != null) {
        entry.exit();
    }
}
```

在 Sentinel 核心 API 的使用方式的代码中，SphU.entry 方法里的参数 /path 代表的是一个资源。如果这个操作被 Sentinel 限流或熔断，那么会抛出 BlockException（FlowException 异常是 BlockException 异常的子类，表示限流异常；DegradeException 异常也是 BlockException 异常的子类，表示降级熔断异常）。如果要处理业务异常，则需要 catch Exception，如果对某些业务异常做降级操作，则需要通过 Tracer.traceEntry 方法进行记录。最后，对于 SphU 返回的 entry，需要进行 exit 释放。

执行代码，在控制台输出 do some business。这时如果给应用添加以下一些限流规则：

```
FlowRule flowRule = new FlowRule();
flowRule.setResource("/path");
flowRule.setCount(0);
flowRule.setGrade(RuleConstant.FLOW_GRADE_QPS);
flowRule.setLimitApp("default");
FlowRuleManager.loadRules(Collections.singletonList(flowRule));
```

再次执行代码，控制状态输出 block by sentinel: FlowException。由于在规则里设置的 QPS 为 0，任何操作都会被限流。

或者给应用添加一些降级熔断规则，代码如下：

```
DegradeRule degradeRule = new DegradeRule();
degradeRule.setGrade(RuleConstant.DEGRADE_GRADE_EXCEPTION_COUNT);
degradeRule.setResource("/path");
degradeRule.setCount(0);
degradeRule.setTimeWindow(10);
DegradeRuleManager.loadRules(Arrays.asList(degradeRule));
```

再次执行代码，控制状态输出 block by sentinel: DegradeException。由于在规则里设置的异常错误数 为 0，0 意味着不发生错误，所以任何操作都会被降级熔断。

从这两个例子可以看到，这些规则都是通过硬编码的方式设置的。数据源的出现就是解决硬编码的问题的，通过加载配置中心的配置来设置对应的规则。比如，可以通过 Sentinel 的 InitFunc SPI 加载 Nacos 配置中心里维护的限流数据：

```java
public class DataSourceInitFunc implements InitFunc {

    @Override
    public void init() throws Exception {
        final String remoteAddress = "localhost";
        final String groupId = "Sentinel:Demo";
        final String dataId = "com.alibaba.csp.sentinel.demo.flow.rule";

        ReadableDataSource<String, List<FlowRule>> flowRuleDataSource = new NacosDataSource<>(remoteAddress, groupId, dataId,
            source -> JSON.parseObject(source, new TypeReference<List<FlowRule>>() {}));
        FlowRuleManager.register2Property(flowRuleDataSource.getProperty());
    }
}
```

这个应用启动并运行完毕后，JVM 并不会自动关闭。这是因为 Sentinel 在本地打开了 8719 端口接收 HTTP 请求，用于对外暴露一些数据，比如，获取/更新规则、metrics 监控数据、基本配置信息、簇点链路等信息。

我们可以使用 curl 命令查看目前设置的熔断规则：

```
curl -XGET http://localhost:8719/getRules?type=degrade
```

返回以下信息：

```
[{
    "count": 0.0,
    "grade": 2,
    "limitApp": "default",
    "minRequestAmount": 5,
```

```
    "resource": "/path",
    "rtSlowRequestAmount": 5,
    "timeWindow": 10
}]
```

查看应用的基本信息：

```
curl -XGET http://localhost:8719/basicInfo
```

返回以下信息：

```
{
    "machine": "jim.local",
    "ip": "192.168.31.232"
}
```

Sentinel 目前在流量控制层支持按照 QPS、并发线程数进行限流。在熔断降级层支持按照响应时间 RT、异常比例、异常数进行熔断降级。在系统保护层支持按照总 QPS、总线程数、最大响应时间 RT、负载、CPU 使用率进行保护。

如图 5-3 所示，Sentinel 的工作原理是通过责任链模式实现的。一条责任链 SlotChain 由多个 Slot 组成。我们可以使用 SPI 定义新的责任链，也可以在责任链上添加新的 Slot。

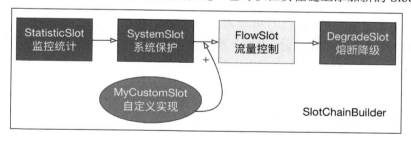

图 5-3

SlotChain 责任链默认的 Slot 链路为：NodeSelectorSlot→ClusterBuilderSlot→LogSlot→StatisticSlot→AuthoritySlot→SystemSlot→FlowSlot→DegradeSlot。各个 Slot 的作用如下：

- NodeSelectorSlot：负责收集资源的路径，并将这些资源的调用路径以树状结构存储起来，用于根据调用路径进行流量控制。
- ClusterBuilderSlot：用于构建资源的 ClusterNode 以及调用来源节点。ClusterNode 保持某个资源运行统计信息（响应时间、QPS、block 数目、线程数、异常数等）以及调用来源统计信息列表。
- LogSlot：异常记录。
- StatisticSlot：统计实时的调用数据。
- AuthoritySlot：黑白名单控制。
- SystemSlot：系统保护控制。
- FlowSlot：流量控制。
- DegradeSlot：熔断降级。

5.4.2　Spring Cloud Alibaba Sentinel

Spring Cloud Alibaba 提供的 Spring Cloud Alibaba Sentinel 是 Sentinel 与 Spring Cloud 体系的整合，该组件为 Sentinel 整合了 Spring MVC、Netflix Zuul、Spring Cloud Gateway、Spring Cloud Circuit Breaker、RestTemplate、OpenFeign、WebFlux 等组件。若想使用 Spring Cloud Alibaba Sentinel，需要引入如下依赖：

```xml
<dependency>
    <groupId>com.alibaba.cloud</groupId>
    <artifactId>spring-cloud-starter-alibaba-sentinel</artifactId>
</dependency>
```

引入 Spring Cloud Alibaba Sentinel 对应的 starter 依赖之后，其内部通过自动化配置类（SentinelCircuitBreakerAutoConfiguration）自动为应用注入一个实现 CircuitBreakerFactory 接口的 SentinelCircuitBreakerFactory。应用代码通过@Autowired 注解进行注入：

```
@Autowired
CircuitBreakerFactory circuitBreakerFactory;
```

引入之后，需要设置降级配置，通过 Customizer 接口自定义配置。示例中对默认的配置进行修改，默认的配置使用异常数作为降级策略，阈值为 3 个，时间窗口为 10s；同时对 ID 为 rt 的 CircuitBreaker 也进行配置，使用响应时间 RT 作为降级策略，阈值为 100ms，时间窗口为 10s：

```
@Bean
public Customizer<SentinelCircuitBreakerFactory> customizer() {
    return factory -> {
        factory.configureDefault(id -> new SentinelConfigBuilder()
```

```
            .resourceName(id)
            .rules(Collections.singletonList(new DegradeRule(id)
                .setGrade(RuleConstant.DEGRADE_GRADE_EXCEPTION_COUNT)
                .setCount(3).setTimeWindow(10)))
            .build());
    factory.configure(builder -> {
        builder
            .rules(Collections.singletonList(new DegradeRule("slow")
                .setGrade(RuleConstant.DEGRADE_GRADE_RT).setCount(100)
                .setTimeWindow(5)));
    }, "rt");
};
```

使用 httpbin response code 为 500 的请求模拟错误调用。调用成功则正常返回；当调用出错时，如果不是被 Sentinel 降级，那么返回 exception occurs with url: {url}，否则返回 degrade by sentinel：

```
public String exp() {
    StringBuilder sb = new StringBuilder();
    CircuitBreaker cb = circuitBreakerFactory.create("temp");
    String url = "https://httpbin.org/status/500";
    for(int index = 0; index < 10; index ++) {
        String httpResult = cb.run(() -> {
            return restTemplate.getForObject(url, String.class);
        }, throwable -> {
            if(throwable instanceof DegradeException) {
                return "degrade by sentinel";
            }
            return "exception occurs with url: " + url;
        });
        sb.append(httpResult).append("<br/>");
    }
    return sb.toString();
}
```

exp 方法内部创建 CircuitBreaker 使用的 ID 为 temp，Customizer 自定义配置内部没有对应的

ID，所以会使用默认的配置。默认配置使用异常数作为降级策略，阈值为 3 个，时间窗口为 10s。每次请求都是一个错误请求，连续发生 3 次被降级，结果输出符合预期：

```
exception occurs with url: https://httpbin.org/status/500
exception occurs with url: https://httpbin.org/status/500
exception occurs with url: https://httpbin.org/status/500
degrade by sentinel
degrade by sentinel
degrade by sentinel
degrade by sentinel
degrade by sentinel
degrade by sentinel
degrade by sentinel
```

使用 httpbin delay 的请求模拟慢调用。调用成功则正常返回；当调用出错时，如果不是被 Sentinel 降级，那么返回 exception occurs with url: {url}，否则返回 degrade by sentinel：

```java
public String rt() {
    StringBuilder sb = new StringBuilder();
    CircuitBreaker cb = circuitBreakerFactory.create("rt");
    String url = "https://httpbin.org/delay/3";
    for(int index = 0; index < 10; index ++) {
        String httpResult = cb.run(() -> {
            Map<String, Object> map = restTemplate.getForObject(url, Map.class);
            return (String) map.get("origin");
        }, throwable -> {
            if(throwable instanceof DegradeException) {
                return "degrade by sentinel";
            }
            return "exception occurs with url: " + url;
        });
        sb.append(httpResult).append("<br/>");
    }
    return sb.toString();
}
```

rt 方法内部创建 CircuitBreaker 使用的 ID 为 rt，Customizer 自定义配置内部有对应的 ID 为

rt 的配置，rt 配置使用响应时间 RT 作为降级策略，阈值为 100ms，时间窗口为 10s。每次都是一个慢请求，都超过了 100ms，并且连续发生 5 次（Sentinel RT 降级默认需要连续触发 5 次），被 Sentinel 降级。结果输出符合预期：

```
127.0.0.1
127.0.0.1
127.0.0.1
127.0.0.1
127.0.0.1
degrade by sentinel
degrade by sentinel
degrade by sentinel
degrade by sentinel
degrade by sentinel
```

如果读者使用 reactive 编程，在引入 webflux-starter 的同时，需要使用 ReactiveCircuitBreakerFactory（自动化配置类 ReactiveSentinelCircuitBreakerAutoConfiguration 自动为应用注入实现 ReactiveCircuitBreakerFactory 接口的 ReactiveSentinelCircuitBreakerFactory 类）代替 CircuitBreakerFactory。同样，ReactiveCircuitBreakerFactory 构造的 ReactiveCircuitBreaker 代替 CircuitBreaker 用来对程序进行保护。

5.4.3　Sentinel 与 OpenFeign 和 RestTemplate

Sentinel 支持 OpenFeign 和 RestTemplate，为调用服务的安全保驾护航。其中，OpenFeign 启用 Sentinel 进行保护时，需要在配置文件里开启（默认关闭），语句如下：

```
feign.sentinel.enabled=true
```

OpenFeign 本身是一个基于接口的声明式的 Rest 客户端，基于接口编程对熔断的支持有天然的优势。比如，一个 EchoService 接口，添加 @FeignClient 注解后底层会生成接口的动态代理，这个代理类其实就是对应的业务处理类。这时可以创建实现 EchoService 接口的 FallbackEchoService 类，这个类内部其实就是熔断降级后需要处理的逻辑。

这里举个例子，InventoryService 接口用于库存方面的操作，BuyService 接口用于购买操作，定义如下：

```java
@FeignClient(name = "inventory-provider")
public interface InventoryService {

    @GetMapping("/save")
    String save();

}

@FeignClient(name = "buy-service", url = "https://httpbin.org/delay/3")
public interface BuyService {

    @GetMapping
    String buy();

}
```

InventoryService 用于模拟请求报错的场景,它对应的服务名 inventory-provider 在注册中心并不存在,所以调用一定会报错。BuyService 用于模拟耗时较长的场景。定义 FallbackService 和 DefaultBusinessService,用于处理熔断降级后的逻辑和系统(业务)异常后的逻辑,代码如下:

```java
class FallbackService implements BuyService, InventoryService {
        @Override
    public String buy() {
        return "buy degrade by sentinel";
    }

    @Override
    public String save() {
        return "save degrade by sentinel";
    }
}

class DefaultBusinessService implements BuyService, InventoryService {
    @Override
    public String buy() {
        return "buy error";
```

```
    }

    @Override
    public String save() {
        return "inventory save error";
    }
}
```

这两个处理异常的类需要通过 @FeignClient 注解里的 fallbackFactory 属性完成，这是一个实现了 feign.hystrix.FallbackFactory 接口的工厂类，基于异常返回相应的处理类。

很明显，针对 Sentinel 抛出的熔断降级 DegradeException 异常需要返回 FallbackService 实例，否则返回 DefaultBusinessService 实例：

```
class FallbackFactory implements feign.hystrix.FallbackFactory {

    private FallbackService fallbackService = new FallbackService();

    private DefaultBusinessService defaultBusinessService = new DefaultBusinessService();

    @Override
    public Object create(Throwable cause) {
        if (cause instanceof BlockException) {
            return fallbackService;
        } else {
            return defaultBusinessService;
        }
    }
}
```

FeignClient 设置 fallbackFactory 属性，代码如下：

```
@FeignClient(name = "inventory-provider", fallbackFactory = FallbackFactory.class)
public interface InventoryService {

    @GetMapping("/save")
```

```
    String save();

}

@FeignClient(name = "buy-service", url = "https://httpbin.org/delay/3",
fallbackFactory = FallbackFactory.class)
public interface BuyService {

    @GetMapping
    String buy();

}
```

在这个例子中，熔断降级生效的前提是配置 Sentinel 相关的数据源，使应用能够加载到对应的降级规则。

Sentinel 对于 RestTemplate 的支持需要在构造 RestTemplate 的时候加上@SentinelRestTemplate 注解：

```
@Bean
@SentinelRestTemplate
public RestTemplate restTemplate() {
    return new RestTemplate();
}
```

使用 httpbin 的 status 和 delay 路径模拟错误请求和耗时调用，再配合 Sentinel 相应的降级规则，可以达到熔断降级的效果，这里不再继续描述。

RestTemplate 对于错误请求的判断通过 ResponseErrorHandler 实现。默认情况下，会使用 DefaultResponseErrorHandler，该 handler 内部认为 Response Code 为 4xx 客户端错误和 5xx 服务器端错误都是一次错误的调用。

Sentinel 对于错误调用也依赖 ResponseErrorHandler 的逻辑。如果一个自定义的 ResponseErrorHandler 只认为 500 才是一个错误调用，那么 404 的调用并不会被熔断：

```
@Bean
@SentinelRestTemplate
```

```java
public RestTemplate restTemplate() {
    RestTemplate template = new RestTemplate();
    template.setErrorHandler(new ResponseErrorHandler() {
        @Override
        public boolean hasError(ClientHttpResponse response) throws IOException {
            return response.getStatusCode() == HttpStatus.INTERNAL_SERVER_ERROR;
        }

        @Override
        public void handleError(ClientHttpResponse response) throws IOException {
            throw new IllegalStateException("illegal status code");
        }
    });
    return template;
}
```

Sentinel 对于 OpenFeign 和 RestTemplate 调用的规则策略类似。

OpenFeign 规则策略如下：

`httpmethod:schema://requesturl`

比如，一个 GET 方法的服务调用 http://my-provider/buy 对应的规则是 GET:http://my-provider/buy。

RestTemplate 规则策略如下：

`httpmethod:schema://host:port/path`

比如，一个 GET 方法的 Rest 调用 https://httpbin.org/status/500 对应的规则是 GET:https://httpbin.org/status/500。

5.4.4 Sentinel 限流与 Dashboard

前面对 Sentinel 的熔断降级进行了讨论，Sentinel 还有一个重大功能——限流。如图 5-4 所示，限流可以对外部流量进行控制，丢弃请求并进行资源保护。

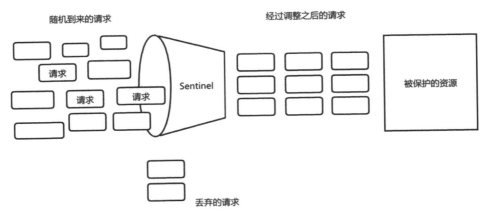

图 5-4

一个典型的流量控制应用场景就是在网关层进行限流。Sentinel 与 Spring Cloud 体系内的网关 Netflix Zuul、Spring Cloud Gateway 都已经进行了整合。限流支持的维度如下：

- QPS 设置。
- 统计时间窗口设置，默认为 1s。
- 来源 IP 限流。
- Host 限流。
- 任意 Header 限流。
- 任意 Cookie 限流。
- 任意 URL Query Parameter 限流。

若要使用 Sentinel 网关限流，只需添加 sentinel gateway 依赖并配置限流规则，就可完成限流操作。

需要添加的 sentinel gateway 依赖如下：

```
<dependency>
    <groupId>com.alibaba.cloud</groupId>
    <artifactId>spring-cloud-alibaba-sentinel-gateway</artifactId>
</dependency>
```

如图 5-5 所示，这是一个满足请求路径中只要存在 name 参数就全部限制访问（QPS 为 0）的限流规则。

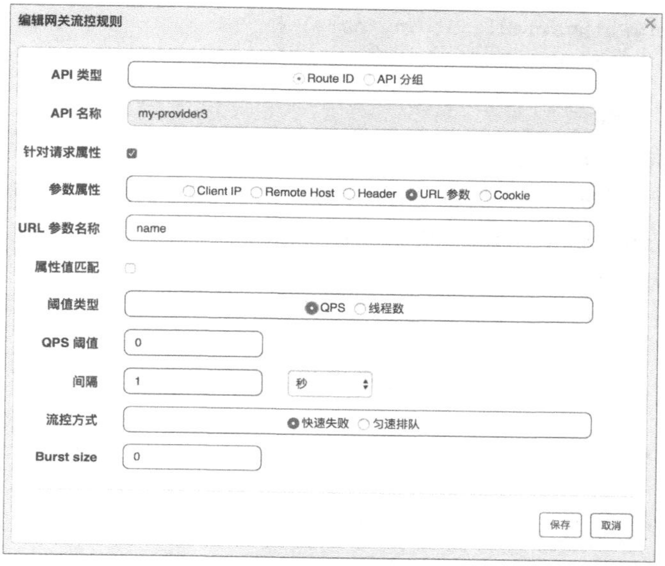

图 5-5

这时访问 http://localhost:8080/s-c-alibaba?name=test（my-provider3 服务对应的 path 为 /s-c-alibaba/**）会被直接拒绝。

Sentinel 除网关相关的支持外，还支持对 Spring MVC 暴露的 RequestMapping 进行流量控制，引入 spring-cloud-starter-alibaba-sentinel 之后会自动为应用添加一个 SentinelWebInterceptor 拦截器，用于拦截所有的请求。SentinelWebInterceptor 会选择 RequestMapping 设置的路径。比如，以下这 3 个 RequestMapping 对应的规则分别是"/springcloud"、"/hello"，以及"/hello/{name}"（通配符 {name} 对 /hello/springcloud 和 /hello/dubbo 都会生效）：

```
@GetMapping("/springcloud")
public String springcloud() {
    return "Spring Cloud";
}

@GetMapping("/hello")
public String hello() {
    return "Hello World";
}

@GetMapping("/hello/{name}")
```

```
public String get(@PathVariable String name) {
    return "Hello: " + name;
}
```

如图 5-6 所示（上下两幅监控图中比较高的都是拒绝 QPS，矮的是通过 QPS），这是对这 3 个资源设置 QPS 分别为 0、1、0 的规则的限流效果（/hello 资源最多只能通过 1 个请求，其余请求会被拒绝。/springcloud 资源不能通过任意请求，所有的请求全部被拒绝）。此时，对应的限流规则列表如图 5-7 所示。

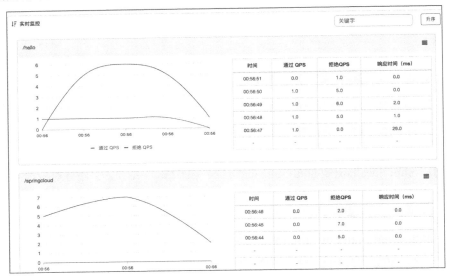

图 5-6

图 5-7

图 5-5~图 5-7 是 Sentinel 对外提供的一个轻量级的开源 Dashboard 控制台，其内部提供机器发现，以及健康情况管理、监控（单机和集群）、规则管理和推送的功能。Dashboard 是一个使用 Spring Boot 开发的应用，如想修改启动端口，可以传入-Dserver.port 参数进行覆盖（默认端口为 8080）。

应用为 Dashboard 进行交互时需要添加 Dashboard 地址配置：

```
spring.cloud.sentinel.transport.dashboard=localhost:9090
```

如图 5-8 和图 5-9 所示，这是新增限流规则和新增熔断降级规则的一些功能截图。

图 5-8　　　　　　　　　　　　　　　　图 5-9

5.4.5　Sentinel 的高级特性

5.4.4 节提到的 Sentinel 限流功能都是直接抛出错误的，这跟配置的限流行为有关，Sentinel 还支持冷启动（让通过的流量缓慢增加，一定时间内到达阈值）、匀速器（让请求匀速通过，使用漏桶算法）限流行为。

Sentinel 还支持多种基于调用关系的限流策略：

- 基于调用方（比如 Dubbo Provider、IP 地址）。
- 基于调用链（调用树控制方向，针对某些链路限流）。
- 基于关联的资源（当两个资源之间具有资源争抢或者依赖关系的时候，这两个资源便有了关联。比如，对数据库同一个字段的读操作和写操作存在争抢，读的速度过高会影响写的速度，写的速度过高会影响读的速度。如果放任读/写操作争抢资源，则争抢本身带来的开销会降低整体的吞吐量。可使用关联限流来避免具有关联关系的资源之间过度争抢）。

下面对基于调用链的限流策略进行介绍。

Sentinel SlotChain 责任链中的 NodeSelectorSlot 内部会构造调用树，为后续在 FlowSlot 里根据调用路径进行流量控制做准备。比如，下面这段代码对应的调用树如图 5-10 所示。

```java
ContextUtil.enter("my-entrance-node");
Entry entryA = null, entryB = null, entryC = null;
try {
    entryA = SphU.entry("resourceA");
    entryB = SphU.entry("resourceB");
    entryC = SphU.entry("resourceC");
} catch (BlockException e) {
    System.err.println("block by Sentinel: " + e.getClass());
} finally {
    if(entryC != null) {
        entryC.exit();
    }
    if(entryB != null) {
        entryB.exit();
    }
    if(entryA != null) {
        entryA.exit();
    }
    ContextUtil.exit();
}
```

又如，下面这段代码对应的调用树如图 5-11 所示。

```java
ContextUtil.enter("my-entrance-node");
Entry entryA = null, entryB = null, entryC = null;
try {
    entryA = SphU.entry("resourceA");
    entryA.exit();
    entryB = SphU.entry("resourceB");
```

```
        entryB.exit();
        entryC = SphU.entry("resourceC");
        int a = 1, b = 2;
        int c = a + b;
        System.out.println(c);
    } catch (BlockException e) {
        System.err.println("block by Sentinel: " + e.getClass());
    } finally {
        if(entryC != null) {
            entryC.exit();
        }
        ContextUtil.exit();
    }
```

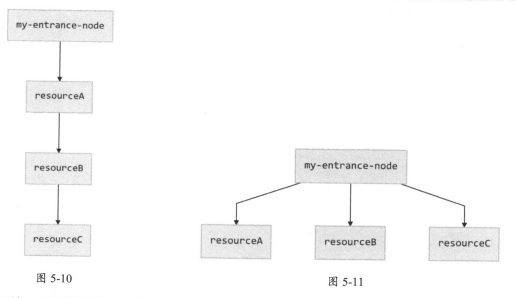

图 5-10   图 5-11

再如，下面这段代码对应的调用树如图 5-12 所示。

```
ContextUtil.enter("my-entrance-node1");
Entry entryA = null;
try {
    entryA = SphU.entry("resourceA");
} catch (BlockException e) {
```

```
        System.err.println("my-entrance-node1 block by Sentinel: " + e.getClass());
    } finally {
        if(entryA != null) {
            entryA.exit();
        }
        ContextUtil.exit();
    }


ContextUtil.enter("my-entrance-node2");
try {
    entryA = SphU.entry("resourceA");
} catch (BlockException e) {
    System.err.println("my-entrance-node2 block by Sentinel: " + e.getClass());
} finally {
    if(entryA != null) {
        entryA.exit();
    }
    ContextUtil.exit();
}
```

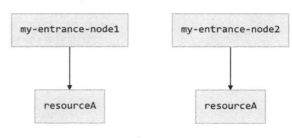

图 5-12

图 5-12 中,入口 my-entrance-node1 和 my-entrance-node2 都调用到了 resourceA。由于 Sentinel 可以基于调用关系进行限流,通过以下限流规则的配置可以限制入口为 my-entrance-node2 的调用:

```
FlowRule rule = new FlowRule();
rule.setGrade(RuleConstant.FLOW_GRADE_QPS);
rule.setCount(0);
```

```
rule.setResource("resourceA");
rule.setRefResource("my-entrance-node2");
rule.setStrategy(RuleConstant.STRATEGY_CHAIN);
rule.setControlBehavior(RuleConstant.CONTROL_BEHAVIOR_DEFAULT);
FlowRuleManager.loadRules(Arrays.asList(rule));
```

上述代码会打印出 my-entrance-node2 block by Sentinel: class com.alibaba.csp.sentinel.slots.block.flow.FlowException。

有关其他高级特性，本书不再具体介绍，读者可自行查看官网上对应的资料。

5.5 Netflix Hystrix

Hystrix 是 Netflix 在 2012 年对外开源的一款熔断降级工具，用于解决分布式系统的延迟和容错问题。官方对 Hystrix 的功能描述如下：

- 通过第三方客户端库（通常通过网络）访问依赖项，对延迟和故障提供保护和控制。
- 在复杂的分布式系统中停止连锁故障。
- 快速失败并迅速恢复。
- 尽可能地回退和优雅地降级。
- 实现近乎实时的监控、警报和操作控制。

这些功能的描述是一个非常经典的熔断器功能描述。

目前 Hystrix 已经处于维护（maintenance）状态，不再继续开发，这个消息是 Netflix 在 2018 年对外宣布的。Hystrix GitHub 页面上也说明了该项目目前处于维护模式。

5.5.1 Hystrix 核心概述

Hystrix 核心 API 的使用方式如下：

```java
public class HelloWorldCommand extends HystrixCommand<String> {

    private final RestTemplate restTemplate = new RestTemplate();
    private final String code;
```

```
    public HelloWorldCommand(String code) {
        super(HystrixCommandGroupKey.Factory.asKey("HelloWorldExample"));
        this.code = code;
    }

    @Override
    protected String run() {
        String url = "http://httpbin.org/status/" + code;
        System.out.println("start to curl: " + url);
        restTemplate.getForObject(url, String.class);
        return "Request success";
    }

    @Override
    protected String getFallback() {
        return "Request failed";
    }
}
```

HystrixCommand 是 Hystrix 对外暴露的抽象类，其内部定义 Hystrix 熔断器的常规操作。两个最重要的方法是 run 和 getFallback，其中，run 方法用于执行业务逻辑，当发生错误或熔断的时候会执行 getFallback 方法。

定义好 HelloWorldCommand 后，使用该类执行以下代码并查看熔断结果：

```
CommandHelloWorld helloWorld1 = new CommandHelloWorld("200");
CommandHelloWorld helloWorld2 = new CommandHelloWorld("500");
System.err.println(helloWorld1.execute() + " and Circuit Breaker is " +
    (helloWorld1.isCircuitBreakerOpen() ? "open" : "closed"));
System.err.println(helloWorld2.execute() + " and Circuit Breaker is " +
    (helloWorld2.isCircuitBreakerOpen() ? "open" : "closed"));
```

控制台输出以下信息：

```
start to curl: http://httpbin.org/status/200
Request success and Circuit Breaker is closed
start to curl: http://httpbin.org/status/500
```

```
Request failed and Circuit Breaker is closed
```

从这个结果可以看到，两次都进行了 run 方法的调用，第 2 次在调用失败的情况下会执行 getFallback 方法。这两次调用的熔断器都处于 Closed 状态。

这个例子与 Sentinel 进行对比，有没有发现少了点什么？没错，就是 Hystrix 熔断的条件是什么，对应的规则或配置应该怎么写？

Hystrix 对应的配置信息通过 HystrixCommand.Setter 完成，比如，以下这段 Setter 配置表示执行超时时间为 1500ms（执行超过 1500ms，就会自动进入 getFallback 方法）：

```
HystrixCommand.Setter.withGroupKey(HystrixCommandGroupKey.Factory.asKey("TimeoutRestExample"))
    .andCommandPropertiesDefaults(
        HystrixCommandProperties.Setter()
            .withExecutionTimeoutInMilliseconds(1500)
)
```

基于这个配置构造 TimeoutRestCommand，并执行以下逻辑：

```
int num = 1;
while (num <= 10) {
    TimeoutRestCommand command = new TimeoutRestCommand(num);
    System.err.println("Execute " + num + ": " + command.execute() + " and Circuit Breaker is " + (command.isCircuitBreakerOpen() ? "open" : "closed"));
    num ++;
}
```

控制台输出以下信息：

```
start to curl: http://httpbin.org/delay/1
Execute 1: Request success and Circuit Breaker is closed
start to curl: http://httpbin.org/delay/2
Execute 2: Request failed and Circuit Breaker is closed
start to curl: http://httpbin.org/delay/3
Execute 3: Request failed and Circuit Breaker is closed
start to curl: http://httpbin.org/delay/4
```

```
Execute 4: Request failed and Circuit Breaker is closed
start to curl: http://httpbin.org/delay/5
Execute 5: Request failed and Circuit Breaker is closed
start to curl: http://httpbin.org/delay/6
Execute 6: Request failed and Circuit Breaker is closed
start to curl: http://httpbin.org/delay/7
Execute 7: Request failed and Circuit Breaker is closed
start to curl: http://httpbin.org/delay/8
Execute 8: Request failed and Circuit Breaker is closed
start to curl: http://httpbin.org/delay/9
Execute 9: Request failed and Circuit Breaker is closed
start to curl: http://httpbin.org/delay/10
Execute 10: Request failed and Circuit Breaker is closed
```

从这个输出可看出，只有 delay 为 1（执行时间小于 1500ms）的执行成功，其他所有的执行由于执行时间大于 1500s，都进入了 getFallback 逻辑。这个例子中，Command 熔断器并没有触发打开的条件。

接下来看一个熔断器打开的例子，使用以下这个 Setter 对应的配置：

```
HystrixCommand.Setter.withGroupKey(HystrixCommandGroupKey.Factory.asKey("CBRestExam
ple"))
    .andCommandPropertiesDefaults(
        HystrixCommandProperties.Setter()
            .withCircuitBreakerErrorThresholdPercentage(10)
            .withCircuitBreakerRequestVolumeThreshold(10)
            .withCircuitBreakerSleepWindowInMilliseconds(3000)
)
```

这个配置表示 10s（可以通过 withMetricsRollingStatisticalWindowInMilliseconds 方法进行修改，默认为 10s）内如果至少有 10 个请求，且其中有 10% 的异常比例，就会触发熔断器进入 Open 状态。3000ms 表示从 Open 状态进入 Half-Open 状态的时间。

基于这个配置构造 CircuitBreakerRestCommand，并执行以下逻辑：

```
num = 1;
while (num <= 15) {
```

```
    CircuitBreakerRestCommand command = new CircuitBreakerRestCommand();
    System.err.println("Execute " + num + ": " + command.execute() + " and Circuit
Breaker is " + (command.isCircuitBreakerOpen() ? "open" : "closed"));
    num ++;
}
Thread.sleep(3000L);
CircuitBreakerRestCommand command = new CircuitBreakerRestCommand("200");
System.err.println("Execute " + num + ": " + command.execute() + " and Circuit
Breaker is " + (command.isCircuitBreakerOpen() ? "open" : "closed"));
```

控制台输出以下信息:

```
start to curl: http://httpbin.org/status/500
Execute 1: Request failed and Circuit Breaker is closed
start to curl: http://httpbin.org/status/500
Execute 2: Request failed and Circuit Breaker is closed
start to curl: http://httpbin.org/status/500
Execute 3: Request failed and Circuit Breaker is closed
start to curl: http://httpbin.org/status/500
Execute 4: Request failed and Circuit Breaker is closed
start to curl: http://httpbin.org/status/500
Execute 5: Request failed and Circuit Breaker is closed
start to curl: http://httpbin.org/status/500
Execute 6: Request failed and Circuit Breaker is closed
start to curl: http://httpbin.org/status/500
Execute 7: Request failed and Circuit Breaker is closed
start to curl: http://httpbin.org/status/500
Execute 8: Request failed and Circuit Breaker is closed
start to curl: http://httpbin.org/status/500
Execute 9: Request failed and Circuit Breaker is closed
start to curl: http://httpbin.org/status/500
Execute 10: Request failed and Circuit Breaker is closed
start to curl: http://httpbin.org/status/500
Execute 11: Request failed and Circuit Breaker is closed
start to curl: http://httpbin.org/status/500
Execute 12: Request failed and Circuit Breaker is open
```

```
Execute 13: Request failed and Circuit Breaker is open
Execute 14: Request failed and Circuit Breaker is open
Execute 15: Request failed and Circuit Breaker is open
start to curl: http://httpbin.org/status/200
Execute 16: Request success and Circuit Breaker is closed
```

从这个输出可看出，前 10 次调用全部报错，异常比率为 100%，大于 10%的触发熔断，后续 5 次调用全部熔断。最后一次调用前休眠（sleep）为 3s，3s 的时间刚好让熔断器进入 Half-Open 状态，最后一次调用成功（如果将 sleep 时间改成 2s，最后一次调用并不会发生，还是会进入熔断操作）。

Hystrix 的运行流程如图 5-13 所示，具体步骤如下：

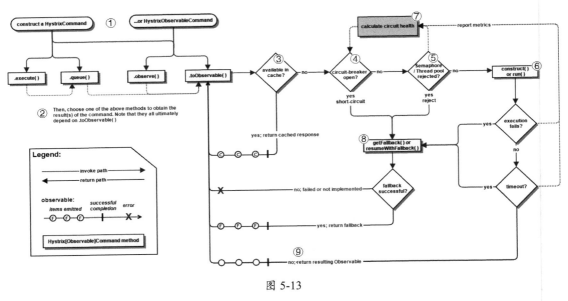

图 5-13

① 构造用于执行业务的 HystrixCommand 或 HystrixObservableCommand。

② 选择调用的方法。比如，HystrixCommand 对应的 execute 或 queue 方法，HystrixObservableCommand 对应的 observe 或 toObservable 方法。

③ 判断缓存池里是否有之前缓存的结果，有则直接返回。

④ 若没有被缓存，则判断熔断器当前是否已打开。

⑤ 若熔断器没有打开，则根据信号量或线程池及当前状态判断是否拒绝。

⑥ 若没有被熔断，则运行业务逻辑。

⑦ 每次调用需要计算新的信号量或线程池消耗情况。

⑧ 熔断器已经打开的情况下调用 fallback 方法返回并结束流程（没有实现 fallback 接口也结束流程）。

⑨ 顺利执行，流程结束。

5.5.2 Spring Cloud Netflix Hystrix

Spring Cloud Netflix 提供 Hystrix 与 Spring Cloud 体系的整合。若想使用它，需要引入以下依赖：

```xml
<dependency>
    <groupId>org.springframework.cloud</groupId>
    <artifactId>spring-cloud-starter-netflix-hystrix</artifactId>
</dependency>
```

由于 Spring Cloud 体系内 Spring Cloud Circuit Breaker、Open Feign 对外暴露的编程模型是一样的，所以底层使用的熔断器用户侧并不关心，唯一需要关心的是熔断器对应的配置。

Spring Cloud Circuit Breaker Sentinel 通过 Sentinel 触发熔断降级。类似的功能可以通过 Hystrix 完成（Hystrix 只支持异常比率降级，本例基于 RT 降级的功能无法完成），只需修改这 3 点即可：

- 将依赖的 com.alibaba.cloud:spring-cloud-starter-alibaba-sentinel 修改成 org.springframework.cloud:spring-cloud-starter-netflix-hystrix。
- Function<Throwable, T> fallback 内将 Sentinel DegradeException 的异常处理改成 Hystrix 对应的异常。
- Customizer 自定义配置相关的处理从 Sentinel 改成 Hystrix 对应的配置即可。

以下这段代码通过 Customizer 配置的 Hystrix 参数与 Sentinel 配置的基于错误数熔断降级的参数能达到相似的效果（1s 内 3 次请求错误率达到 100%）：

```java
@Bean
public Customizer<HystrixCircuitBreakerFactory> customizer() {
    return factory -> {
```

```
        factory.configureDefault(id -> HystrixCommand.Setter.withGroupKey(
            HystrixCommandGroupKey.Factory.asKey(id))
            .andCommandPropertiesDefaults(
                HystrixCommandProperties.Setter()
                    .withMetricsRollingStatisticalWindowInMilliseconds(1000)
                    .withCircuitBreakerRequestVolumeThreshold(3)
                    .withCircuitBreakerErrorThresholdPercentage(100)
                    .withCircuitBreakerSleepWindowInMilliseconds(10000)
                    .withExecutionTimeoutInMilliseconds(5000)
            )
        );
    };
}
```

同样，OpenFeign 通过 Sentinel 触发熔断降级。类似的功能可以通过 Hystrix 完成，只需修改这 3 点即可：

- 将依赖的 com.alibaba.cloud:spring-cloud-starter-alibaba-sentinel 修改成 org.springframework.cloud:spring-cloud-starter-netflix-hystrix。
- FallbackFactory 内将 Sentinel DegradeException 的异常处理改成 Hystrix 对应的异常。
- 配置文件将 sentinel 对应的配置改成 hystrix 配置。

以下这段配置内容与 Sentinel 配置的基于错误数熔断降级的参数能达到相似的效果（1s 内 3 次请求错误率达到 100%）：

```
feign.hystrix.enabled=true

hystrix.command.default.execution.isolation.thread.timeoutInMilliseconds=1000
hystrix.command.default.metrics.rollingStats.timeInMilliseconds=1000
hystrix.command.default.circuitBreaker.requestVolumeThreshold=3
hystrix.command.default.circuitBreaker.errorThresholdPercentage=100
hystrix.command.default.circuitBreaker.sleepWindowInMilliseconds=5000
```

如果开发者只使用 Spring MVC 开发，可以在 Controller 或 Service 对外暴露的方法上加上 @HystrixCommand 注解：

```java
@HystrixCommand(groupKey = "ControllerGroup", fallbackMethod = "fallback")
@GetMapping("/exp")
public String exp() {
    return new RestTemplate().getForObject("https://httpbin.org/status/500", String.class);
}

public String fallback() {
    return "Hystrix fallback";
}
```

@HystrixCommand 注解内部提供了 commandProperties 属性，用于设置降级相关的规则（Sentinel 也提供了@SentinelResource 注解，用于完成同样的事情）。

5.5.3 Hystrix 限流与 Dashboard

Hystrix 提供了两种隔离策略：线程池隔离和信号量隔离，其中，最常用的是线程池隔离，这也是默认的隔离策略。如果使用信号量隔离，可以用来完成限流操作。比如，这个 HystrixController：

```java
@RestController
class HystrixController {

    @HystrixCommand(commandKey = "Hello", groupKey = "HelloGroup", fallbackMethod = "fallback")
    @GetMapping("/hello")
    public String hello() {
        return "Hello World";
    }

    public String fallback(Throwable throwable) {
        return "Hystrix fallback";
    }
}
```

配置 QPS 数：

```
hystrix.command.Hello.execution.isolation.semaphore.maxConcurrentRequests=0
hystrix.command.Hello.execution.isolation.strategy=SEMAPHORE
hystrix.command.Hello.execution.timeout.enabled=false
```

访问/hello，全部返回 Hystrix fallback，并且这个 throwable 是一个 RuntimeException，且 messge 为 "could not acquire a semaphore for execution"。

网关组件 Netflix Zuul 和 Spring Cloud Gateway 默认都集成了 Hystrix。

如图 5-14 所示，这是在 Zuul 上使用 Hystrix 限流和降级的数据统计（Zuul 自动集成 Hystrix，每次调用都使用 Hystrix 进行保护）。

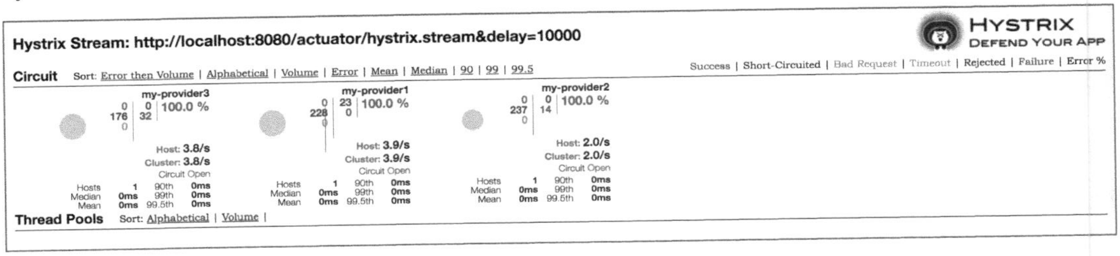

图 5-14

在图 5-14 中，my-provder1 对应的配置是 QPS 为 0，数字 23 表示 Rejected 的请求数，228 表示熔断的请求。my-provider2 对应的配置是 10s 内 15 次请求，达到 50% 的异常比例进入熔断降级状态，数字 14 表示失败调用，237 表示熔断的请求。my-provider3 对应的配置是 10s 内 30 次请求，达到 50% 的异常比例进入熔断降级状态，数字 32 表示失败调用，176 表示熔断的请求。

同时，这 3 个服务的熔断器状态都已经打开。在 Zuul 对应的路由配置信息中，服务 ID 会作为 Hystrix 的 CommandKey，比如，如下配置中 my-provider1 作为服务 ID，其对应的 hystrix 配置也使用了 my-provider1 作为 CommandKey：

```
zuul.routes.my-provider1.path=/dubbo/**
zuul.routes.my-provider1.service-id=my-provider1
hystrix.command.my-provider1.execution.isolation.semaphore.maxConcurrentRequests=0
hystrix.command.my-provider1.execution.isolation.strategy=SEMAPHORE
hystrix.command.my-provider1.execution.timeout.enabled=false
```

如图 5-15 所示，这是在 Spring Cloud Gateway 上使用 Hystrix 限流和降级的数据统计。

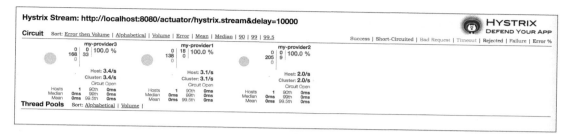

图 5-15

图 5-15 中，my-provider1、my-provider2 和 my-provider3 对应的配置与在 Zuul 上使用 Hystrix 是一样的：

```
@Bean
public RouteLocator hystrixRouteLocator(RouteLocatorBuilder builder) {
    return builder.routes()
            .route(r -> r.path("/dubbo/**")
                    .filters(f -> f.hystrix(new Consumer<Config>() {
                        @Override
                        public void accept(Config config) {
                            config.setSetter(HystrixObservableCommand.Setter
                                    .withGroupKey(HystrixCommandGroupKey.Factory.asKey("SCGHystrix"))
                                    .andCommandKey(HystrixCommandKey.Factory.asKey("my-provider1"))
                                    .andCommandPropertiesDefaults(
                                            HystrixCommandProperties.Setter().withExecutionIsolationSemaphoreMaxConcurrentRequests(
                                                    0).withExecutionIsolationStrategy(
                                                    ExecutionIsolationStrategy.SEMAPHORE).withExecutionTimeoutEnabled(false)));
                        }
                    }).stripPrefix(1))
                    .uri("lb://my-provider1")
            )
            .route(r -> r.path("/springcloud/**")
                    .filters(f -> f.hystrix(new Consumer<Config>() {
                        @Override
                        public void accept(Config config) {
```

```java
                        config.setSetter(HystrixObservableCommand.Setter
                            .withGroupKey(HystrixCommandGroupKey.Factory.asKey("SCGHystrix"))
                            .andCommandKey(HystrixCommandKey.Factory.asKey("my-provider2"))
                            .andCommandPropertiesDefaults(
                                HystrixCommandProperties.Setter().
withExecutionTimeoutInMilliseconds(5000)
                                    .withMetricsRollingStatisticalWindowInMilliseconds(10000)
                                    .withCircuitBreakerRequestVolumeThreshold(15)
                                    .withCircuitBreakerErrorThresholdPercentage(50)
                                    .withCircuitBreakerSleepWindowInMilliseconds(5000)));
                    }
                }).stripPrefix(1))
                .uri("lb://my-provider2")
        ).route(r -> r.path("/s-c-alibaba/**")
            .filters(f -> f.hystrix(new Consumer<Config>() {
                @Override
                public void accept(Config config) {
                    config.setSetter(HystrixObservableCommand.Setter
                        .withGroupKey(HystrixCommandGroupKey.Factory.asKey("SCGHystrix"))
                        .andCommandKey(HystrixCommandKey.Factory.asKey("my-provider3"))
                        .andCommandPropertiesDefaults(
                            HystrixCommandProperties.Setter().
withExecutionTimeoutInMilliseconds(5000)
                                .withMetricsRollingStatisticalWindowInMilliseconds(10000)
                                .withCircuitBreakerRequestVolumeThreshold(30)
                                .withCircuitBreakerErrorThresholdPercentage(50)
                                .withCircuitBreakerSleepWindowInMilliseconds(5000)));
                    }
                }).stripPrefix(1))
                .uri("lb://my-provider3")
        )
        .build();
}
```

从 Dashboard 的统计数据上看，达到了限流熔断的效果。

5.5.4 Hystrix 的高级特性

Hystrix 有以下 3 种比较重要的配置。

- CommandKey：一个 CommandKey 代表一个 HystrixCommand，用于监控、熔断、metrics 统计、缓存等功能。比如，Dashboard 上显示的每个监控项都是一个 CommandKey。
- GroupKey：用于完成同一个目的的 HystrixCommand 可以使用相同的 GroupKey。可以聚合同一个 GroupKey 下所有的 HystrixCommand，用于报警、统计。
- ThreadPoolkey：一个 ThreadPoolkey 代表一个线程池，用于监控、metrics 统计、缓存等功能。如果不设置，默认使用 GroupKey。

使用同一个 CommandKey 会共享同一份 Hystrix 配置，使用同一个 ThreadPoolkey 会共享同一份线程池。

下面这段 Hystrix 代码中，使用了相同的 CommandKey 的方法是 hello1 和 hello2，这两个方法共享 QPS 为 0 的限流策略。singleThread 和 singleThread2 方法共享相同的 ThreadPoolKey，它们使用相同的线程池（如果不设置 ThreadPoolKey，默认会使用 GroupKey；若不设置 GroupKey，则会使用类名作为 GroupKey）：

```
@RestController
class HystrixController {

    @HystrixCommand(commandKey = "hello", fallbackMethod = "fallback",
        commandProperties = {
        @HystrixProperty(name = "execution.isolation.strategy", value = "SEMAPHORE"),
        @HystrixProperty(name = "execution.isolation.semaphore.maxConcurrentRequests", value = "0"),
        @HystrixProperty(name = "execution.isolation.thread.timeoutInMilliseconds", value = "5000")
    })
    @GetMapping("/hello1")
    public String hello1() {
        return new RestTemplate().getForObject("https://httpbin.org/status/200",
String.class);
    }
```

```java
    @HystrixCommand(commandKey = "hello", fallbackMethod = "fallback")
    @GetMapping("/hello2")
    public String hello2() {
        return new RestTemplate().getForObject("https://httpbin.org/status/200",
String.class);
    }

    @HystrixCommand(threadPoolKey = "singleThread",fallbackMethod = "fallback",
        threadPoolProperties = {
            @HystrixProperty(name = "coreSize", value = "1")
    })
    @GetMapping("/singleThread")
    public String singleThread() {
        // 只会打印同一个线程名
        System.out.println("--" + Thread.currentThread().getName());
        return new RestTemplate().getForObject("https://httpbin.org/status/200",
String.class);
    }

    @HystrixCommand(threadPoolKey = "singleThread",fallbackMethod = "fallback")
    @GetMapping("/singleThread2")
    public String singleThread2() {
        // 只会打印同一个线程名
        System.out.println("--" + Thread.currentThread().getName());
        return new RestTemplate().getForObject("https://httpbin.org/status/200",
String.class);
    }

    @HystrixCommand(fallbackMethod = "fallback")
    @GetMapping("/multipleThread")
    public String multipleThread() {
        // 最多 10 个线程名
        System.out.println("--" + Thread.currentThread().getName());
        return new RestTemplate().getForObject("https://httpbin.org/status/200",
String.class);
    }
```

```java
public String fallback(Throwable throwable) {
    return "Hystrix fallback: " + throwable.getMessage();
}
```
}

Hystrix Dashboard 会统计各个线程池的状态，如图 5-16 所示。

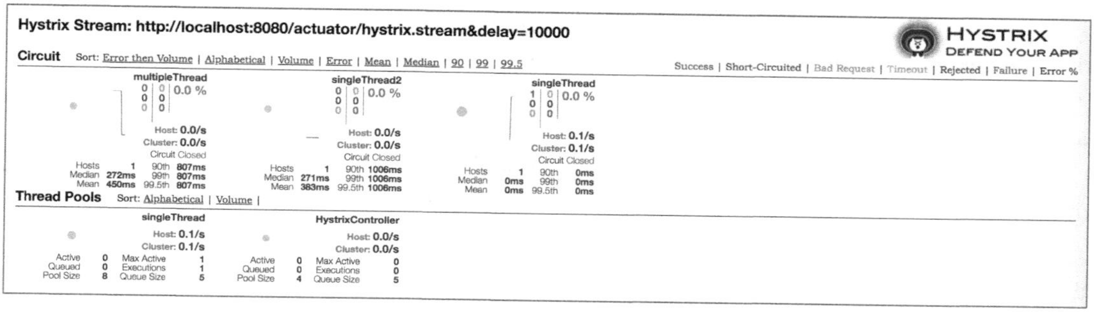

图 5-16

Hystrix wiki 对外提供的各项配置说明如表 5-1 所示。

表 5-1

配 置 项	作 用	默 认 值
execution.isolation.strategy	隔离策略：线程池隔离（THREAD）和信号量隔离（SEMAPHORE）	线程池（THREAD）
execution.isolation.thread.timeoutInMilliseconds	执行超时时间，超时后进入 fallback 逻辑	1000（ms）
execution.timeout.enabled	执行超时时间是否生效。开启后超时配置才会生效	true
execution.isolation.thread.interruptOnTimeout	HystrixCommand 执行超时后是否可以被打断	true
execution.isolation.thread.interruptOnCancel	HystrixCommand 执行被取消时是否可以被打断	True
execution.isolation.semaphore.maxConcurrentRequests	使用信号量隔离策略后最大 QPS 数	10（QPS）
fallback.isolation.semaphore.maxConcurrentRequests	HystrixCommand.getFallback() 方法最大请求数	10（请求数）

续表

配 置 项	作 用	默 认 值
fallback.enabled	HystrixCommand.getFallback() 方法是否生效	true
circuitBreaker.enabled	熔断功能是否开启	true
circuitBreaker.requestVolumeThreshold	时间窗口内最小能触发熔断的请求数（如果时间窗口内没有达到最小请求数，即使异常比率达标，也不会触发熔断效果）	20（个）
circuitBreaker.sleepWindowInMilliseconds	熔断器从 Open 状态进入 Hafl-Open 状态的时间	5000（ms）
circuitBreaker.errorThresholdPercentage	异常比率。时间窗口内最小能触发熔断的请求数中错误请求数占比	50（百分比）
circuitBreaker.forceOpen	熔断器是否强制打开，如打开，所有的请求都会被拒绝	false
circuitBreaker.forceClosed	熔断器是否强制关闭，如关闭，达到异常比率的请求还是可以通过	false
metrics.rollingStats.timeInMilliseconds	请求统计时间窗口	10000（ms）
metrics.rollingStats.numBuckets	时间窗口内桶的个数。默认时间窗口是 10s，10s 内分 10 个桶，每个桶记录着调用成功（Success）、失败（Failure）、超时（Timeout）和拒绝（Rejection）次数	10（个）
metrics.rollingPercentile.enabled	是否统计响应时间百分比。Dashboard 会展示这些百分比统计数据	true
metrics.rollingPercentile.timeInMilliseconds	统计响应时间百分比时的时间窗口	60000（ms）
metrics.rollingPercentile.numBuckets	统计响应时间百分比时滑动窗口要划分的桶个数	6（个）
metrics.rollingPercentile.bucketSize	统计响应时间百分比时每个滑动窗口桶内保存的最大请求数，桶内的请求超出这个值后，会覆盖最前面保存的数据	100（个）

续表

配 置 项	作 用	默 认 值
metrics.healthSnapshot.intervalInMilliseconds	用来设置采集影响断路器状态的健康快照（请求的成功、错误百分比）的间隔等待时间	500（ms）
requestCache.enabled	将请求结果缓存，下一个具有相同 key 的请求将直接从缓存中取出结果，减少请求开销。缓存的 key 由 HystrixCommand.getCacheKey 方法获取	true
requestLog.enabled	每次 HystrixCommand 执行都会被日志记录	true
coreSize	使用线程池隔离策略时核心线程数	10（个）
maximumSize	线程池最大线程数	10（个）
maxQueueSize	线程池队列长度	-1（使用 SynchronousQueue 队列，否则使用 LinkedBlockingQueue）
queueSizeRejectionThreshold	队列拒绝阈值，如果设置该值，线程池内的队列长度无效（maxQueueSize 参数等于-1，该配置无效）	5（队列长度）
keepAliveTimeMinutes	核心线程数小于最大线程数时，那些多出来的线程的存活时间	1min
allowMaximumSizeToDivergeFromCoreSize	设置为 true 时，当最大线程数比核心线程数小时，最大线程数会使用核心线程数	false

5.6　Resilience4j

Resilience4j 是 Netflix Hystrix 宣布 Hystrix 进入维护模式时自己推荐代替的熔断器。Resilience4j 声明它的设计是受 Hystrix 的启发，并且是为 Java 8 和函数式编程设计的。官方对 Resilience4j 的功能描述如下：

- Retry：重试（有些错误是临时的，很短的时间内会自愈，可以通过重试去解决）。
- Circuit Breaker：熔断器。
- Rate Limiter：限流。
- Time Limiter：限制执行时间。
- Bulkhead：限制并发执行次数。
- Cache：缓存一些类似请求的结果。
- Fallback：熔断后自定义操作。

5.6.1 Resilience4j 体验

Resilience4j Rate Limiter 的使用方式如下：

```
RateLimiterConfig config = RateLimiterConfig.custom()
    .timeoutDuration(Duration.ofMillis(100))
    .limitRefreshPeriod(Duration.ofSeconds(1))
    .limitForPeriod(1)
    .build();
RateLimiter rateLimiter = RateLimiter.of("httpbin", config);

RestTemplate restTemplate = new RestTemplate();

Supplier<String> limitSupplier = RateLimiter
    .decorateSupplier(rateLimiter, () -> {
        return restTemplate.
            getForEntity("http://httpbin.org/status/200", String.class).
            getStatusCode().toString();
    });

ExecutorService executorService = Executors.newFixedThreadPool(5);
executorService.submit(new Runnable() {
    @Override
    public void run() {
        Try<String> execute = Try.ofSupplier(limitSupplier);
        System.out.println(execute.isSuccess() + ": " + (execute.isSuccess() ?
execute.get() : execute.getCause()));
```

```java
        }
    });
    executorService.submit(new Runnable() {
        @Override
        public void run() {
            Try<String> execute = Try.ofSupplier(limitSupplier);
            System.out.println(execute.isSuccess() + ": " + (execute.isSuccess() ? execute.get() : execute.getCause()));
        }
    });
    executorService.submit(new Runnable() {
        @Override
        public void run() {
            Try<String> execute = Try.ofSupplier(limitSupplier);
            System.out.println(execute.isSuccess() + ": " + (execute.isSuccess() ? execute.get() : execute.getCause()));
        }
    });
    try {
        Thread.sleep(10000L);
    } catch (InterruptedException e) {
        e.printStackTrace();
    }
    executorService.shutdown();
```

控制台输出以下信息：

```
true: 200 OK
false: io.github.resilience4j.ratelimiter.RequestNotPermitted: RateLimiter 'httpbin' does not permit further calls
false: io.github.resilience4j.ratelimiter.RequestNotPermitted: RateLimiter 'httpbin' does not permit further calls
```

Resilience4j 限流算法封装在 RateLimiter 接口内，其对应的两个实现 AtomicRateLimiter 和 SemaphoreBasedRateLimiter 分别使用令牌桶算法和固定并发数限流。默认情况下，RateLimiter 提供的静态方法 of 内部使用的都是令牌桶算法实现 AtomicRateLimiter。

RateLimiterConfig 表示限流功能对应的配置类，比如，limitForPeriod 属性表示对应时间内可以通过的流量个数；limitRefreshPeriod 表示限流生效的时间（limitRefreshPeriod=1s，limitForPeriod=1，表示 1s 内只允许通过 1 个请求）；timeoutDuration 表示等待获取允许通过的超时时间，如果超过该时间依然被限流，则会抛出异常，否则会继续执行。

示例中，由于 1s 内只能有一个请求通过，所以 3 个请求中只有 1 个正常返回。

Resilience4j Circuit Breaker 的使用方式如下：

```
CircuitBreaker circuitBreaker = CircuitBreaker.of("httpbin",
        CircuitBreakerConfig.custom().failureRateThreshold(50).
minimumNumberOfCalls(2).build());

RestTemplate restTemplate = new RestTemplate();

Supplier<String> decoratedSupplier1 = CircuitBreaker
    .decorateSupplier(circuitBreaker, () -> {
        return restTemplate.
            getForEntity("http://httpbin.org/status/500", String.class).
            getStatusCode().toString();
    });

Supplier<String> decoratedSupplier2 = CircuitBreaker
    .decorateSupplier(circuitBreaker, () -> {
        return restTemplate.
            getForEntity("http://httpbin.org/status/200", String.class).
            getStatusCode().toString();
    });

String result1 = Try.ofSupplier(decoratedSupplier1)
    .recover(throwable -> "Hello from Recovery1" + throwable.getMessage()).get();

System.out.println(result1);

String result2 = Try.ofSupplier(decoratedSupplier2)
    .recover(throwable -> "Hello from Recovery2" + throwable.getMessage()).get();
```

```
System.out.println(result2);

String result3 = Try.ofSupplier(decoratedSupplier1)
    .recover(throwable -> "Hello from Recovery3" + throwable.getMessage()).get();

System.out.println(result3);
```

控制台输出以下信息：

```
Hello from Recovery1: 500 INTERNAL SERVER ERROR
200 OK
Hello from Recovery3CircuitBreaker 'httpbin' is OPEN and does not permit further calls
```

CircuitBreakerConfig 表示 Circuit Breaker 功能对应的配置类，minimumNumberOfCalls 对应的参数表示最少调用次数会触发熔断，failureRateThreshold 表示错误比例。

上述 Resilience4j Circuit Breaker 示例的前面两次调用中，一次成功，另一次失败（decoratedSupplier1 返回的 Response Code 为 500 会报错，decoratedSupplier2 返回的 Response Code 为 200 则正常执行）。两次调用满足最少调用次数，且失败 1 次也满足 50%的失败比例，达到熔断状态。第三次调用虽然会成功，但由于处于熔断状态，所以直接返回 Hello from Recovery3。

Resilience4j Bulkheader 的使用方式如下：

```
BulkheadConfig config = BulkheadConfig.custom()
    .maxConcurrentCalls(1)
    .maxWaitDuration(Duration.ofMillis(200))
    .build();

Bulkhead bulkhead = Bulkhead.of("httpbin", config);

RestTemplate restTemplate = new RestTemplate();

Supplier<String> decoratedSupplier = Bulkhead
    .decorateSupplier(bulkhead, () -> {
        return restTemplate.
```

```
                getForEntity("http://httpbin.org/delay/2", String.class).
                getStatusCode().toString();
        });

ExecutorService executorService = Executors.newFixedThreadPool(5);
executorService.submit(new Runnable() {
    @Override
    public void run() {
        Try<String> execute = Try.ofSupplier(decoratedSupplier);
        System.out.println(execute.isSuccess() + ": " + (execute.isSuccess() ?
execute.get() : execute.getCause()));
    }
});
executorService.submit(new Runnable() {
    @Override
    public void run() {
        Try<String> execute = Try.ofSupplier(decoratedSupplier);
        System.out.println(execute.isSuccess() + ": " + (execute.isSuccess() ?
execute.get() : execute.getCause()));
    }
});
executorService.submit(new Runnable() {
    @Override
    public void run() {
        Try<String> execute = Try.ofSupplier(decoratedSupplier);
        System.out.println(execute.isSuccess() + ": " + (execute.isSuccess() ?
execute.get() : execute.getCause()));
    }
});

executorService.shutdown();
```

控制台输出以下信息：

```
true: 200 OK
false: io.github.resilience4j.bulkhead.BulkheadFullException: Bulkhead 'httpbin' is
full and does not permit further calls
```

```
false: io.github.resilience4j.bulkhead.BulkheadFullException: Bulkhead 'httpbin' is
full and does not permit further calls
```

BulkheadConfig 表示 Bulkhead 功能对应的配置类，maxConcurrentCalls 对应的参数表示最大的并发调用，maxWaitDuration 表示线程在没有资源（并发调用次数已满）的情况下最大等待时间。

本例中使用线程池调用 httpbin 的 delay 服务。由于最大并发数只有 1，所以后面两次调用都抛出了异常。

5.6.2　Spring Cloud Resilience4j

Resilience4j 与 Spring Cloud 体系已经进行了整合，若想使用它，应引入以下依赖：

```xml
<dependency>
    <groupId>org.springframework.cloud</groupId>
    <artifactId>spring-cloud-starter-circuitbreaker-resilience4j</artifactId>
</dependency>
```

同样，由于编程模型一致，Spring Cloud Circuit Breaker Resilience4j 的使用只需关心配置即可。

以下这段代码通过 Customizer 配置的 Resilience4J 参数与 Sentinel 和 Hystrix 配置的基于错误数熔断降级的参数，能达到相似的效果（最少 3 次请求错误率达到 100% 触发熔断）：

```java
@Bean
public Customizer<Resilience4JCircuitBreakerFactory> customizer() {
    return factory -> {
        factory.configureDefault(id -> {
            Resilience4JCircuitBreakerConfiguration configuration
                = new Resilience4JConfigBuilder.Resilience4JCircuitBreakerConfiguration();
            configuration.setCircuitBreakerConfig(
                    CircuitBreakerConfig.custom().minimumNumberOfCalls(3).failureRateThreshold(100).build()
            );
            configuration.setTimeLimiterConfig(TimeLimiterConfig.ofDefaults());
            return configuration;
```

```
            }
        );
    };
}
```

Resilience4j 与 OpenFeign 的集成需要加上 io.github.resilience4j:resilience4j-feign 依赖。这个集成并不像 Sentinel 和 Hystrix 那样简单，它不能使用@FeignClient 注解里的 fallback 属性，需要使用 FeignDecorators，再配合 Resilience4jFeign 的 builder 静态方法构造 Resilience4jFeign.Builder，或者不使用 Resilience4jFeign.Builder，直接使用 Resilience4j 提供的注解。

Resilience4jFeign 的使用方式如下：

```
@Bean
@Scope("prototype")
@ConditionalOnMissingBean
public Feign.Builder resilience4jBuilder() {
    CircuitBreaker circuitBreaker = CircuitBreaker.of("my-service",
CircuitBreakerConfig.custom().minimumNumberOfCalls(3).failureRateThreshold(100).build());
    RateLimiter rateLimiter = RateLimiter.ofDefaults("my-service");
    FeignDecorators decorators = FeignDecorators.builder()
        .withRateLimiter(rateLimiter)
        .withCircuitBreaker(circuitBreaker)
        .withFallbackFactory(exception -> new MyFallback(exception))
        .build();

    return Resilience4jFeign.builder(decorators);
}
```

使用 FeignDecorators 设置 RateLimiter 和 CircuitBreaker，再配合使用 FallbackFactory 判断异常类型进行处理。

这种方式对所有的 FeignClient 接口都设置一样的 RateLimiter、CircuitBreaker 和 FallbackFactory。如果需要对不同的 FeignClient 设置不一样的配置，则需要单独对每个 FeignClient 进行处理。

使用 Resilience4j 提供的注解方式如下：

```java
@FeignClient(name = "inventory-provider")
public interface InventoryService {

    @GetMapping("/save")
    @CircuitBreaker(name = "inventory", fallbackMethod = "fallbackSave")
    String save();

    default String fallbackSave(Throwable cause) {
        if (cause instanceof CallNotPermittedException) {
            return "fallback by r4j";
        }
        return "biz error: " + cause.getMessage();
    }

}
```

使用 @CircuitBreaker 修饰 FeignClient 接口里定义的方法，被熔断后调用 fallbackMethod 方法。@CircuitBreaker 注解有一个 name 属性，表示对应的是哪个 CircuitBreaker，可以通过配置文件指定（以下配置文件内的 inventory 表示 Circuit Breaker 的 name）：

```
resilience4j.circuitbreaker.backends.inventory.minimum-number-of-calls=3
resilience4j.circuitbreaker.backends.inventory.failure-rate-threshold=100
```

或者通过注入 CircuitBreakerRegistry 进行 CircuitBreaker 的构造：

```java
@Autowired
CircuitBreakerRegistry circuitBreakerRegistry;

@Bean
CommandLineRunner runner() {
    return args -> {
        circuitBreakerRegistry.circuitBreaker("inventory", CircuitBreakerConfig.custom().minimumNumberOfCalls(3)
                .failureRateThreshold(100).build());
    };
}
```

5.6.3 Resilience4j 的高级特性

Resilience4j 提供了 Registry 接口，用于 Resilience4j 各个功能点的配置。比如，BulkheadRegistry、CircuitBreakerRegistry、RateLimiterRegistry、TimeLimiterRegistry 等都是实现了 Registry 接口的子接口。它们针对 Bulkhead、CircuitBreaker、RateLimiter、TimeLimiter 这些功能提供配置的操作。

默认情况下，这些接口只有基于内存的实现，比如 InMemoryCircuitBreakerRegistry、InMemoryRateLimiterRegistry、InMemoryBulkheadRegistry 等。这些基于内存的实现会在对应的自动化配置类中被构造。

我们可以基于一些配置中心去实现 Registry 接口，以达到修改配置中心内的配置信息，从而让 Resilience4j 对应的策略动态生效。

Resilience4j 对外提供的 Circuit Breaker 各项配置说明如下：

- failureRateThreshold：异常调用比例。默认为 50%，服务调用的失败比例大于或等于这个比例后，就会进入 Open 状态。
- slowCallRateThreshold：慢调用比例。默认为 100%，调用时间超过 slowCallDurationThreshold 配置的时间算是慢调用。服务调用的慢调用比例大于或等于这个比例时，就会进入 Open 状态。
- slowCallDurationThreshold：慢调用时间阈值。默认为 60000ms，服务调用时间超过这个阈值时，认为是一次慢调用。
- permittedNumberOfCallsInHalfOpenState：Half-Open 状态下允许继续调用的次数。默认为 10 次。
- slidingWindowType：滑动窗口类型。默认为基于次数的类型（COUNTBASED），还有一种是基于时间的类型（TIMEBASED）。如果是基于次数的类型，那么最近的 slidingWindowSize 次调用会被统计和记录，否则最近的 slidingWindowSize 内会统计和记录时间内的调用。
- slidingWindowSize：根据 slidingWindowType 滑动窗口类型充当不同的作用。基于次数的类型记录最近 slidingWindowSize 次调用，基于时间的类型记录最近 slidingWindowSize 的调用。
- minimumNumberOfCalls：触发熔断生效的最小调用次数。默认为 10 次，如果配置的失败比例是 10 次，但只有 9 次调用且全部失败，由于没有达到最小调用次数的条件，这个场景并不会触发熔断。

- waitDurationInOpenState：从 Open 到 Half-Open 的等待时间。默认为 60000ms。
- automaticTransitionFromOpenToHalfOpenEnabled：在不需要额外条件的情况下是否可以自动从 Open 切换到 Half-Open 状态。默认是 false。
- recordExceptions：需要明确的异常列表，这些异常会增加异常比例。
- ignoreExceptions：需要忽略的异常列表，这些异常不会对异常比例的统计有影响。
- recordException：一个 Predicate 函数，参数是 Throwable，用于判断哪些异常是错误的调用。默认情况下，所有的异常都是错误的调用。
- ignoreException：一个 Predicate 函数，参数是 Throwable，用于忽略不需要进入异常统计的异常。默认情况下，所有的异常都需要统计。

下面这段代码中，遍历 10 次调用全部发送异常，没有被熔断，因为熔断的最小调用次数是 10 次，10 次调用刚好触发条件，没有第 11 次调用，导致没看到熔断效果：

```
CircuitBreaker circuitBreaker = CircuitBreaker.of("httpbin",
    CircuitBreakerConfig.custom().minimumNumberOfCalls(10).failureRateThreshold(20).
build());

RestTemplate restTemplate = new RestTemplate();
for (int i = 0; i < 10; i++) {
    String result = Try.ofSupplier(CircuitBreaker
        .decorateSupplier(circuitBreaker, () -> {
            return restTemplate.
                getForEntity("http://httpbin.org/status/500", String.class).
                getStatusCode().toString();
    })).recover(throwable -> "fallback: " + throwable.getMessage()).get();
    System.out.println(result);
}
```

如果在构造 CircuitBreaker 的时候加上 ignoreExceptions(HttpServerErrorException.InternalServerError.class)代码，表示不对 HttpServerErrorException.InternalServerError 异常进行错误率统计，所以循环中，即使大于 10 次调用，也不会触发熔断。同样的效果也可以通过 ignoreException 方法使用 Predicate 函数完成：

```
ignoreException( throwable -> {
    if(throwable instanceof HttpServerErrorException.InternalServerError) {
```

```
        return true;
    }
    return false;
})
```

下面这段代码设置最小调用次数为 2 次，慢调用比例为 100%，超过 2s 算慢调用的熔断效果。代码里遍历的 5 次调用中，前面 2 次成功，后面 3 次被熔断：

```
CircuitBreaker circuitBreaker = CircuitBreaker.of("httpbin",
CircuitBreakerConfig.custom().
    minimumNumberOfCalls(2).
    slowCallRateThreshold(100).
    slowCallDurationThreshold(Duration.ofSeconds(2)).
    build());

RestTemplate restTemplate = new RestTemplate();
for (int i = 0; i < 5; i++) {
String result = Try.ofSupplier(CircuitBreaker
    .decorateSupplier(circuitBreaker, () -> {
        return restTemplate.
            getForEntity("http://httpbin.org/delay/3", String.class).
            getStatusCode().toString();
    })).recover(throwable -> "fallback: " + throwable.getMessage()).get();
System.out.println(result);
}
```

下面这段代码设置从 Open 到 Half-Open 的时间为 10s，出现熔断后 sleep 为 10s，发现进入 Half-Open 状态熔断失效：

```
CircuitBreaker circuitBreaker = CircuitBreaker.of("httpbin",
    CircuitBreakerConfig.custom().
        minimumNumberOfCalls(4).
        failureRateThreshold(100).
        waitDurationInOpenState(Duration.ofSeconds(10)). // 10s 后进入 Half-Open 状态
        build());
```

```java
RestTemplate restTemplate = new RestTemplate();
for (int i = 0; i < 6; i++) {
    String result = Try.ofSupplier(CircuitBreaker
        .decorateSupplier(circuitBreaker, () -> {
            return restTemplate.
                getForEntity("http://httpbin.org/status/500", String.class).
                getStatusCode().toString();
        })).recover(throwable -> "fallback: " + throwable.getMessage()).get();
    System.out.println(result);
}

try {
    Thread.sleep(10000L);
} catch (InterruptedException e) {
    e.printStackTrace();
}

for (int i = 0; i < 3; i++) {
    String result = Try.ofSupplier(CircuitBreaker
        .decorateSupplier(circuitBreaker, () -> {
            return restTemplate.
                getForEntity("http://httpbin.org/status/500", String.class).
                getStatusCode().toString();
        })).recover(throwable -> "fallback: " + throwable.getMessage()).get();
    System.out.println(result);
}
```

如表 5-2 所示，给出了 3 种熔断器（Sentinel、Hystrix 和 Resilience4j）的对比（来自 Sentinel GitHub Wiki）。

表 5-2

名 称	Sentinel	Hystrix	Resilience4j
隔离策略	信号量隔离（并发线程数限流）	线程池隔离/信号量隔离	信号量隔离
熔断降级策略	基于响应时间、异常比率、异常数等	异常比率模式、超时熔断	基于异常比率、响应时间

名称	Sentinel	Hystrix	Resilience4j
实时统计实现	滑动窗口（LeapArray）	滑动窗口（基于 RxJava）	Ring Bit Buffer
动态规则配置	支持多种配置源	支持多种数据源	有限支持
扩展性	丰富的 SPI 扩展接口	插件的形式	接口的形式
基于注解的支持	支持	支持	支持
限流	基于 QPS，支持基于调用关系的限流	有限的支持	Rate Limiter
集群流量控制	支持	不支持	不支持
流量整形	支持预热模式、匀速排队模式等多种复杂场景	不支持	简单的 Rate Limiter 模式
系统自适应保护	支持	不支持	不支持
控制台	提供开箱即用的控制台，可配置规则、查看秒级监控、机器发现等	简单的监控检查	不提供控制台，可对接其他监控系统
多语言支持	Java/C++	Java	Java
开源社区状态	活跃	停止维护	较活跃

5.7 案例：使用 Sentinel 保护应用，防止服务雪崩

服务雪崩指的是因下游服务的不可用拖累上游，导致这个上游服务不可用，进而影响这个上游服务的上游也不可用，最终引起所有的链路不可用。这个现象被称为服务雪崩。

比如，某个电商系统里订单服务依赖下游的配送服务，配送服务又依赖下游的第三方短信服务。某一天，第三方的短信服务出现了问题，导致请求的响应时间达到 10s，短信服务响应时间过长，导致上游配送服务的响应时间也跟着变长甚至崩溃（请求积压，无法释放链接），配送服务响应时间过长又引起订单服务响应时间过长。这时内部所有的链路（订单服务和配送服务）均不可用，整个过程如图 5-17 所示。

针对上述问题，可以利用熔断器解决。如图 5-18 所示，在识别到第三方短信服务响应慢的情况下，会触发熔断机制，直接通过调用本地的容错方法返回错误信息，而不会再去调用下游的短信服务。

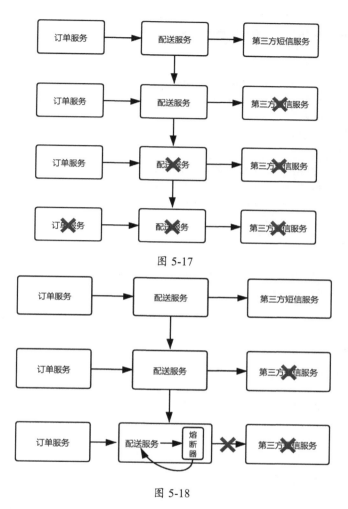

图 5-17

图 5-18

下面使用程序来模拟这个场景。

（1）编写 sms-service 第三方服务的代码

首先编写 sms-service 第三方服务的代码。这里前 1000 次调用都是正常调用，后面所有的调用模拟第三方服务出现响应慢的问题：

```
@RestController
class SMSController {
```

```java
private AtomicInteger count = new AtomicInteger();

@GetMapping("/send")
public String send(String orderId, int delaySecs) {
    int num = count.addAndGet(1);
    if(num >= 1000) {
        if(delaySecs > 0) {
            try {
                Thread.sleep(1000 * delaySecs);
            } catch (InterruptedException e) {
                return "Interrupted: " + e.getMessage();
            }
        }
    }
    System.out.println(orderId + " send successfully");
    return "success";
}
```

（2）编写 delivery-service 配送服务

然后编写 delivery-service 配送服务。通过 OpenFeign 调用短信服务，出错情况下使用 FallbackFactory 处理：

```java
@RestController
class DeliveryController {

    @Autowired
    private SMSService smsService;

    @GetMapping("/delivery")
    public String delivery(String orderId) {
        String smsResult = smsService.send(orderId, 5);
        if(smsResult.equalsIgnoreCase("sms service has some problems")) {
            return "delivery " + orderId + " failed: " + smsResult;
        } else {
            return "delivery " + orderId + " success";
```

```java
        }
    }
}

@Bean
public FallbackFactory fallbackFactory() {
    return new FallbackFactory();
}

@FeignClient(name = "sms-service", fallbackFactory = FallbackFactory.class)
public interface SMSService {

    @GetMapping("/send")
    String send(@RequestParam("orderId") String orderId, @RequestParam("delaySecs") int delaySecs);

}

class FallbackInventoryService implements SMSService {

    @Override
    public String send(String orderId, int delaySecs) {
        return "sms service has some problems";
    }
}

class FallbackFactory implements feign.hystrix.FallbackFactory {

    private FallbackInventoryService fallbackService = new FallbackInventoryService();

    @Override
    public Object create(Throwable cause) {
        return fallbackService;
    }
}
```

注意，配送服务对应的应用启动时需要设置 JVM 内存参数，这样效果比较直观：

```
-Xmx64m -Xms64m -Xss1024k
```

（3）通过 wrk 模拟大流量场景调用配送服务

通过 wrk 模拟大流量场景调用配送服务的语句如下：

```
./wrk -c 500 -t 10 -d 10 http://localhost:8081/delivery?orderId=aaaa
```

wrk 执行结束后，发现配送服务发生异常，出现内存溢出（Out Of Memory，缩写为 OOM）（这是因为下游短信服务响应过慢，导致配送服务积压请求，从而消耗完内存）：

```
Exception in thread "http-nio-8081-Acceptor"
Exception: java.lang.OutOfMemoryError thrown from the UncaughtExceptionHandler in thread "http-nio-8081-Acceptor"
Exception in thread "Catalina-utility-1" java.lang.OutOfMemoryError: GC overhead limit exceeded
2020-07-10 00:24:43.328 ERROR 6279 --- [81-ClientPoller]
org.apache.tomcat.util.net.NioEndpoint    : Error processing poller event

java.lang.OutOfMemoryError: GC overhead limit exceeded
```

这时配送服务的上游订单服务也会受到影响。接下来引入 com.alibaba.cloud:spring-cloud-starter-alibaba-sentinel 依赖并打开 Feign 的开关 feign.sentinel.enabled=true，再配置资源的熔断策略：

```
[
  {
    "resource": "GET:http://sms-service/send",
    "count": 20,
    "grade": 0,
    "timeWindow": 30
  }
]
```

重新使用 wrk 模拟大流量场景，结束后发现程序运行正常，并没有出现内存溢出，因为下游响应过慢导致配送服务触发了熔断，不进行短信服务的调用。

第 6 章
Spring 生态消息驱动

本章将介绍 Spring 生态内消息体系。整个 Spring 生态中与消息有关的 3 个项目分别是 Spring Framework 里的 spring-messaging 模块、Spring Integration 和 Spring Cloud Stream 项目。

Spring Framework 内部的 spring-messaging 模块是定义消息编程模型的基础模块，内部定义了如消息（Message）接口、MessageChannel（消息通道）接口、MessageHandler（消息处理器）接口、@MessageMapping、@Header 等 WebSocket 相关的注解等诸多内容。

Spring Integration 在 spring-messaging 的基础上根据 Enterprise Integration Patterns（企业集成模式）内消息部分的功能抽象了更多消息的概念，比如：MessageDispatcher（消息分发器）、Transformer（消息转换器）、Aggregator（消息聚合器）等内容。

Spring Cloud Stream 在 Spring Integration 的基础上提出了 Binder、Binding 等概念让开发者能够通过 Bean 的注入及相关注解，就能轻松地完成消息的发送和订阅来进行业务开发。

6.1 消息中间件概述

假设有如图 6-1 所示的场景：Producer 实时给 Consumer 发送数据，Consumer 进行消费。

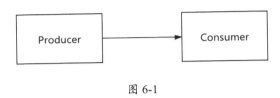

图 6-1

图 6-1 的模型有如下一些弊端：

- Producer 生产数据的频率如果比 Consumer 消费数据的频率高，会导致消息一直堆积在 Consumer 上消费不完，最终可能会出现内存溢出。
- 如果 Consumer 挂掉，堆积在 Consumer 上的数据会丢失。
- Producer 和 Consumer 两者强耦合。

消息中间件出现后，会在两者之间加一层 Topic（或者叫 Broker），如图 6-2 所示，Producer 发送消息（数据）到 Topic 中，Consumer 再从 Topic 中订阅消息。

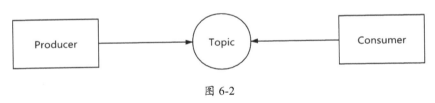

图 6-2

图 6-2 的模型提供了以下两大能力。

- 异步解耦：在没有消息中间件的情况下，Producer 应用会给 Consumer 应用发送数据，每次发送需要 Consumer 的正常响应，两者强依赖。消息中间件的出现在两者之间产生了一个 Topic（或叫 Broker）的概念，Producer 只需要往 Topic 上发消息，Consumer 只需要在 Topic 上订阅消息即可，以达到两个应用解耦的效果。发送和订阅操作也可以是异步操作。这时如果新的应用 Order 也需处理 Producer 发送的数据，直接在 Topic 上订阅消息即可，不需要在 Producer 里再写一遍新的发送逻辑。
- 削峰填谷：如果 Consumer 实例数量少或者配置不高，会遇到消息堆积处理不完的场景，甚至导致 Consumer 出现内存溢出。引入消息中间件后可以将消息堆积在 Topic 上，Consumer 视情况进行消费，从而达到削峰填谷的作用。

看到这里，有的读者可能会想：这个 Topic 似乎可以使用一个数据库来代替？答案是肯定的，但是需要对数据库做以下设计来解决这些场景：

- Producer 产生的数据在业务方不仅只有 Consumer 这一个消费者消费，比如，有一个库存应用 Inventory 也需要订阅 Topic 数据，这时需要维护一个 Topic 订阅关系的表。
- 顺序消费的场景每次获取数据需要进行排序。
- Topic 可以设置 offset（偏移），从某个位偏移量开始重新消费数据，这时需要维护一个 offset 位移量的表。

一个完整的消息中间件当然不只有异步解耦、削峰填谷功能，它还提供顺序收发、分布式事务一致性、大数据分析等功能。一个完整的消息中间件也不单单只是这一个 Topic(Broker)节点，它还需要提供高可用功能，当 Topic 挂掉时，确保消息不丢失。

目前业界主流的消息中间件有 Apache Kafka、Apache RocketMQ、RabbitMQ、Apache ActiveMQ。每个 MQ 在设计层面上大同小异，但是在对外的 API 上却有比较大的区别。

比如 Apache RocketMQ 消息发送/接收代码如下：

```
// 消息发送
DefaultMQProducer producer = new DefaultMQProducer(PRODUCE_RGROUP);
producer.setNamesrvAddr("127.0.0.1:9876");
producer.start();
Message message = new Message("my_topic", "Apache RocketMQ Message".getBytes());
SendResult sendResult = producer.send(message);
// 消息接收
consumer = new DefaultMQPushConsumer(CONSUMER_GROUP);
consumer.setNamesrvAddr("127.0.0.1:9876");
consumer.subscribe("my_topic", "*");
consumer.registerMessageListener((MessageListenerOrderly) (msgs, context) -> {
    MessageExt msg = msgs.get(0);
    consume(msg);
    return ConsumeOrderlyStatus.SUCCESS;
});
consumer.start();
```

Apache Kafka 消息发送/接收代码如下：

```java
// 消息发送
Properties kafkaProps = new Properties();
kafkaProps.put("bootstrap.servers", "localhost:9092");
kafkaProps.put("key.serializer",
"org.apache.kafka.common.serialization.StringSerializer");
kafkaProps.put("value.serializer",
"org.apache.kafka.common.serialization.StringSerializer");
KafkaProducer producer = new KafkaProducer<String, String>(kafkaProps);
ProducerRecord<String, String> record = new ProducerRecord<String, String>("my_topic",
        "Apache Kafka Message");
producer.send(record);
// 消息接收
Properties props = new Properties();
props.put("bootstrap.servers", "localhost:9092");
props.put("group.id", "my_group");
props.put("key.deserializer",
"org.apache.kafka.common.serialization.StringDeserializer");
props.put("value.deserializer","org.apache.kafka.common.serialization.StringDeserializer");
KafkaConsumer<String, String> consumer = new KafkaConsumer<String, String>(props);
consumer.subscribe(Collections.singletonList("my_topic"));
while (true) {
    ConsumerRecords<String, String> records = consumer.poll(100);
    for (ConsumerRecord<String, String> record : records) {
        consume(msg);
        consumer.commitSync();
    }
}
```

从 Apache RocketMQ 和 Apache Kafka 对消息的发送和接收代码来看，这两者的 API 完全不一致。

有没有一套框架能够对外提供统一的 API 来解决消息的发送和接收？类似如下代码：

```java
// 消息发送
Message msg = new Message("Hello Message");
```

```
Broker broker = new Broker("my_topic");
broker.send(msg);
// 消息接收
@MessageListener("my_topic")
public void receiveMsg(Message msg) {
    consume(msg);
}
```

这也是本章要讲到的知识点 Spring Cloud Stream。它屏蔽了底层消息中间件的实现细节，对外提供统一的 API，只需几行代码，即可完成消息的发送和接收，用于构建高度可扩展的事件驱动系统。

想要了解 Spring Cloud Stream，首先需要了解 Spring 整个体系对消息的支持，这涉及两个模块：Spring Messaging 和 Spring Integration。

6.2 Spring 与消息

spring-messaging 是 Spring Framework 框架 4.0 推出的一个子模块，其内部包含了 Spring Integration 项目中关键的消息抽象，比如 Message、MessageChannel 和 MessageHandler，用来作为基于消息的应用的基础模块。spring-messaging 内部还包含了一组注解，比如 @MessageMapping、@Payload 和 @Header，用于映射消息与方法之间的关系（类似 Spring MVC 请求和方法的映射编程模型）。

接下来将介绍这些抽象接口和注解。

6.2.1 消息编程模型的统一

spring-messaging 模块对消息的编程模型进行了统一，不论是 Apache RocketMQ 的 Message，还是 Apache Kafka 的 ProducerRecord，在 spring-messaging 中被统一称为 org.springframework.messaging.Message 接口：

```java
public interface Message<T> {
    T getPayload();

    MessageHeaders getHeaders();
}
```

Message 接口有两个方法，分别是 getPayload 和 getHeaders，用于获取消息体和消息头。这也意味着一个消息（Message）由 Header 和 Payload 组成。

Payload 是一个泛型，意味着消息体可以存放任意数据类型。

Header 是一个 MessageHeaders 类型的消息头，MessageHeaders 是一个实现了 java.util.Map<String, Object> 接口的类，内部对数据的操作做了一些限制（MessageHeaders 是一个 Immutable 类型的对象，不能随意对其中的元数据做修改），k-v 形式的消息头可以设置任意类型的 value。

消息可以通过 MessageBuilder 构造，MessageBuilder 对外暴露了一些静态方法，如 withPayload，createMessage 用于消息的构造：

```
MessageBuilder.withPayload("custom payload").setHeader("k", "v").build();
MessageBuilder.createMessage("custom payload", new
MessageHeaders(Collections.singletonMap("k", "v")));
```

消息（Message）是一个接口，具体的实现类有以下几种：

- GenericMessage：普通消息，这是一个不可变（immutable）的消息，无法新增、修改和删除 Header 中的数据。
- ErrorMessage：错误消息，如果 Payload 是一个 Throwable 异常，那么对应的消息就是错误消息。
- MutableMessage：可变消息，跟 GenericMessage 的区别是可以新增、修改、删除 Header 中的数据。

6.2.2 消息的发送和订阅

有了消息之后，需要把消息发送到 Topic(Broker) 里，这个 Topic 对应的编程模型是消息通道（MessageChannel），代码如下：

```
public interface MessageChannel {
    default boolean send(Message<?> message) {
        return send(message, INDEFINITE_TIMEOUT);
    }
```

```
    boolean send(Message<?> message, long timeout);
}
```

如图 6-3 所示，调用 MessageChannel 的 send 方法可以将消息发送到这个 MessageChannel 中。

图 6-3

消息在消息通道中是怎么被消费的呢？可以通过 MessageChannel 的子接口 PollableChannel 去完成：

```
public interface PollableChannel extends MessageChannel {
    @Nullable
    Message<?> receive();

    @Nullable
    Message<?> receive(long timeout);
}
```

PollableChannel 是一种以"拉"的方式获取消息的消息通道，可以调用 receive 方法去拉消息通道内的消息。

消息通道还有另外一个子类 SubscribableChannel 可以通过订阅的方式获取消息：

```
public interface SubscribableChannel extends MessageChannel {
    boolean subscribe(MessageHandler handler);

    boolean unsubscribe(MessageHandler handler);
}
```

如图 6-4 所示，需要通过 MessageHandler 去消费订阅到的消息。

```
public interface MessageHandler {

    void handleMessage(Message<?> message) throws MessagingException;

}
```

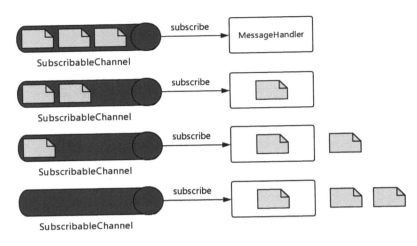

图 6-4

消息通道拦截器（ChannelInterceptor）用于在 MessageChannel 发送消息前、发送消息后、发送消息完成时（发生异常也算完成）进行拦截，这是对 SubscribableChannel 类型的处理。如果是 PollableChannel 类型的消息通道，会在消息接收前、接收后、接收完成时（发生异常也算完成）进行拦截。

无论是 SubscribableChannel 还是 PollableChannel，都可以在拦截的时候对消息内部的属性做一定修改，或者决定是否过滤消息。

以上这些内容都是 spring-messaging 对消息的一些概念的定义，接下来简单看两段代码来理解 SubscribeMessageChannel 和 PollableChannel 的概念。

我们自定义继承 AbstractSubscribableChannel 抽象类（AbstractSubscribableChannel 实现了 SubscribableChannel 接口）的消息通道 MySubscribableChannel。该消息通道订阅消息的时候会根据内部的 MessageHandler 列表随机选择一个进行消费：

```java
public class MySubscribableChannel extends AbstractSubscribableChannel {

    private Random random = new Random();

    @Override
    protected boolean sendInternal(Message<?> message, long timeout) {
        if (message == null || CollectionUtils.isEmpty(getSubscribers())) {
```

```
            return false;
        }
        Iterator<MessageHandler> iter = getSubscribers().iterator();
        int index = 0, targetIndex = random.nextInt(getSubscribers().size());
        while(iter.hasNext()) {
            MessageHandler handler = iter.next();
            if(index == targetIndex) {
                handler.handleMessage(message);
                return true;
            }
            index ++;
        }
        return false;
    }
}
```

使用 MySubscribeChannel 完成消息的发送/消费,代码如下:

```
AtomicInteger successCount = new AtomicInteger(0);
AtomicInteger failCount = new AtomicInteger(0);      // ①
MySubscribeChannel channel = new MySubscribeChannel();   // ②

channel.addInterceptor(new ChannelInterceptor() {   // ③
    @Override
    public Message<?> preSend(Message<?> message, MessageChannel channel) {
        String ignoreKey = "ignore";
        if (message.getHeaders().containsKey(ignoreKey) &&
            message.getHeaders().get(ignoreKey, Boolean.class)) {   // ④
            return null;
        }
        return message;
    }

    @Override
    public void postSend(Message<?> message, MessageChannel channel, boolean sent) {
```

```java
        }

        @Override
        public void afterSendCompletion(Message<?> message, MessageChannel channel,
boolean sent, Exception ex) {
            if (sent) {   // ⑤
                successCount.incrementAndGet();
            } else {
                failCount.incrementAndGet();
            }
        }
});

channel.send(MessageBuilder.withPayload("custom payload1").setHeader("k1",
"v1").build());   // ⑥
channel.subscribe(msg -> {
    System.out.println("[" + Thread.currentThread().getName() + "] handler1 receive:
" + msg);
});
channel.subscribe(msg -> {
    System.out.println("[" + Thread.currentThread().getName() + "] handler2 receive:
" + msg);
});
channel.subscribe(msg -> {   // ⑦
    System.out.println("[" + Thread.currentThread().getName() + "] handler3 receive:
" + msg);
});
channel.send(MessageBuilder.withPayload("custom payload2").setHeader("k2", "v2").
build());
channel.send(MessageBuilder
    .createMessage("custom payload3", new
MessageHeaders(Collections.singletonMap("ignore", true))));   // ⑧
System.out.println("successCount: " + successCount.get() + ", failCount: " +
failCount.get());
}
```

上述代码中：

① 定义两个变量，用于统计消息是否发送成功。

② 定义 MySubscribableChannel 这个 SubscribableChannel，用于消息的发送。

③ 为 MySubscribableChannel 添加匿名 ChannelInterceptor 拦截器。

④ 拦截器的 preSend 方法会判断是否过滤消息，会过滤掉 HEADER 中存在 key 为 ignore、value 为 true 的键值对。

⑤ 拦截器的 afterSendCompletion 方法会根据 sent 结果统计消息是否发送成功。

⑥ 发送一个 payload 为 custom payload1，HEADER 中存在 key 为 k1、value 为 v1 的键值对的消息。这时由于还没有 MessageHandler 订阅，而且 MySubscribableChannel 对于没有 MessageHandler 订阅的场景会忽略。本次发送无任何反馈，send 方法返回值是 false。

⑦ 分别使用 3 个匿名的 MessageHandler 对 MySubscribableChannel 进行订阅。

⑧ 分别发送对应的消息。Payload 为 payload2 的消息会被正常消费，Payload 为 payload3 的消息由于 HEADER 中存在 key 为 ignore、value 为 true 的消息头而被过滤。

控制台最终输出以下信息：

```
[main] handler3 receive: GenericMessage [payload=custom payload2, headers={k2=v2, id=9eca0b7d-29a9-427b-00a5-18fcef784bdf, timestamp=1583248428254}]
successCount: 1, failCount: 1
```

从输出消息的内容来说，消息构造之后会自动添加 key 分别为 id 和 timestamp 的消息头。id 对应的值是一个 UUID，timestamp 是消息创建时对应的时间戳。

这里的 handler3 有可能是 handler1 或 handler2，因为 MySubscribableChannel 内部进行了随机选择。

程序一共发送 3 条消息，第一条消息由于在还没有 MessageHandler 订阅的情况下进行发送，直接被忽略。后续两条消息中，一条被过滤，另一条被成功处理，所以最终的输出包含了 1 条处理成功和 1 条处理失败的信息。

接下来，我们体验以拉的方式消费消息。自定义 MyPollableChannel，这是一个实现 PollableChannel 接口的消息通道，消息的存储和获取分别使用阻塞队列完成：

```java
public class MyPollableChannel implements PollableChannel {

    private BlockingQueue<Message> queue = new ArrayBlockingQueue<>(1000);

    @Override
    public Message<?> receive() {
        return queue.poll();
    }

    @Override
    public Message<?> receive(long timeout) {
        try {
            return queue.poll(timeout, TimeUnit.MILLISECONDS);
        } catch (InterruptedException e) {
            e.printStackTrace();
        }
        return null;
    }

    @Override
    public boolean send(Message<?> message, long timeout) {
        return queue.add(message);
    }
}
```

使用 MyPollableChannel 完成消息的发送/消费：

```
MyPollableChannel channel = new MyPollableChannel();   // ①
channel.send(MessageBuilder.withPayload("custom payload1").setHeader("k1",
"v1").build());
channel.send(MessageBuilder.withPayload("custom payload2").setHeader("k2",
"v2").build());
channel.send(MessageBuilder
    .createMessage("custom payload3", new
MessageHeaders(Collections.singletonMap("ignore", true))));   // ②
```

```
System.out.println(channel.receive());
System.out.println(channel.receive());
System.out.println(channel.receive());
System.out.println(channel.receive());   // ③
```

上述代码中：

① 构造 MyPollableChannel 这个 MyPollableChannel 类型的消息通道。

② 往 MyPollableChannel 消息通道内发送 3 条消息。

③ 调用 4 次 receive 方法从 MyPollableChannel 获取消息并打印内容。

控制台最终输出以下信息：

```
GenericMessage [payload=custom payload1, headers={k1=v1, id=53546078-b36a-f4f0-
211a-3166b302e816, timestamp=1583252190939}]
GenericMessage [payload=custom payload2, headers={k2=v2, id=f5942e8a-3a8a-d09f-
d235-83ee6fac7e36, timestamp=1583252190939}]
GenericMessage [payload=custom payload3, headers={ignore=true, id=922ec49a-2a0f-
3d45-3a9b-cf543a9a8ad6, timestamp=1583252190940}]
null
```

我们发现，通过 MyPollableChannel 的 send 方法发送出去的 3 条消息，在 MyPollableChannel 的 receive 方法内依次被获取到。因为一共只发送了 3 条消息，第 4 个调用 receive 方法会返回 null。

读者看了这两个例子后可能会问：这两段代码似乎只看到了一些消息的概念，并没有看到如何使用消息中间件和这些概念的整合。这些内容后续在讲解 Spring Cloud Stream 的时候会进行分析。

6.2.3 WebSocket

WebSocket 是一种与 HTTP 协议类似的网络通信协议，它可以应用于 HTTP 协议不能覆盖的一些场景。

HTTP 协议存在以下问题：

- 单向，只能由客户端发起通信。

- Request/Reponse 通信模型。客户端发起通信，服务器端响应结果。
- 无状态。
- 半双工（Half-Duplex）协议。

WebSocket 协议出现的目的是在浏览器和服务器之间建立一个不受限的双向通信的通道，这个双向通道可以让服务器主动给浏览器发送消息。

1. Spring 与 WebSocket

spring-messaging 内部提供了一套与 spring-web 非常相似的编程模型在 WebSocket 场景中使用。

spring-web 对外暴露了以下编程模型：

- @Controller 或 @RestController 注解，表示这是一个控制器。
- @RequestMapping 修饰方法，表示这是对外暴露的一个请求映射。
- @Header 解析 HTTP Header 并设置到方法参数中。
- @ResponseBody 方法返回值写到 Response Body 中。
- @RequestParam 解析 HTTP 参数并设置到方法参数中。
- @ExceptionHandler HTTP 异常处理器。
- @PathVariable 解析 Restful 风格的 HTTP 路径。
- HttpMessageConverter 接口用于 Request 参数和 Response 返回内容的转换。比如，@RequestParam 注解对应参数的具体值是从 HTTP 参数被 HttpMessageConverter 转换而成的，比如@ResponseBody 注解对应的返回值由 HttpMessageConverter 转换成 Response Body。
- HandlerMethodArgumentResolver 方法参数解析器。@RequestParam 和@Header 等注解对应的 HTTP 内容如何被解析成参数是由该接口完成的。
- HandlerMethodReturnValueHandler 方法返回值处理器。@ResponseBody 注解、ModelAndView、HttpEntity 等类型的返回值内容如何被解析成 Response Body 由该接口完成。

spring-messaging 针对这些模型都有类似的接口或注解：

- @Controller 或 @RestController 注解，表示这是一个控制器。
- @MessageMapping 修饰方法，表示这是对外暴露的一个消息映射。
- @Header 解析 Message Header 并设置到方法参数中。

- @SendTo/@SendToUser 方法返回值被转换成 Message 消息，并发送到指定的目的地。
- @Payload 解析消息体并设置到对应的方法参数中。
- @MessageExceptionHandler 消息异常处理器。
- @DestinationVariable 解析消息对应的路径。
- MessageConverter 接口用于 Message 和 Object 的内容转换。比如，String 类型的参数的具体值是从 Message Payload 中的字节数组被 MessageConverter 转换而成的，比如，Map 类型的返回值由 MessageConverter 转换 Message 得到。Map 中的 header key 读取消息头，payload key 读取消息体。
- HandlerMethodArgumentResolver 方法参数解析器。@Payload 和 @DestinationVariable 等注解对应的消息内容如何被解析成参数是由该接口完成的。
- HandlerMethodReturnValueHandler 方法返回值处理器。@SendTo、@SendToUser 注解修饰的返回值内容如何被解析成新的目的地是由该接口完成的。

2. SockJS 和 SMOTP 协议

WebSocket 协议需要浏览器支持，如果一些浏览器不支持，但又需要 WebSocket 功能，该怎么办？可以通过 HTTP 轮询（客户端定时给服务器发送请求）的方式模拟 WebSocket。作为一个开发者，肯定不想关心哪些浏览器支持或不支持 WebSocket。有没有一种框架可以屏蔽浏览器，对外提供一套 API 来实现 WebSocket 效果？

答案是肯定的。这套框架就是 SockJS，SockJS 屏蔽了底层不兼容性，优先使用原生 WebSocket，如果还不支持，会自动降为轮询的方式。

WebSocket 协议定义了两种类型的消息：文本（text）和二进制（binary），但是它们的内容格式并没有定义。这意味着服务器端想要跟客户端通信，两者必须要统一消息的规范。

Simple (Streaming) Text Orientated Messaging Protocol 简称为 STOMP 协议，中文名称是简单（流）文本定向消息协议，可以作为 WebSocket 消息内容协议。STOMP 协议是消息队列的一种协议，类似于 AMQP 和 JMS，目前 RabbitMQ 和 ActiveMQ 都已经支持 STOMP。

STOMP 协议是基于帧（frame）的协议，帧的格式如下：

```
COMMAND
header1:value1
header2:value2

Body^@
```

上述帧格式中：

- COMMAND 命令可以是 SEND（发送）、SUBSCRIBE（订阅）、UNSUBSCRIBE（取消订阅）或 MESSAGE（广播）等。
- HEADER 是 k-v 形式的数据格式。比如，可以设置目的地。
- Body 可以是任意形式的文本。

比如，下面这段帧表示客户端发送消息到 /queue/trace，内容是 json 数组：

```
SEND
destination:/queue/trade
content-type:application/json
content-length:44

{"action":"BUY","ticker":"MMM","shares",44}^@
```

6.2.4 案例：使用 spring-messaging 处理 WebSocket

spring-messaging 内部提供的消息映射模型可以用来处理 STOMP 协议内容，下列代码中，WebSocketController 内部暴露了一些 MessageMapping 映射：

```java
@Controller
class WebSocketController {

    @Autowired
    private SimpMessageSendingOperations messagingTemplate;

    @MessageMapping("/subscribe")   // ①
    public void subscribe() {
        messagingTemplate.convertAndSend("/topic/tom", "jerry");
        messagingTemplate.convertAndSend("/topic/jerry", "tom");
    }

    @MessageMapping("/payload")    // ②
    public void payload(@Payload User user, @Header(value = "content-type") String contentType) {
```

```java
        System.out.println("payload: " + user);
        System.out.println("header content-type: " + contentType);
    }

    @MessageMapping("/path/{var}")    // ③
    public void path(@DestinationVariable String var, Message message) {
        System.out.println("receive: " + message);
    }

    @MessageMapping("/message")    // ④
    public void message(String msg) {
        if (msg.contains("input1")) {
            messagingTemplate.convertAndSend("/topic/messages1", msg);
        } else if (msg.contains("input2")) {
            messagingTemplate.convertAndSend("/topic/messages2", msg);
        } else if (msg.contains("input3")) {
            messagingTemplate.convertAndSend("/topic/messages3", msg);
        } else {
            throw new IllegalStateException("unknown msg");
        }
    }

    @MessageExceptionHandler    // ⑤
    @SendTo("/topic/error")
    public String handleException(Throwable exception) {
        return exception.getMessage();
    }
}
```

图 6-5、图 6-6 所示是一个 WebSocket 前端交互页面，具体的按钮意义如下。

① 单击 subscribe 按钮会给/app/subscribe 发送消息，服务器端 subscribe 分发往/topic/tom 和/topic/jerry 分别发送 Payload 为 jerry 和 tom 的消息。客户端监听并设置 Jerry 和 Tom 对应 label 的值。

② 单击 payload 按钮会给/app/payload 发送消息，服务器端 payload 方法使用@Payload 和 @Header 注解解析客户端发送的消息。

③ 单击 DestinationVariable 按钮会给 /app/path/hi 发送消息，服务器端 path 分发使用 @DestinationVariable 解析客户端发送的 topic，会从 HEADER 里取 key 为 `DestinationVariable-MethodArgumentResolver.templateVariables` 的对象，这个对象里 key 为参数名，表示最终的 path 变量，这个例子中 key 为 `var`。

④ 单击 Send Message 按钮会给 /app/message Topic 发送消息，服务器端 message 分发接收客户端的消息，判断消息内容后再分发给客户端不同的 Topic。结果 input1、input2 和 input3 分别订阅到 Topic，再设置内容。

⑤ 通过@MessageExceptionHandler 处理服务器端异常的情况，这里直接返回异常的 message，并发送给客户端的 /topic/error。

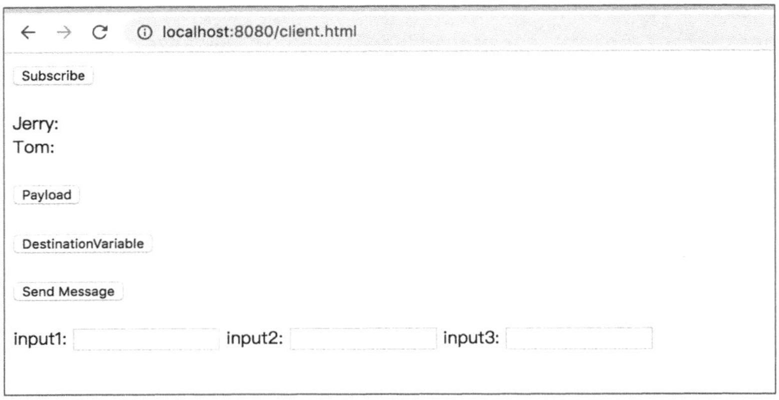

图 6-5

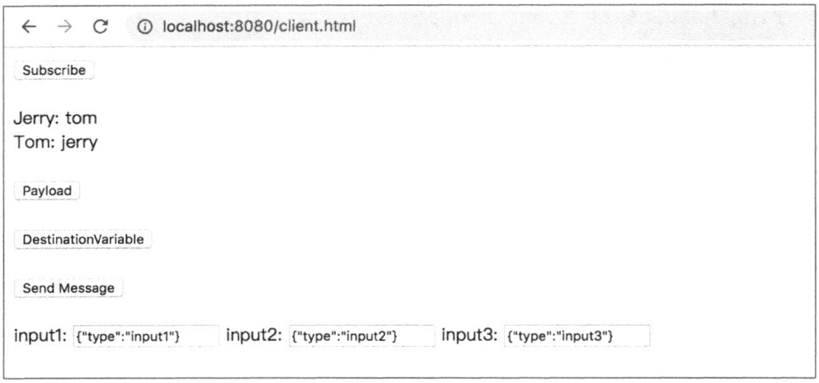

图 6-6

说明：本例中为什么客户端所有的 Topic 都多了 /app 前缀？这是因为服务器端设置了全局的目的地前缀（destination prefix）为/app。

6.3 Spring Integration

Spring Integration 项目是 Spring 对 Enterprise Integration Patterns（企业集成模式）内容的实现，这些实现扩展了原有的 Spring 消息编程模型。

如图 6-7 所示，EIP 内部提出了很多概念，比如消息的过滤、聚合、点对点发送、发布/订阅模式、消息转换、TCP/UDP、JMS、E-mail 等内容。spring-messaging 模块仅仅是对 Spring Integration 项目中消息部分关键的内容进行抽象和简单实现，Spring Integration 项目是真正复杂的实现。

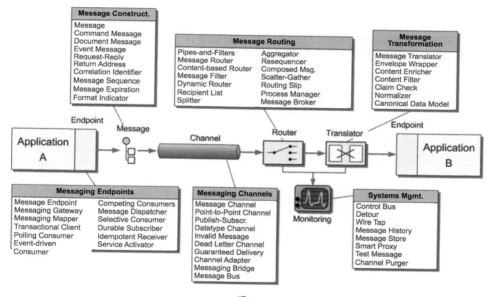

图 6-7

Spring Integration 描述了从应用 A（Application A）发送消息到应用 B（Application B）接收消息的整个过程：应用 A 通过 Endpoint 生产消息，消息发送到 MessageChannel 中，再获取 MessageChannel 中的消息，根据 Router（路由）信息决定往哪个 MessageChannel 发送消息，该过程中可能会通过 Translator 对消息内容做一些修改（路由和 Translator 消息转换过程会有专门的监控系统进行监控），最后 MessageChannel 中的消息被 Endpoint 解析，然后被应用 B 消费。

在这个过程中，Endpoint 端点可以有 MessageEndpoint、Polling Consumer、MessageDispatcher 等多种实现；MessageChannel 消息通道也有多种方式，比如，发布/订阅模式（Publish-Subscr）、点对点模式（Point-to-Point）；消息路由（Router）过程中可以对消息进行分割（Splitter）、聚合（Aggregator）；消息转换（Translator）过程中也可以对内容进行转换，去掉不重要的信息（Content Filter）；最后通过 Endpoint 端点被应用 B 消费。

6.3.1　Spring Integration 核心组件概述

下面具体介绍 Spring Integration 内部的各个核心组件。

① MessageDispatcher：消息分发器。消息分发给 MessageHandler 的策略接口。如图 6-8 所示，比如，有 3 个 MessageHandler 处理器，消息转发器的作用是决定怎么分发给这些 MessageHandler。BroadcastingDispatcher 的实现是广播模式，每次消费分发给所有的 MessageHandler。UnicastingDispatcher 的实现是单播模式，每次消费只能让一个 MessageHandler 处理，单播模式分发给哪个 MessageHandler，涉及负载均衡策略 LoadBalancingStrategy（消息发送者发送了 3 条消息（消息体内容分别是 1、2、3），这 3 条消息被消息接收者接收。接收者使用 MessageDispatcher（消息分发器）处理消息，消息体内容为 1 的消息被上面的消息处理器消费，消息体内容为 2 的消息被中间的消息处理器消费，消息体内容为 3 的消息被下面的消息处理器消费）。

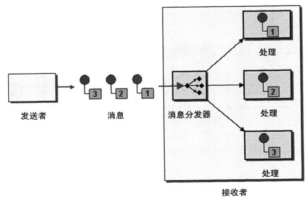

图 6-8

② LoadBalancingStrategy：单播 UnicastingDispatcher 模式下分发 MessageHandler 的负载

均衡策略，默认只有轮询实现：RoundRobinLoadBalancingStrategy。如果使用其他算法，可自定义。

③ Transformer：消息转换器。可以把 Message A 转换成 Message B，在转换过程中，比如可以过滤消息内 HEADER 或 Payload 信息，仅仅保留重要部分。如图 6-9 所示，消息转换器将消息原本比较丰富的内容进行了删减。

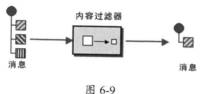

图 6-9

④ MessageSelector：消息选择器。如图 6-10 所示，决定是否选择（accept）消息，与 MessageFilter 配合进行消息的过滤（有 3 条消息发送到了消息通道里，消息通道内的 MessageFilter（消息过滤器）对消息进行过滤后发送到新的消息通道里，消费者从新的消息通道中消费这两条消息）。

图 6-10

⑤ MessageRouter：消息路由。如图 6-11 所示，获取 MessageChannel 中的消息，并根据不同的条件发送给其他不同的 MessageChannel（有 1 条消息发送到消息通道里，消息通道内的消息路由订阅通道里的消息，再根据内部的逻辑把这个消息发送到另外一个消息通道里）。

⑥ Aggregator：消息聚合器。如图 6-12 所示，把一组消息根据一些条件聚合成一条消息（3 条消息经过消息聚合器后，聚合器内部把这 3 条消息聚合成 1 条消息。这个被聚合消息的消息体内包含之前单独的 3 条消息的消息体内容）。

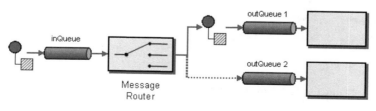

图 6-11

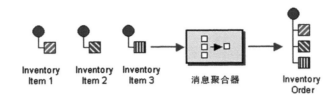

图 6-12

⑦ Splitter：消息分割器。如图 6-13 所示，把一条消息根据一些条件分割成多条消息（1 条消息经过消息分割器后，分割器内部把这 1 条消息分割成 3 条消息，这 3 条消息的消息体是之前单独的 1 条消息的消息体内容分割的内容）。

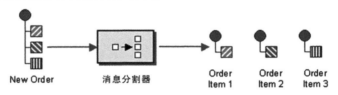

图 6-13

⑧ ChannelAdapter：通道适配器。如图 6-14 所示，这是应用跟 MessageChannel 之间的桥梁（发送方的应用可以根据消息端点将消息发送到 MessageChannel，接收方的应用可以根据消息端点读取 MessageChannel 上的消息。ChannelAdapter 分为 OutboundChannelAdapter 和 InboundChannelAdapter。OutboundChannelAdapter 表示 MessageChannel 上的消息发送到应用上，InboundChannelAdapter 表示读取应用上的消息并发送到 MessageChannel 上）。

⑨ PollingConsumer：消息轮询消费者。如图 6-15 所示，消息发送方发送消息后，消息消费方以轮询的方式拉取消息消费（Sender 表示消息发送者，它发送的消息会在 Receiver（消息接收者）内部被一直轮询拉取）。

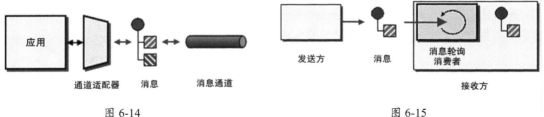

图 6-14　　　　　　　　　　　　　　图 6-15

⑩ MessagingGateway：消息网关。如图 6-16 所示，以 HTTP 网关的形式将消息的操作暴露出去，用户通过 Rest 请求就可与消息系统进行交互（左右两个区域分别是两个应用，它们

的消息网关暴露了 HTTP 端点，应用内部通过 HTTP 端点发送消息，消息被发送到消息中间件内，应用内部也可以 HTTP 端点来接收消息，这个消息是从消息中间件内订阅而来的）。

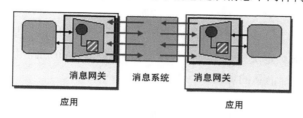

图 6-16

6.3.2 Spring Integration 核心组件的使用

Spring Integration 内部简化了这些组件的使用方式。比如，@Filter 注解可以用来过滤消息（所有的消息中，只要消息体内包含关键字"keywords"，这些消息就会被过滤掉）：

```
@Filter(inputChannel = "input", discardChannel = "discard", outputChannel = "output")
public boolean receiveByFilter(String receiveMsg) {
    if(receiveMsg.contains("keywords")) {
        return true;
    }
    return false;
}
```

过滤掉的消息会被发送到名称为 discard 的 MessageChannel 中，没被过滤的消息则被发送到名称为 output 的 MessageChannel 中。

@Transformer 注解用于消息的转换，会删除 HEADER 中 key 为 secret 的内容：

```
@Transformer(inputChannel = "input", outputChannel = "output")
public Message receiveByTransformer(Message message) {
    message.getHeaders().remove("secret");
    return message;
}
```

@Splitter 注解用于消息的分割，会把一个消息内的 Payload 内容根据 "-" 分割成多个消息：

```
@Splitter(inputChannel = "input", outputChannel = "output")
public String[] receiveBySplitter(String receiveMsg) {
    return receiveMsg.split("-");
}
```

@MessagingGateway 定义一个消息网关接口，可以在 Controller 中注入并进行调用，底层通过动态代理的方式进行消息的发送：

```
@MessagingGateway(name = "testGateway", defaultRequestChannel = "input")
public interface TestGateway {
    @Gateway(requestChannel = "input", replyTimeout = 2, requestTimeout = 200)
    String order(OrderMsg orderMsg);
}
```

@Poller 注解用于消息轮询消费，跟 ChannelAdapter 合作进行消息的消费（每次最多拉取 1 条消息，间隔 10s 操作一次）：

```
@Bean
@InboundChannelAdapter(value = "input", poller = @Poller(fixedDelay = "10000",
maxMessagesPerPoll = "1"))
public MessageSource<OrderMsg> orderMessageSource() {
    return () -> {
        return MessageBuilder.withPayload(new OrderMsg(randomGoods,
random.nextInt(5))).build();
    };
}
```

Spring Integration 内部提供了以下一些新的 MessageChannel 实现：

- DirectChannel：单播模式的消息通道，默认使用轮询负载均衡策略。
- ExecutorChannel：基于线程池的单播模式的消息通道，默认使用轮询负载均衡策略。
- PublishSubscribeChannel：基于线程池的广播模式的消息通道。
- QueueChannel：PollableChannel 接口的实现类，基于队列和信号量完成。
- PriorityChannel：PollableChannel 接口的实现类，基于优先队列。

若要使用 Spring Integration，可以加上 `spring-boot-starter-integration` 依赖，该依赖是 Spring Boot 和 Spring Integration 整合的 Starter。

spring-messaging 模块和 Spring Integration 项目对消息的统一做了抽象。spring-integration-amqp（RabbitMQ）和 spring-integration-kafka（Kafka）这些项目是针对不同消息中间件的适配。比如，spring-integration-kafka 内部的 KafkaInboundGateway 就是对 MessagingGateway 的适配，KafkaMessageSource 是对 PollingConsumer 的适配，KafkaMessageDrivenChannelAdapter 是对 ChannelAdapter 的适配。同样，这些功能在 spring-integration-amqp 模块中对应的类是 AmqpInboundGateway、AmqpMessageSource 和 AmqpInboundChannelAdapter。

通过这两个 ChannelAdapter，可以直接从各消息中间件上读取消息，并发送到 MessageChannel 上。

Spring Cloud Alibaba 的 Spring Cloud Stream RocketMQ Binder 模块也提供了 RocketMQ-InboundChannelAdapter，可以直接从 RocketMQ 上读取消息并发送到 MessageChannel 上。

6.4 Spring Cloud Stream

官方对 Spring Cloud Stream（简称 SCS）的定义是：一个为与消息中间件连接的微服务构建事件驱动的框架。

SCS 在 Spring Integration 项目的基础上再进行了一些封装，并提出一些新的概念，让开发者能够更简单地使用这套消息编程模型。如图 6-17 所示，这是 Spring Messaging、Spring Integration 和 Spring Cloud Stream 三者之间的关系。

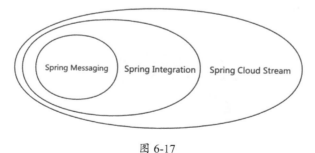

图 6-17

6.4.1 使用 Spring Cloud Stream 发送和接收消息

下面是一段使用 SCS 完成事件驱动的编码（@StreamListener 类似于@EventListener 进行消息接收）：

```
@StreamListener("input")
@SendTo("output")
public String receive(String msg) {
    return msg.toUpperCase();
}
```

receive 方法通过 @StreamListener 方法读取消息中间件对应的 Topic 的消息，再通过 @SendTo 把方法的返回值写入消息中间件对应的 Topic 的队列中。读取的 Topic 和写入的 topic 是哪个呢？这两个 Topic 是通过 Binding 配置的：

```
spring.cloud.stream.bindings.input.destination=input-topic
spring.cloud.stream.bindings.input.group=test-group
spring.cloud.stream.bindings.output.destination=output-topic
```

这段配置表示 input 这个 Binding 对应的 Topic 和 group 分别是 input-topic 和 test-group，output 这个 Binding 对应的 Topic 是 output-topic。

这个 Binding 是谁创建的呢？是消息中间件提供的 Binder 创建的，比如 Kafka Binder、RabbitMQ Binder、RocketMQ Binder。

如图 6-18 所示，这是 Spring Cloud Stream 的编程模型。通过 RabbitMQ Binder 构建 Input Binding，用于读取 RabbitMQ 上的消息，将 Payload 内容转成大写，再通过 Kafka Binder 构建的 Output Binding 写入 Kafka 中。图 6-18 中，中间 4 行非常简单的代码就可以完成从 RabbitMQ 读取消息，再写入 Kafka 的动作。

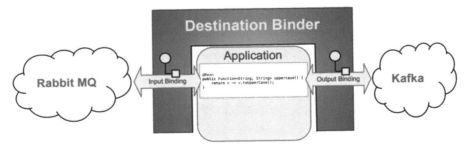

图 6-18

下列代码是使用 SCS 以最简单的方式完成消息的发送和接收的：

```
@SpringBootApplication
@EnableBinding({Source.class, Sink.class})   // ①
public class SCSApplication {

    public static void main(String[] args) {
        new SpringApplicationBuilder().sources(SCSApplication.class)
            .web(WebApplicationType.NONE).run(args);
    }

    @Autowired
    Source source;   // ②

    @Bean
    public CommandLineRunner runner() {
        return (args) -> {
            source.output().send(MessageBuilder.withPayload("custom payload").setHeader("k1", "v1").build());   // ③
        };
    }

    @StreamListener(Sink.INPUT)   // ④
    @SendTo(Source.OUTPUT)   // ⑤
    public String receive(String msg) {
        return msg.toUpperCase();
    }
}
```

上述代码中：

① 使用 @EnableBinding 注解，注解中有两个参数 Source 和 Sink，它们都是接口。Source 接口内部有个 MessageChannel 类型返回值的 output 方法，被 @Output 注解修饰，表示这是一个 Output Binding；Sink 接口内部有个 SubscribableChannel 类型返回值的 intput 方法，被 @Input 注解修饰，表示这是一个 Input Binding。@EnableBinding 注解会针对这两个接口生成动态代理。

② 注入 @EnableBinding 注解对 Source 接口生成的动态代理。

③ 使用 @EnableBinding 注解对 Source 接口生成的动态代理内部的 MessageChannel 发送一条消息。最终消息会被发送到消息中间件对应的 Topic 里。

④ @StreamListener 注解订阅@EnableBinding 注解对 Sink 接口生成的动态代理内部的 SubscribableChannel 中的消息，这里会订阅到消息中间件对应的 Topic 和 group。

⑤ 消息处理结果发送到 @EnableBinding 注解对于 Source 接口生成的动态代理内部的 MessageChannel。最终消息会被发送到消息中间件对应的 Topic 里。

@EnableBinding 注解里的 Source 和 Sink 是两个接口，定义如下：

```
public interface Source {

    String OUTPUT = "output";

    @Output(Source.OUTPUT)
    MessageChannel output();

}

public interface Sink {

    String INPUT = "input";

    @Input(Sink.INPUT)
    SubscribableChannel input();

}
```

6.4.2 理解 Binder 和 Binding

Binder 用于绑定一个对象（app interface）到一个逻辑名字（logical name）上。这个名字用来识别消息的消费者或生产者，被绑定的对象可以是一个队列、一个 ChannelAdapter、一个 MessageChannel 或者一个 Spring Bean。

Binder 接口对外暴露了两个方法，分别是 bindConsumer 和 bindProducer。其中，Consumer 除名字和绑定对象外，还需要 group 及对应的配置信息；Producer 同样也需要对应的配置信息。

Binder 接口的定义如下：

```
public interface Binder<T, C extends ConsumerProperties, P extends ProducerProperties> {

    Binding<T> bindConsumer(String name, String group, T inboundBindTarget,
            C consumerProperties);

    Binding<T> bindProducer(String name, T outboundBindTarget, P
producerProperties);

}
```

Binding 是应用与消息中间件之间的桥梁，由 Binder 构造。Binding 分 Producer Binding 和 Consumer Binding。Producer Binding 理解为应用发出的消息（output）和消息中间件之间的桥梁，Consumer Binding 理解为消息中间件和应用对消息的处理（input）之间的桥梁。

如图 6-19 和图 6-20 所示，分别是 Producer Binding（也称 Output Binding）和 Consumer Binding（也称 Input Binding）的解释。

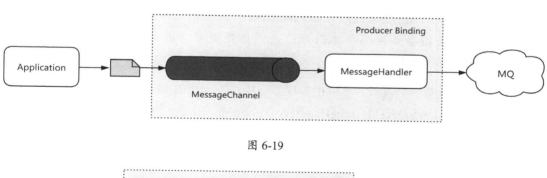

图 6-19

图 6-20

6.4.3 深入理解 Spring Cloud Stream

在 6.4.1 节中，我们以最简单的编码完成了消息的发送和接收，接下来对整个过程进行分析，理解 SCS 底层到底做了什么事情。

1. 消息的发送

消息发送的代码片段如下：

```
source.output().send(MessageBuilder.withPayload("custom payload").setHeader("k1",
"v1").build());
```

从上述代码得到以下两个疑问：

- 发送的明明是 Spring Message，存储到 Kafka、RabbitMQ 或 RocketMQ 中却变成了这些消息中间件的消息模型。这是怎么发生的？
- 往 MessageChannel 中发消息时，消息会存储到 MessageChannel 中，存储到 MessageChannel 也表示存储到内存中。为什么最终变成了存储到消息中间件中？

这两个行为在 Producer Binding 中完成。Producer Binding 首先会通过 SendingHandler 这个 MessageHandler 去订阅消息发送的 MessageChannel，SendingHandler 内部会根据配置信息决定是否使用固定的消息头对消息体做修改，然后把处理完的消息委托给 Binder 构造的 MessageHandler 继续处理。

由于 Binder 是一个接口，不同的实现类会构造不一样的 MessageHandler，比如 Kafka Binder 构造 KafkaProducerMessageHandler；RabbitMQ Binder 构造 AmqpOutboundEndpoint；RocketMQ Binder 构造 RocketMQMessageHandler；RocketMQ Binder 构造 RocketMQMessage-Handler。

这些 Binder 构造的 MessageHandler 内部做了以下两件事情：

- 把 Spring Message 转换成对应 MQ 的 Message Model。
- 转换后的 Message Model 发送到 MQ broker 中。

Spring Cloud Stream 消息发送过程的总结如图 6-21 所示。

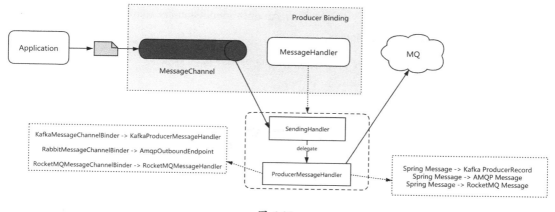

图 6-21

2. 消息的接收

消息接收的代码片段如下：

```
@StreamListener(Sink.INPUT)
public String receive(String msg) {
    return msg.toUpperCase();
}
```

从上述代码得到以下两个疑问：

- 订阅的明明是消息中间件 Message Model，代码中通过 @StreamListener 收到的只是一个 String。这期间发生了什么？
- 消息中间件的订阅方式不一样，为什么这里可以统一用 @StreamListener 进行订阅？

这两个行为在 Consumer Binding 中完成。Consumer Binding 首先会根据 Binder 得到 ChannelAdapter，利用 ChannelAdapter 读取消息中间件上的消息，再把这些消息转换成 Spring Message，最后发送到 MessageChannel 中。

之后通过 @StreamListener 注解读取这个 MessageChannel 上的消息，@StreamListener 注解底层实际上是构造了一个 MessageHandler，用 MessageHandler 去订阅 MessageChannel 上的注解。同样，其他像 @ServiceActivator、@Transformer、@Filter 等注解底层也会构造 MessageHandler，去订阅 MessageChannel 上的消息。

不同的 Binder 构造不一样的 ChannelAdapter，比如 Kafka Binder 构造 KafkaMessage-

DrivenChannelAdapter；RabbitMQ Binder 构造 AmqpInboundChannelAdapter；RocketMQ Binder 构造 RocketMQInboundChannelAdapter。

如图 6-22 所示，这是 Spring Cloud Stream 消息接收过程的总结。

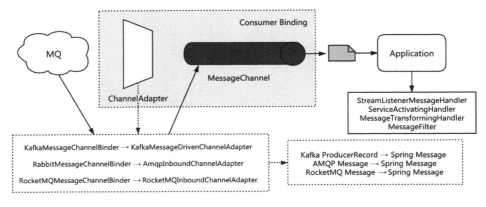

图 6-22

6.4.4 Spring Cloud Stream 的高级特性

Spring Cloud Stream 在统一消息编程模型的基础上，还提供了以下 11 个高级特性。

1. 重试

Consumer 端可以配置重试次数，当消息消费失败的时候会进行重试。

底层使用 Spring Retry 去重试，重试次数可自定义配置：

```
spring.cloud.stream.bindings.<channelName>.consumer.maxAttempts=3
```

默认重试次数为 3，配置大于 1 时才会生效。

2. 消息发送失败的处理

Producer 发送消息出错的情况下，可以配置错误处理，将错误信息发送给对应 ID 的 MessageChannel。

消息发送失败的场景下，会将消息发送到一个 MessageChannel。这个 MessageChannel 会取 ApplicationContext 中 name 为 "topic.errors"（Topic 就是配置的 destination）的 Bean，如果找不到，就会自动构造一个 PublishSubscribeChannel（广播）。然后使用 BridgeHandler 订阅这个

MessageChannel，同时再设置 ApplicationContext 中 name 为 errorChannel 的 PublishSubscribeChannel 消息通道为 BridgeHandler 的 outputChannel。

由于 name 为 errorChannel 的 MessageChannel 是一个全局的错误消息的消息通道，笔者建议使用 "topic.errors" 这个消息通道，并且设置发送到广播模式的 PublishSubscribeChannel 消息通道中（若不手动设置，SCS 默认情况下会自动构造一个 name 为 "topic.errors" 的 DirectChannel）。如果是单播模式的 DirectChannel，会导致 BridgeHandler 和自定义的 MessageHandler 同时订阅这个 MessageChannel 产生不符合预期的结果。

```
spring.cloud.stream.bindings.output.destination=test-output

// 定义 "topic.errors" 消息通道为 PublishSubscribeChannel
@Bean("test-output.errors")
MessageChannel testOutputErrorChannel() {
    return new PublishSubscribeChannel();
}


@Service
class ErrorProduceService {

    // 订阅 "topic.errors" 消息通道，处理错误消息
    @ServiceActivator(inputChannel = "test-output.errors")
    public void receiveProduceError(Message receiveMsg) {
        System.out.println("receive error msg: " + receiveMsg);
    }

}
```

消息发送失败的处理逻辑默认是关闭的，需要加上如下配置打开：

```
spring.cloud.stream.bindings.<channelName>.producer.errorChannelEnabled=true
```

3. 消费错误处理

Consumer 消费消息出错的情况下，可以配置错误处理，将错误信息发送给对应 ID 的 MessageChannel。

消息错误处理与生产错误处理大致相同。错误的 MessageChannel 对应的 name 为 "topic.group.errors"，还会加上多个 MessageHandler 订阅的一些判断，使用 ErrorMessage-Strategy 创建错误消息等内容。

笔者建议直接使用 "topic.group.errors" 这个消息通道，并且设置发送到单播模式的 DirectChannel 消息通道中（使用@ServiceActivator 注解接收会直接构成 DirectChannel），这样会确保只会被唯一的一个订阅了 "topic.group.errors" 的 MessageHandler 处理（Spring Cloud Stream 对于消息消费失败处理做过一些优化），否则可能会被多个 MessageHandler 处理，导致出现一些意想不到的结果。

```
spring.cloud.stream.bindings.input.destination=test-input
spring.cloud.stream.bindings.input.group=test-input-group

@Service
class ErrorConsumeService {

    @StreamListener(Sink.INPUT)
    public void receive(String receiveMsg) {
        throw new RuntimeException("Oops");
    }

    @ServiceActivator(inputChannel = "test-input.test-input-group.errors")
    public void receiveConsumeError(Message receiveMsg) {
        System.out.println("receive error msg: " + receiveMsg);
    }

}
```

4. 自定义 MessageChannel 类型

默认情况下，Output Binding 对应的 MessageChannel 和 Input Binding 对应的 SubscribableChannel 会被构造成 DirectChannel。SCS 提供了 BindingTargetFactory 接口进行扩展，比如，可以扩展构造 PublishSubscribeChannel 这种广播类型的 MessageChannel。

默认情况下，BindingTargetFactory 接口只有两个实现类，分别是 SubscribableChannel-BindingTargetFactory 和 MessageSourceBindingTargetFactory。

- SubscribableChannelBindingTargetFactory：针对 Input Binding 和 Output Binding 都会构造成 DirectWithAttributesChannel 类型的 MessageChannel（一种带 HashMap 属性的 DirectChannel）。
- MessageSourceBindingTargetFactory：不支持 Output Binding，Input Binding 会构造成 DefaultPollableMessageSource。DefaultPollableMessageSource 内部维护着 MessageSource 属性，该属性用于拉取消息。

5. Endpoint 端点

SCS 提供了 BindingsEndpoint，可以获取 Binding 信息或对 Binding 的生命周期进行修改，比如 start、stop、pause 或 resume。

BindingsEndpoint 的 ID 是 bindings，对外暴露了以下 3 个操作：

- 修改 Binding 状态，可以改成 STARTED、STOPPED、PAUSED 和 RESUMED，对应 Binding 接口的 4 个操作。
- 查询单个 Binding 的状态信息。
- 查询所有 Binding 的状态信息。

BindingsEndpoint 的源码如下：

```
@Endpoint(id = "bindings")
public class BindingsEndpoint {

    ...

    @WriteOperation
    public void changeState(@Selector String name, State state) {
        Binding<?> binding = BindingsEndpoint.this.locateBinding(name);
        if (binding != null) {
            switch (state) {
            case STARTED:
                binding.start();
                break;
            case STOPPED:
                binding.stop();
                break;
```

```java
            case PAUSED:
                binding.pause();
                break;
            case RESUMED:
                binding.resume();
                break;
            default:
                break;
            }
        }
    }

    @ReadOperation
    public List<?> queryStates() {
        List<Binding<?>> bindings = new ArrayList<>(gatherInputBindings());
        bindings.addAll(gatherOutputBindings());
        return this.objectMapper.convertValue(bindings, List.class);
    }

    @ReadOperation
    public Binding<?> queryState(@Selector String name) {
        Assert.notNull(name, "'name' must not be null");
        return this.locateBinding(name);
    }

    ...

}
```

6. Metrics 指标

该功能自动与 micrometer 集成进行 Metrics 统计，可以通过前缀"spring.cloud.stream.metrics"进行相关配置，配置项 `spring.cloud.stream.bindings.applicationMetrics.destination` 会构造 MetersPublisherBinding，将应用相关的 metrics 发送到 MQ 中。

7. Serverless

默认与 Spring Cloud Function 集成。

可以使用 Function 处理消息，代码如下：

```java
@SpringBootApplication
@EnableBinding(Processor.class)
public class FunctionApplication {

    public static void main(String[] args) {
        new SpringApplicationBuilder(FunctionApplication.class)
                .web(WebApplicationType.NONE).run(args);
    }

    @Bean
    public Function<String, String> uppercase() {
        return x -> x.toUpperCase();
    }

    @Bean
    public Function<String, String> addprefix() {
        return x -> "prefix-" + x;
    }
}
```

配置文件需要加上 function 配置：

```
spring.cloud.stream.function.definition=uppercase|addprefix
```

8. Partition 统一

SCS 统一 Partition 相关的设置，可以屏蔽不同 MQ Partition 的设置。

Producer Binding 提供的 ProducerProperties 提供了一些 Partition 相关的配置：

- partitionKeyExpression：partition key 提取表达式。
- partitionKeyExtractorName：是一个实现 PartitionKeyExtractorStrategy 接口的 Bean name。PartitionKeyExtractorStrategy 是一个根据 Message 获取 partiton key 的接口。如果两者都配置，优先取 partitionKeyExtractorName 配置对应的 Bean。
- partitionSelectorName：是一个实现 PartitionSelectorStrategy 接口的 Bean name。

PartitionSelectorStrategy 接口是一个根据 partition key 决定选择哪个 partition 的接口。
- partitionSelectorExpression：partition 选择表达式，会根据表达式和 partition key 得到最终的 partition。如果两者都配置，优先根据 partitionSelectorExpression 表达式解析 partition。
- partitionCount：partition 个数。该属性不一定会生效，Kafka Binder 和 RocketMQ Binder 会使用 topic 上的 partition 个数覆盖该属性。

Partition 分区功能生效的前提是 partitionKeyExpression 或 partitionKeyExtractorName，这两个配置至少存在一个且必须是一个 output channel。

Partition 功能的生效通过 MessageConverterConfigurer$PartitioningInterceptor 这个 ChannelInterceptor 拦截完成，会在 preSend 过程中对消息 HEADER 进行修改：

```java
public final class PartitioningInterceptor implements ChannelInterceptor {
    ...

    @Override
    public Message<?> preSend(Message<?> message, MessageChannel channel) {
        if (!message.getHeaders().containsKey(BinderHeaders.PARTITION_OVERRIDE)) { // ①
            int partition = this.partitionHandler.determinePartition(message);
            return MessageConverterConfigurer.this.messageBuilderFactory
                    .fromMessage(message)
                    .setHeader(BinderHeaders.PARTITION_HEADER, partition).build(); // ②
        }
        else {
            return MessageConverterConfigurer.this.messageBuilderFactory
                    .fromMessage(message)
                    .setHeader(BinderHeaders.PARTITION_HEADER,
                            message.getHeaders()
                                    .get(BinderHeaders.PARTITION_OVERRIDE))
                    .removeHeader(BinderHeaders.PARTITION_OVERRIDE).build();   // ③
        }
    }
}
```

```
}
public class PartitionHandler {

    ...

    public int determinePartition(Message<?> message) {
        Object key = extractKey(message);    // ④

        int partition;
        if (this.producerProperties.getPartitionSelectorExpression() != null) {
            partition = this.producerProperties.getPartitionSelectorExpression()
                    .getValue(this.evaluationContext, key, Integer.class);    // ⑤
        }
        else {
            partition = this.partitionSelectorStrategy.selectPartition(key,
                    this.partitionCount);    // ⑥
        }
        return Math.abs(partition % this.partitionCount);    // ⑦
    }

    private Object extractKey(Message<?> message) {
        Object key = invokeKeyExtractor(message);
        if (key == null && this.producerProperties.getPartitionKeyExpression() != null) {
            key = this.producerProperties.getPartitionKeyExpression()
                    .getValue(this.evaluationContext, message);
        }
        Assert.notNull(key, "Partition key cannot be null");

        return key;
    }

    ...

}
```

上述代码中：

① 若 Message Header 中不存在 key 为 scst_partitionOverride，则通过 PartitionHandler 获取 partition 分区。获取的步骤参考代码中④～⑦处的代码。

② PartitioningInterceptor 在原先 Message 的基础上，Message Header 添加 key 为 scst_partition，value 为 PartitionHandler 得到的 partition 分区数据。

③ Message Header 中若存在 key 为 scstpartitionOverride，PartitioningInterceptor 在原先 Message 的基础上，Message Header 添加 key 为 scstpartition，value 为原先 Message Header 上 key 为 scstpartitionOverride 的数据，并且删除 header 中 key 为 scstpartitionOverride 的数据。

④ 找出 ApplicationContext 中 PartitionKeyExtractorStrategy 类型的 Bean（如果配置了 partitionKeyExtractorName，根据 Bean name 寻找，否则找出唯一的 PartitionKeyExtractorStrategy 类型的 Bean），找到后调用 extractKey 方法获取 partition key。如果 partition key 为空，基于消息内容解析 partitionKeyExpression 表达式属性得到 partition key。

⑤ 如果配置的 partitionSelectorExpression 表达式不为空，根据 partition key 解析表达式得到分区位置。

⑥ 如果配置的 partitionSelectorExpression 表达式为空，找出 ApplicationContext 中 PartitionSelectorStrategy 类型的 Bean（如果配置了 partitionSelectorName，根据 Bean name 寻找，否则找出唯一的 PartitionSelectorStrategy 类型的 Bean）。调用 PartitionSelectorStrategy 的 selectPartition，得到分区位置。

⑦ 最终的 partiton 需要对 partitionCount 取余，再取这个结果的绝对值。

关于 partition key 的获取，下面这两段代码配置的效果是一致的：

```
spring.cloud.stream.bindings.output.producer.partition-key-expression=headers['partitionKey']
spring.cloud.stream.bindings.output.producer.partition-key-extractor-name=myKeyExtractor

public class MyPartitionKeyExtractor implements PartitionKeyExtractorStrategy {
    @Override
    public Object extractKey(Message<?> message) {
        return message.getHeaders().get("partitionKey");
    }
}
```

```java
@Bean
MyPartitionKeyExtractor myKeyExtractor() {
    return new MyPartitionKeyExtractor();
}
```

关于 partition selector 的获取，下列这两段代码配置的效果是一致的：

```
spring.cloud.stream.bindings.output.producer.partition-selector-expression=id
spring.cloud.stream.bindings.output.producer.partition-selector-name=mySelector
```

```java
public class MyPartitionSelector implements PartitionSelectorStrategy {
    @Override
    public int selectPartition(Object key, int partitionCount) {
        return ((User) key).getId();
    }
}
```

```java
@Bean
MyPartitionSelector mySelector() {
    return new MyPartitionSelector();
}
```

9. Polling Consumer

实现 MessageSource 进行 polling 操作的 Consumer。

普通的 Pub/Sub 模式需要定义 SubscribableChannel 类型的返回值，Polling Consumer 需要定义 PollableMessageSource 类型的返回值：

```java
public interface MySink {

    String INPUT = "input";

    @Input(INPUT)
    PollableMessageSource input();
}
```

10. 支持多个 Binder 同时使用

支持多个 Binder 同时使用，在配置 Binding 的时候需要指定对应的 Binder。

配置全局默认的 Binder：

```
spring.cloud.stream.default-binder=rocketmq
```

配置各个 Binder 内部的配置信息：

```
spring.cloud.stream.binders.rocketmq.environment.xx=xx
spring.cloud.stream.binders.rocketmq.type=rocketmq
```

配置 Binding 对应的 Binder：

```
spring.cloud.stream.bindings.<channelName>.binder=kafka
spring.cloud.stream.bindings.<channelName>.binder=rocketmq
spring.cloud.stream.bindings.<channelName>.binder=rabbit
```

11. 建立事件机制

比如，新建 BindingCreatedEvent 事件，用户的应用就可以监听该事件在创建 Input Binding 或 Output Binding 时做的业务相关的处理。

第 7 章
消息总线

本章将介绍 Spring Cloud 体系内的 Spring Cloud Bus（消息总线）。Spring Cloud Bus 定义了 RemoteApplicationEvent 远程事件，远程事件的定位是跨 JVM 发送/接收事件。

远程事件的发送/接收底层依赖 Spring Cloud Stream 消息的发送。其中，远程事件的发送相当于消息发送到指定的 Topic，远程事件的接收相当于订阅指定 topic 的消息。

7.1　消息总线概述

在第 4.4.4 节曾提到一个问题：在分布式系统中，如果一个服务对应 100 个应用实例，那么在这 100 个节点上手动刷新配置显然是不合理的。本节介绍的消息总线可以解决这个问题。

Enterprise Integration Patterns（企业集成模式）对消息总线的定义为：消息总线是一种消息传递基础结构，它允许不同的系统通过一组共享的接口（消息总线）进行通信。如图 7-1 所示，3 个应用分别表示 3 个系统，它们连接到消息总线上，当某个系统发送消息到消息总线上时，其他系统可以接收到这个消息并做相应的处理（图 7-1 中右上角的系统发送消息到消息总线后，另外两个系统包括右上角的系统自身都会从消息总线收到右上角系统发送的消息）。

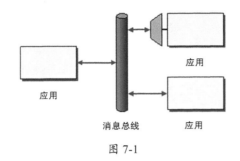

图 7-1

从消息总线的定义可知，这个功能完全可以用消息队列来完成：应用 A 发送消息到指定的 Topic，应用 B、C、D、E 订阅这个 Topic，收到消息后完成对应的业务处理。

这个解释功能在实现上没有问题，不过有以下几个问题需要解决：

- 消息发送/订阅不同的消息队列 MQ 实现方式不一样，如何统一？
- 开发者并不想了解消息队列，只想添加监听器监听变化并做相应的业务处理。
- 应用 A 发送消息，并不想让所有监听 Topic 的应用都进行业务处理，只想让指定的应用做处理。

Spring Cloud Bus 是 Spring Cloud 对于消息总线定义对外提供的组件，它的出现解决了上述 3 个问题，下面我们来分析这 3 个问题是如何被解决的，并介绍 Spring Cloud Bus 的原理。

7.2 深入理解 Spring Cloud Bus

在深入介绍 Spring Cloud Bus 之前，回答 7.1 节提出的 3 个问题：

- 通过 Spring Cloud Stream 统一发送消息。
- 定义 RemoteApplicationEvent 远程事件，用于屏蔽消息发送/接收细节，开发者只需接收远程事件，与 Spring 原生事件机制完美整合。
- RemoteApplicationEvent 远程事件定义源服务（originService）和目标服务（destinationService），用于过滤事件接收方。

7.2.1 Spring Cloud Bus 的使用

我们先来尝试使用 Spring Cloud Bus 的 Demo，然后深入分析其原理。

Spring Cloud Bus 的 Demo 定义了 3 个应用：node1、node2 和 node3。这 3 个应用各自发送事件，并查看其他应用的接收情况。

node1、node2 和 node3 应用的代码相同，具体如下：

```
@SpringBootApplication
@RemoteApplicationEventScan(basePackages = "deep.in.spring.cloud")  //①
public class SCBNode1 {

    public static void main(String[] args) {
        SpringApplication.run(SCBNode1.class, args);
    }

    @Autowired
    ApplicationContext applicationContext;

    @Value("${spring.cloud.bus.id}")  // ②
    String originService;

    @RestController
    class BusController {

        @GetMapping("/event")
        String event(
            @RequestBody User user,
            @RequestParam(required = false) String destination) {  // ③
            applicationContext.publishEvent(new CustomEvent(this, user, originService, destination));  // ④
            return "ok";
        }

    }

    @Service
    class EventReceiver {
```

```
        @EventListener
        public void receive(CustomEvent event) {   // ⑤
            System.out.println("receive: " + event.getUser());
        }

    }

}

public class CustomEvent extends RemoteApplicationEvent {   // ⑥

    private User user;

    public CustomEvent() {
    }

    public CustomEvent(Object source, User user, String originService, String destinationService) {   // ⑦
        super(source, originService, destinationService);
        this.user = user;
    }

    public User getUser() {
        return user;
    }

    public void setUser(User user) {
        this.user = user;
    }
}
```

上述代码中：

① 使用 @RemoteApplicationEventScan 注解扫描远程事件。如果自定义的远程事件不使用该注解进行扫描，这些事件会被识别成 UnknownRemoteApplicationEvent。

② 读取配置文件 spring.cloud.bus.id 配置项的配置，该配置用于表示这次远程事件的来源。Spring Cloud Bus 在发送远程事件的时候会判断事件来源，从而确定是否进行消息发送。

③ 定义一个 Controller，对外暴露的 Rest 接口需要一个 destination 参数和 Request Body 内容。destination 表示事件的目的地，比如，node1 只给 node3 发送远程事件是通过 destination 的设置完成的。

④ 使用 Spring 的事件机制发送 CustomEvent 事件，这是一个远程事件。

⑤ 通过 @EventListener 接收 CustomEvent 事件。

⑥ CustomEvent 是一个远程事件，继承 Spring Cloud Bus 对外提供的 RemoteApplicationEvent 远程事件。

⑦ CustomEvent 的构造方法可以传入事件的来源，并自定义对象、来源和目的地。

node1、node2 和 node3 这 3 个应用唯一的区别是配置文件，分别如下：

```
spring.application.name=scb-node1
spring.cloud.bus.id=scb-node1
server.port=8080
```

node2 和 node3 对应的 spring.application.name 和 spring.cloud.bus.id 配置为 node2 和 node3，且端口为 8081 和 8082。

访问 node1 应用对外暴露的 Rest 接口：

```
curl --header "Content-Type:application/json" -XGET http://localhost:8080/event\?destination\=scb-node3 -d '{"id":1, "name": "deep in spring cloud"}'
```

destination 传入了 scb-node3，表示 node1 应用发送的远程事件只被 node3 接收，调用成功之后，node3 控制台和 node1 控制台分别打印如下内容：

```
node1 -> receive: User{id=1, name='deep in spring cloud'}
node3 -> receive: User{id=1, name='deep in spring cloud'}
```

访问 node2 应用对外暴露的 Rest 接口：

```
curl --header "Content-Type:application/json" -XGET http://localhost:8081/event -d '{"id":2, "name": "hello world"}'
```

destination 参数没有传递，表示 node2 应用发送的远程事件会被消息总线上的所有应用接收（node1 和 node3），调用成功之后，node1、node2 和 node3 控制台分别打印如下内容：

```
node1 -> receive: User{id=2, name='hello world'}
node2 -> receive: User{id=2, name='hello world'}
node3 -> receive: User{id=2, name='hello world'}
```

总结这个 Demo 的关键信息：

- 无任何消息相关的处理，所有的交互通过事件完成。
- RemoteApplicationEvent 远程事件与 Spring 原生事件机制完美整合，发送 RemoteApplicationEvent 事件可以让消息总线上的其他节点接收。
- RemoteApplicationEvent 远程事件的发送可以指定 destination（目的地），让消息总线上的部分节点接收。

7.2.2 Spring Cloud Bus 的原理

Spring Cloud Bus 强依赖 Spring Cloud Stream。目前 Spring Cloud Bus 的实现 Spring Cloud AMQP Bus、Spring Cloud Kafka Bus 或 Spring Cloud RocketMQ Bus 底层强依赖 Spring Cloud Stream Rabbit Binder、Spring Cloud Stream Kafka Binder 以及 Spring Cloud Stream RocketMQ Binder。如果想更换消息总线的实现，只需修改对应的依赖即可，代码层面没有任何变化。

Spring Cloud Bus 在初始化的时候会通过 Spring Cloud Stream 提供的 @EnableBinding 注解构造 Spring Cloud Bus 对应的 Binding（Spring Cloud Bus 依赖 Spring Cloud Stream，意味着如果 MQ 没有对应的 Spring Cloud Stream，则无法使用 Spring Cloud Bus 组件）：

```
...
@EnableBinding(SpringCloudBusClient.class)
...
public class BusAutoConfiguration
```

SpringCloudBusClient 接口内部会构造 Ouput Binding 和 Input Binding，代码如下：

```
public interface SpringCloudBusClient {

    String INPUT = "springCloudBusInput";
```

```
    String OUTPUT = "springCloudBusOutput";

    @Output(SpringCloudBusClient.OUTPUT)
    MessageChannel springCloudBusOutput();

    @Input(SpringCloudBusClient.INPUT)
    SubscribableChannel springCloudBusInput();
}
```

Output Binding 的作用是在接收 RemoteApplicationEvent 远程事件的时候，将事件封装成消息通过 Spring Cloud Stream 发送到 MQ。

Input Binding 的作用是读取 MQ 上的消息，把消息封装成事件，再通过 Spring 的事件机制将事件发送出去，消息总线上的应用通过 EventListener 接收这个事件。

Spring Cloud Bus 远程事件的发送/接收流程如图 7-2 所示。

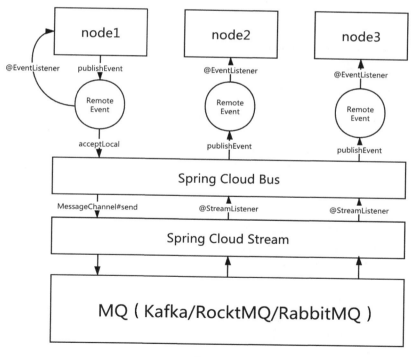

图 7-2

下列步骤是对图 7-2 的详细解释。

（1）node1 应用通过 Spring 事件机制提供的 ApplicationEventPublisher#publishEvent 方法发送事件。

（2）远程事件 RemoteApplicationEvent 继承 Spring 提供的事件父类 ApplicationEvent，表示可以通过 Spring 事件机制提供的 @EventListener 注解进行接收。

（3）Spring Cloud Bus 内部专门使用 @EventListener 注解接收远程事件（BusAutoConfiguration#acceptLocal）。

（4）Spring Cloud Bus 接收到远程事件之后会通过 Spring Cloud Stream 提供的 Output Binding 将远程事件封装成 Message 进行消息发送（Message Payload 为远程事件）。

（5）其他应用（node2 和 node3）也使用了 Spring Cloud Bus，并且订阅了同一个 Topic（默认情况下，使用 Spring Cloud Bus 会发送/订阅 springCloudBus 这个 Topic 消息），Spring Cloud Bus 内部会通过 Spring Cloud Stream 提供的 @StreamListener 注解接收消息（BusAutoConfiguration#acceptRemote）。

（6）其他应用（node2 和 node3）接收消息之后会判断远程事件的目的地（destination）是否与自身的 ID 一致。如果一致，通过 Spring 的事件机制发送该远程事件（ApplicationEventPublisher#publishEvent）。

（7）其他应用（node2 和 node3）通过 @EventListener 注解接收远程事件并做相应的业务处理。

从图 7-2 可以看到：Spring Cloud Stream 暴露的 Output Binding 和 Input Binding 必须读取 MQ 上的同一个 Topic。默认情况下，使用 springCloudBus 这个 Topic，可以通过 Binding 配置进行覆盖：

```
spring.cloud.stream.bindings.springCloudBusOutput.destination=myTopic
spring.cloud.stream.bindings.springCloudBusInput.destination=myTopic
spring.cloud.stream.bindings.springCloudBusInput.group=myGroup
```

7.2.3 Spring Cloud Bus 事件

Spring Cloud Bus 自身提供了以下一些远程事件。

（1）EnvironmentChangeRemoteApplicationEvent：配置信息修改远程事件。

Spring Cloud Bus 对外提供了 ID 为 "bus-env" 的 EnvironmentBusEndpoint，用于修改当前应用的配置信息。EnvironmentBusEndpoint 内部会发送 EnvironmentChangeRemoteApplicationEvent 远程事件。比如，调用 EnvironmentBusEndpoint，添加 key 为 book、value 是 deepinspringcloud 的配置项：

```
curl --header "Content-Type:application/json" -XPOST
'http://localhost:8080/actuator/bus-env?name=book&value=deepinspringcloud'
```

同样，Spring Cloud Bus 内部提供 EnvironmentChangeListener 监听器，用于 EnvironmentChangeRemoteApplicationEvent 远程事件的接收：

```java
public class EnvironmentChangeListener
        implements ApplicationListener<EnvironmentChangeRemoteApplicationEvent> {

    private static Log log = LogFactory.getLog(EnvironmentChangeListener.class);

    @Autowired
    private EnvironmentManager env;

    @Override
    public void onApplicationEvent(EnvironmentChangeRemoteApplicationEvent event) {
        Map<String, String> values = event.getValues();
        log.info("Received remote environment change request. Keys/values to update"
                + values);
        for (Map.Entry<String, String> entry : values.entrySet()) {
            this.env.setProperty(entry.getKey(), entry.getValue());
        }
    }
}
```

node1 应用发送 EnvironmentChangeRemoteApplicationEvent 事件，自身通过 EnvironmentChangeListener 监听器修改配置。node2 应用和 node3 应用监听远程事件后再进行发送，再次通过 EnvironmentChangeListener 监听修改配置，从而达到消息总线上的任意一个应用修改了配置，都会使总线上的所有应用配置全部生效的效果。

（2）AckRemoteApplicationEvent：远程事件发送成功确认事件。

远程事件发送出去后，需要机制确认该事件是否发送成功。比如，node1 发送了远程事件，node2 和 node3 接收到事件后会发送 AckRemoteApplicationEvent 确认自己收到远程事件。

（3）RefreshRemoteApplicationEvent：配置刷新远程事件。

RefreshListener 内部接收该事件并调用 ContextRefresher 进行全局配置的刷新。有了该接口，就能完美解决分布式系统下手动调用 N 个节点进行配置刷新的问题。只需在消息总线的任意一个入口发送 RefreshRemoteApplicationEvent 事件，消息总线上的所有实例就能全部刷新配置。

那么怎么发送 RefreshRemoteApplicationEvent 远程事件呢？可以通过 Spring Cloud Bus 对外暴露的 ID 为"bus-refresh"的 RefreshBusEndpoint 来实现。

RefreshBusEndpoint 的调用可以指定 destination（node1 只要求 node3 刷新全局配置）：

```
curl -XPOST 'http://localhost:8080/actuator/bus-refresh/scb-node3'
```

（4）UnknownRemoteApplicationEvent：未知远程事件。

7.2.1 节的应用如果删除了@RemoteApplicationEventScan 注解，发出的 CustomEvent 远程事件会被解析成 UnknownRemoteApplicationEvent。这是因为@RemoteApplicationEventScan 注解声明后，会构造一个 BusJacksonMessageConverter 这个 MessageConverter，它内部会在 jackson 内部注册一个能识别这些 package 内部的远程事件的 module，在发送消息的时候会对消息 Payload 进行解析。如果没有注册 module，解析失败后会自动转成 UnknownRemote-ApplicationEvent（未知远程事件）。

Spring Cloud Bus 还提供了消息发送追踪的功能，可以追踪远程事件的发送；当远程事件发送后会发送一个 SentApplicationEvent 本地事件。比如，CustomEvent 远程事件发送后，可以通过 @EventListener 监听 SentApplicationEvent 事件（SentApplicationEvent 对外暴露了 getType 方法，用于获取远程事件的类型）。消息发送追踪功能默认关闭，可以通过 `spring.cloud.bus.trace.enabled=true` 配置打开。

如表 7-1 所示，这是 Spring Cloud Bus 所有功能对外提供的配置项。

表 7-1

配 置 项	作 用	默 认 值
spring.cloud.bus.enabled	是否启用 Spring Cloud Bus 自动化配置	true

续表

配 置 项	作 用	默 认 值
spring.cloud.bus.env.enabled	是否构造 EnvironmentChangeListener。该监听器用于监听 EnvironmentChange-RemoteApplicationEvent 远程事件，用于同一消息总线上所有实例配置的新增/更新	true
spring.cloud.bus.refresh.enabled	是否构造 RefreshListener。该监听器用于监听 RefreshRemoteApplicationEvent 远程事件，用于同一消息总线上所有实例配置的动态刷新	true
spring.cloud.bus.ack.enabled	AckRemoteApplicationEvent 事件开关，用于确认远程事件发送是否成功	true
spring.cloud.bus.ack.destinationService	发送 AckRemoteApplicationEvent 远程事件对应的 destination（目的地）	null（默认发送给所有的 destination）
spring.cloud.bus.ack.destinationService	发送 AckRemoteApplicationEvent 远程事件对应的 destination（目的地）	null（默认发送给所有的 destination）
spring.cloud.bus.trace.enabled	是否开启消息发送追踪功能	False
spring.cloud.bus.destination	消息发送/接收对应的 Topic	springCloudBus
spring.cloud.bus.id	应用程序唯一标识符	如果不设置，默认读取表达式 ${vcap.application.name:${spring.application.name:application}}:${vcap.application.instanceindex:${spring.application.index:${local.server.port:${server.port:0}}}}:${vcap.application.instanceid:${random.value}} 的值

7.2.4 Spring Cloud Bus 源码分析

Spring Cloud Bus 几乎所有的核心代码都在 BusAutoConfiguration 自动化配置类里，下列代码是其中一段对远程事件发送/接收的分析：

```java
@EventListener(classes = RemoteApplicationEvent.class)   // ①
public void acceptLocal(RemoteApplicationEvent event) {
    if (this.serviceMatcher.isFromSelf(event)
            && !(event instanceof AckRemoteApplicationEvent)) {  // ②

this.cloudBusOutboundChannel.send(MessageBuilder.withPayload(event).build());  // ③
    }
}

@StreamListener(SpringCloudBusClient.INPUT)   // ④
public void acceptRemote(RemoteApplicationEvent event) {
    if (event instanceof AckRemoteApplicationEvent) {
        if (this.bus.getTrace().isEnabled()
&& !this.serviceMatcher.isFromSelf(event)
                && this.applicationEventPublisher != null) {  // ⑤
            this.applicationEventPublisher.publishEvent(event);
        }
        return;
    }
    if (this.serviceMatcher.isForSelf(event)
            && this.applicationEventPublisher != null) {  // ⑥
        if (!this.serviceMatcher.isFromSelf(event)) {  // ⑦
            this.applicationEventPublisher.publishEvent(event);
        }
        if (this.bus.getAck().isEnabled()) {  // ⑧
            AckRemoteApplicationEvent ack = new AckRemoteApplicationEvent(this,
                    this.serviceMatcher.getServiceId(),
                    this.bus.getAck().getDestinationService(),
                    event.getDestinationService(), event.getId(), event.getClass());
            this.cloudBusOutboundChannel
                    .send(MessageBuilder.withPayload(ack).build());
            this.applicationEventPublisher.publishEvent(ack);
        }
    }
    if (this.bus.getTrace().isEnabled() && this.applicationEventPublisher != null) {  // ⑨
        this.applicationEventPublisher.publishEvent(new SentApplicationEvent(this,
```

```
            event.getOriginService(), event.getDestinationService(),
            event.getId(), event.getClass()));
    }
}
```

上述代码中：

① 利用 Spring 事件的监听机制监听本地所有的 RemoteApplicationEvent 远程事件（比如 bus-env 会在本地发送 EnvironmentChangeRemoteApplicationEvent 事件，bus-refresh 会在本地发送 RefreshRemoteApplicationEvent 事件。这些事件在这里都会被监听到）。

② 判断本地接收到的事件不是 AckRemoteApplicationEvent 远程确认事件（否则会出现死循环，一直接收消息、发送消息……），以及该事件是应用自身发送出去的（事件发送方是应用自身），如果都满足，则执行③处的代码。

③ 构造 Message 并将该远程事件作为 Payload，然后使用 Spring Cloud Stream 构造的 Binding name 为 springCloudBusOutput 的 MessageChannel 将消息发送到 broker。

④ @StreamListener 注解消费 Spring Cloud Stream 构造的 Binding name 为 springCloudBusInput 的 MessageChannel，接收的消息为远程消息。

⑤ 如果该远程事件是 AckRemoteApplicationEvent 远程确认事件，并且应用开启了消息追踪 trace 开关，同时该远程事件不是应用自身发送的（事件发送方不是应用自身，表示事件是其他应用发送过来的），那么本地发送 AckRemoteApplicationEvent 远程确认事件，表示应用确认收到了其他应用发送过来的远程事件。流程结束。

⑥ 如果该远程事件是其他应用发送给应用自身的（事件的接收方是应用自身），那么执行⑦、⑧处的代码，否则执行⑨处的代码。

⑦ 该远程事件若不是应用自身发送（事件发送方不是应用自身）的，将该事件以本地的方式发送出去。应用自身一开始已经在本地被对应的消息接收方处理了，无须再次发送。

⑧ 如果开启了 AckRemoteApplicationEvent 远程确认事件的开关，构造 AckRemoteApplication-Event 事件，并在远程和本地都发送该事件（本地发送是因为⑤处的代码没有进行本地 AckRemoteApplicationEvent 事件的发送，也就是自身应用对自身应用确认；远程发送是为了告诉其他应用，自身应用收到了消息）。

⑨ 如果开启了消息追踪 Trace 的开关，本地构造并发送 SentApplicationEvent 事件。

Spring Cloud Bus 内部代码的逻辑执行流程如图 7-3 所示。

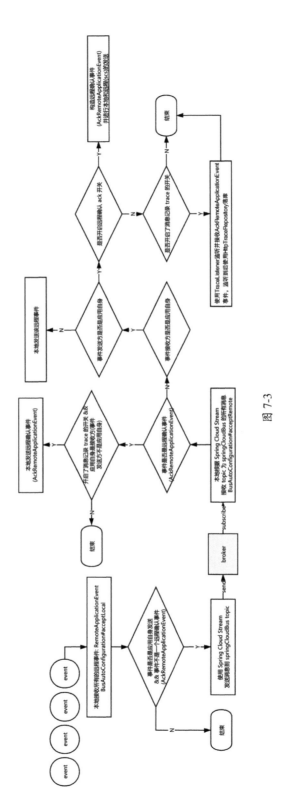

图 7-3

7.3　案例：使用 Spring Cloud Bus 完成多节点配置动态刷新

Spring Cloud Bus 与配置相关的操作可以通过 EnvironmentBusEndpoint 和 RefreshBusEndpoint 完成。

调用 node1 暴露的 EnvironmentBusEndpoint：

```
curl --header "Content-Type:application/json" -XPOST
'http://localhost:8080/actuator/bus-env?name=bookname&value=deepinspringcloud'
```

发送完毕后，进入 node1、node2 和 node3 的 EnvironmentEndpoint，会发现多了一个 name 为 manager 的 PropertySource：

```
{
    ...
    "propertySources": [{
        "name": "manager",
        "properties": {
            "bookname": {
                "value": "deepinspringcloud"
            }
        }
    }
    ...
    ]
}
```

修改 node1、node2 和 node3 的 application.properies 文件（如果使用 Spring Cloud Config Server 和 Git，只需在 Git 仓库上修改配置），加上 book.category=springcloud 配置，调用 node2 暴露的 RefreshBusEndpoint：

```
curl -XPOST 'http://localhost:8081/actuator/bus-refresh'
```

发送完毕后，进入 node1、node2 和 node3 的 EnvironmentEndpoint，会发现多了一个 name 为 book.category 的配置，且值为 springcloud。

可以看到，Spring Cloud Bus 的引入完美地解决了分布式系统场景下多实例手动刷新全局配置的问题。

如图 7-4 所示，这是引入 Spring Cloud Bus 后配置刷新的整个过程。

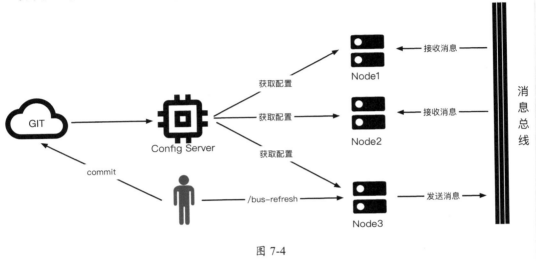

图 7-4

开发者也可以自定义远程事件来处理自身的业务逻辑。

第 8 章

Spring Cloud Data Flow

本章将介绍 Spring Cloud Data Flow 项目，该项目涉及如下模块知识。

- Spring Cloud Stream：流处理任务使用到的框架，使用消息队列作为中间存储模块。
- Spring Cloud Task：批处理使用到的框架。使用 CommandLineRunner 或 ApplicationRunner 表示子任务的执行。Spring Cloud Data Flow 内部提供了 composedtaskrunner-task 这个 Task 类型的应用，用于任务的组合。
- Spring Cloud Batch：批处理使用到的另一个框架。Spring Cloud Task 是一个功能比较粗的框架，而 Spring Batch 在批处理上定义了很多编程模型。
- Spring Cloud Deployer：定义部署 SPI。无论是批处理还是流处理，都需要部署在对应的平台上。
- Spring Cloud Skipper：发现 Spring Boot 应用并管理其生命周期的工具。Spring Cloud Data Flow 上应用的注册、发现都是基于 Skipper 完成的。
- Spring Cloud Data Flow Server：整合以上多个模块的项目。

- Spring Cloud Data Flow Dashboard/Spring Cloud Data Flow Shell：Data Flow Server 对应的客户端。可以在 Dashboard UI 界面或者 Shell 完成。

在介绍上述模块之前，先介绍通过信用卡反欺诈系统和统计 GitHub 仓库的各项数据这两个流处理和批处理任务来体验 Spring Cloud Data Flow 的使用。

最后，针对上述模块的原理和使用进行深入讲解。

8.1 批处理/流处理概述

Spring Cloud Data Flow 是一个能对微服务应用进行批处理和流处理的编排和管理的工具，支持把批处理和流处理任务部署在本地、Cloud Foundry 或 Kubernetes 上。

1. 批处理

批处理是指在没有交互（interaction）或中断（interruption）的情况下，处理一批有限数量的数据。实现批处理的应用可以被称为临时（ephemeral）或短暂（short-lived）的应用。

批处理一个典型的使用场景是账单数据的录入，如图 8-1 所示，中间的账单程序根据左边用户购买数据生成对应的账单，再把账单数据入库到右边的数据表里。

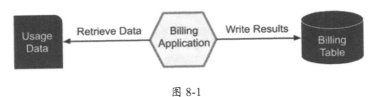

图 8-1

在使用批处理的过程中有很多问题需要解决，比如，任务状态记录（什么时间点开始、什么时间点结束）、异常状态记录、运行状态记录、批处理框架等。这些功能在 Spring 体系中都有框架支持，可以通过 Spring Batch 完成批任务的处理或通过 Spring Cloud Task 完成任务调度的记录。

Spring Cloud Data Flow 可以对这些批处理任务进行编排，如图 8-2 所示。

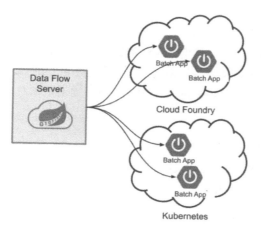

图 8-2

2. 流处理

流处理是指在无交互且无中断（without interaction or interruption）情况下对数据进行无限制的处理。实现流处理的应用可以被称为运行长久的（long-lived）应用。

流处理的一个典型使用场景是实时数据分析或录入，如图 8-3 所示，HTTP 服务对外暴露的信息可以获取一些数据，消息中间件对数据进行二次处理，最终使用 JDBC 写入数据库。

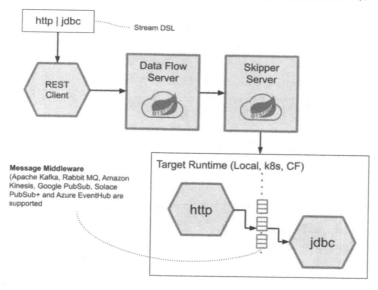

图 8-3

Spring Cloud Data Flow 对流处理的应用定义了以下 3 种类型。

- Source：消息发送者，会将消息发送到目的地。上述示例中的 HTTP 就是一个 Source。
- Sink：消息订阅者，从目的地消费消息。上述示例中的 jdbc 就是一个 Sink。
- Processor：Source 和 Sink 的集合体。从目的地消费消息，并将处理好的结果发送到另外一个目的地。

这 3 种类型都可以被注册到 Spring Cloud Data Flow 中，Spring Cloud Data Flow 可以对这些流处理应用进行编排。Spring Cloud Data Flow 是如何记录这些应用并对它们进行编排的呢？这需要了解它内部的一些核心组件。

Spring Cloud Data Flow 的核心组件如图 8-4 所示。

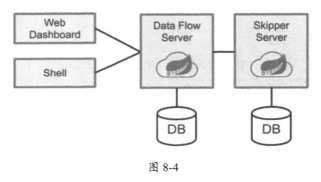

图 8-4

在图 8-4 中：

- Data Flow Server：Spring Cloud Data Flow 服务器端，对外提供了一个 Dashboard，具有注册应用、查看正在运行的应用、查看/创建流处理任务、查看/创建批处理任务等功能。
- Skipper Server：Spring Cloud Skipper Server，用于在多个云平台上发现 Spring Boot 应用并管理其生命周期的工具。
- Spring Cloud DataFlow Shell：Data Flow Server 对外提供了一些 Rest API 用于各种操作，比如：http://localhost:9393/streams/definitions 可以查看流处理的定义。Spring Cloud DataFlow Shell 定义了一套 Shell 脚本，使用这些 Shell 脚本也可以达到与 Data Flow 交互的目的。比如，注册一个 sink: app register --name my-sink-app --type sink -uri maven://org.springframework.cloud:my-sink-app:jar:0.1.0.BUILD-SNAPSHOT。

默认情况下，Data Flow Server 和 Skipper Server 会使用 H2 内存数据库存储数据。可以通过启动参数修改数据源配置的方式使用其他数据库，如 MySQL、Oracle、DB2 或 SQL Server。比如，Data Flow Server 使用 MySQl 数据库：

```
java -jar spring-cloud-dataflow-server/target/spring-cloud-dataflow-server-
2.2.2.RELEASE.jar \
    --spring.datasource.url=jdbc:mysql://localhost:3306/scdfs \
    --spring.datasource.username= \
    --spring.datasource.password= \
    --spring.datasource.driver-class-name=org.mariadb.jdbc.Driver
```

再如，Skipper Server 使用 Oracle 数据库：

```
java -jar spring-cloud-skipper-server-2.3.0.RELEASE.jar \
    --spring.datasource.url=jdbc:oracle:thin:@localhost:1521/scss \
    --spring.datasource.username= \
    --spring.datasource.password= \
    --spring.datasource.driver-class-name=oracle.jdbc.OracleDriver
```

8.2 流处理案例：信用卡反欺诈系统

本节介绍模拟的信用卡反欺诈系统。银行客户会在不同的银行办理信用卡，每张信用卡都有自己的额度，如果客户所有信用卡的消费额度超过指定额度，说明这个客户目前的资金比较紧张，后期可能导致无法偿还的问题出现。

本例假设信用卡指定的总额度不超过 2000 元。如果客户最近两次消费都超过这个额度，就会进入黑名单，不允许再消费。

这个系统主要由 Source 和 Sink 两个应用组成。Source 应用模拟信用卡消费记录的产生，每 2s 定时产生一条随机的信用卡消费记录。Sink 应用读取消费记录，并判断是否存在欺诈行为。

编写 Source 应用代码，生成信用卡消费记录：

```
@SpringBootApplication
@EnableBinding(Source.class)
```

```java
@EnableScheduling
public class CreditCardSourceApplication {

    private final Logger logger =
LoggerFactory.getLogger(CreditCardSourceApplication.class);

    public static Random random = new Random();

    public static void main(String[] args) {
        SpringApplication.run(CreditCardSourceApplication.class, args);
    }

    @Autowired
    private Source source;

    @Scheduled(fixedDelay = 2000)
    public void sendMsg() {
        int cost = random.nextInt(2000);
        String cardType = CreditCardRecord.cardTypes.get(
            String.valueOf(random.nextInt(CreditCardRecord.cardTypes.size())));
        String user = CreditCardRecord.users.get(String.valueOf(random.nextInt(CreditCardRecord.users.size())));
        CreditCardRecord record = new CreditCardRecord(user, new BigDecimal(cost), cardType);
        logger.info("credit card cost record: " + record);
        source.output().send(MessageBuilder.withPayload(record).build());
    }
}
```

通过@Scheduled 触发调度器每 2s 生成一条 CreditCardRecord 消费记录，通过 Spring Cloud Stream 写入到 MQ 中。

编写 Sink 应用代码进行信用卡欺诈检测：

```java
@SpringBootApplication
@EnableBinding(Sink.class)
```

```java
public class CreditCardSinkApplication {

    private final Logger logger =
LoggerFactory.getLogger(CreditCardSinkApplication.class);

    private Map<String, BigDecimal> lastCostInfo = new HashMap<>();
    private Set<String> blackList = new HashSet<>();

    private BigDecimal warningMoney = new BigDecimal(2000);

    public static void main(String[] args) {
        SpringApplication.run(CreditCardSinkApplication.class, args);
    }

    @StreamListener(Sink.INPUT)
    public void receive(CreditCardRecord record) {
        if (blackList.contains(record.getUser())) {
            logger.info(record.getUser() + " now is in black list");
            return;
        }
        logger.info(record.getUser() + " cost " + record.getCost() + " with " + record.getCardType());
        if (lastCostInfo.containsKey(record.getUser())) {
            BigDecimal recentlyCostMoney = lastCostInfo.get(record.getUser()).add(record.getCost());
            if (recentlyCostMoney.compareTo(warningMoney) >= 0) {
                logger.warn(record.getUser() + " recently cost " + recentlyCostMoney + ", go to black list");
                blackList.add(record.getUser()); // 超过 2000，加入黑名单
                return;
            }
        }
        lastCostInfo.put(record.getUser(), record.getCost());
    }
}
```

从 MQ 中读取信用卡消费记录，如果不同的信用卡连续两次消费总和超过 2000 元，客户就被列入黑名单，后续所有的信用卡消费就会失败。

接下来使用 Spring Cloud Data Flow 对这两个应用的流处理任务进行编排。

1. 注册并添加 Source 和 Sink 应用

注册 Source 和 Sink 应用，如图 8-5 所示。

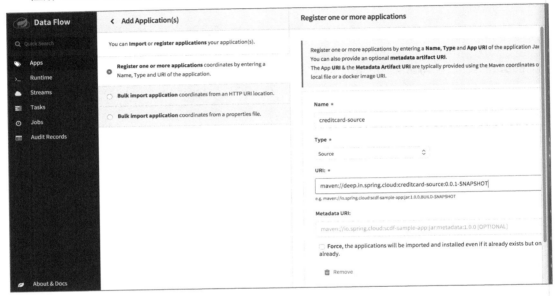

图 8-5

应用添加成功之后，Apps 应用列表页会显示刚才添加的 Source 和 Sink 应用，如图 8-6 所示。

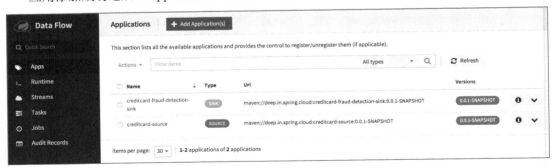

图 8-6

2. 创建 Stream 流处理任务

创建 Stream 流处理任务，如图 8-7 所示。

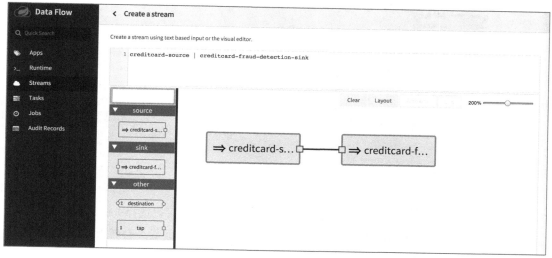

图 8-7

Dashboard 创建流处理任务的方式有以下两种：

- 拖曳界面上的 Source、Sink 或 Processor 应用并编排成任务。
- 在输入框内使用 DSL 描述任务编排：creditcard-source | creditcard-fraud-detection-sink。这段 DSL 表示 creditcard-source 产生的数据会放入 MQ，然后被 creditcard-fraud-detection-sink 消费。

流处理任务添加成功之后。Streams 流处理列表页会显示刚才添加的流处理任务（这时流处理任务还未开始部署，状态 Status 显示为 UNDEPLOYED），如图 8-8 所示。

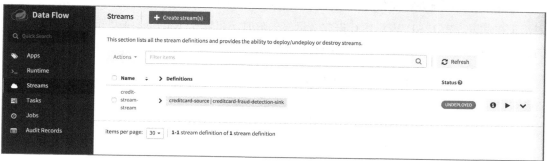

图 8-8

3. 执行流处理任务

单击 Streams 列表页上对应流处理的部署按钮进行部署，如图 8-9 所示。

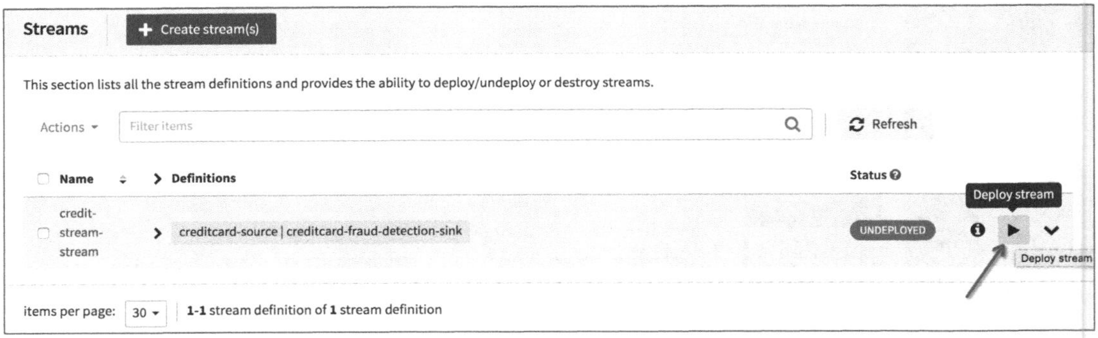

图 8-9

部署的时候需要填写一些配置。比如，运行平台 platform，可以选择 local 或 Kubernetes；Generic Deployer 实例配置信息；应用配置等信息。这个配置界面最终会转换成一段配置文件，并在 Freetext 页显示，我们也可以绕过界面配置，直接通过 Freetext 页面配置：

```
app.creditcard-source.spring.cloud.stream.bindings.output.destination=scdf-creditcard
app.creditcard-fraud-detection-sink.spring.cloud.stream.bindings.input.destination=scdf-creditcard
app.creditcard-fraud-detection-sink.spring.cloud.stream.bindings.input.group=scdf-group
```

通过 Freetext 页面的配置，也会自动映射到 Builder 页面，如图 8-10 所示。

4. 查看流处理任务执行状态

创建 Streams 流处理任务后，Streams 列表页会显示当前流处理任务的状态。DEPLOYING 表示正在部署，FAILED 表示运行失败（运行失败时，可以在 Runtime 菜单下查看是哪个应用的运行出了问题），DEPLOYED 表示已经部署完，正在运行。

流处理任务运行失败的界面如图 8-11 所示。

第 8 章 Spring Cloud Data Flow

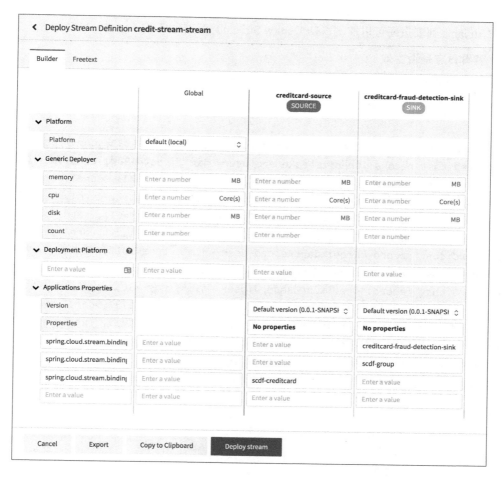

图 8-10

图 8-11

在 Streams 列表页还可以取消（UnDeploy Stream）流处理任务。

5. 查看日志是否生效

进入 Runtime 菜单，查看 Source 和 Sink 应用对应的目录工作位置，并查看日志。

Source 日志如下：

```
credit card cost record: CreditCardRecord{user='tom', cost=1036, cardType='ABC'}
credit card cost record: CreditCardRecord{user='jim', cost=1900, cardType='CMBC'}
credit card cost record: CreditCardRecord{user='jim', cost=233, cardType='CCB'}
credit card cost record: CreditCardRecord{user='jim', cost=172, cardType='CMB'}
credit card cost record: CreditCardRecord{user='jerry', cost=611, cardType='CMB'}
credit card cost record: CreditCardRecord{user='jim', cost=1221, cardType='CMB'}
credit card cost record: CreditCardRecord{user='jim', cost=1017, cardType='ABC'}
credit card cost record: CreditCardRecord{user='jerry', cost=1691, cardType='CMBC'}
credit card cost record: CreditCardRecord{user='jim', cost=38, cardType='ABC'}
credit card cost record: CreditCardRecord{user='jim', cost=1137, cardType='CCB'}
credit card cost record: CreditCardRecord{user='tom', cost=1719, cardType='CMBC'}
credit card cost record: CreditCardRecord{user='jim', cost=546, cardType='CMB'}
credit card cost record: CreditCardRecord{user='jim', cost=1661, cardType='ABC'}
credit card cost record: CreditCardRecord{user='jerry', cost=850, cardType='BCM'}
```

Sink 日志如下：

```
tom cost 1036 with ABC
jim cost 1900 with CMBC
jim cost 233 with CCB
jim recently cost 2133, go to black list
jim now is in black list
jerry cost 611 with CMB
jim now is in black list
jim now is in black list
jerry cost 1691 with CMBC
jerry recently cost 2302, go to black list
jim now is in black list
jim now is in black list
tom cost 1719 with CMBC
```

```
tom recently cost 2755, go to black list
jim now is in black list
jim now is in black list
jerry now is in black list
```

结果符合预期，3 个用户消费金额超过 2000 元后进入黑名单，后续所有的消费全部失败。

这个模拟的反信用卡欺诈系统只用到了 Source 和 Sink 类型的应用，如果中间需要对信用卡消费记录进行修改，可以引入 Processor 类型的应用进行处理。

8.3 批处理案例：统计 GitHub 仓库的各项指标数据

GitHub 仓库数据存储在本地文本文件里，批处理任务会读取文件里的 GitHub 仓库信息，请求 GitHub Open API 获取仓库的 commit 提交次数、issue 个数、fork 个数和 releases 次数。

编写 Task 应用代码，具体如下：

```
@SpringBootApplication
@EnableTask
public class SpringCloudTaskSimpleApplication {

    private final Logger logger =
LoggerFactory.getLogger(SpringCloudTaskSimpleApplication.class);

    public static void main(String[] args) {
        SpringApplication.run(SpringCloudTaskSimpleApplication.class, args);
    }

    @Bean
    RestTemplate restTemplate() {
        return new RestTemplate();
    }

    @Autowired
    private DataSource dataSource;
```

```java
@Autowired
private RestTemplate restTemplate;

@Value("${sleep.time:10000}")
private Long sleepTime;

private String[] fileLocation = new String[] {
    //"spring-cloud/spring-cloud-netflix",
    "alibaba/spring-cloud-alibaba"
};

@Bean
@Order(50)
public CommandLineRunner commandLineRunner() {
    return args -> {
        logger.info("sleepTime: " + sleepTime);
        JdbcTemplate jdbcTemplate = new JdbcTemplate(dataSource);
        jdbcTemplate.execute("CREATE TABLE IF NOT EXISTS " +
            "GITHUB_REPO ( " +
            "repo_name varchar(50), commits int, " +
            "issues int, forks int, releases int)");
    };
}

@Bean
@Order(100)
CommandLineRunner runner() {
    return (args) -> {
        JdbcTemplate jdbcTemplate = new JdbcTemplate(dataSource);
        Arrays.stream(fileLocation).forEach(githubRepo -> {
            int releases = calculate(
                "https://api.github.com/repos/" + githubRepo + "/releases?page={page}&per_page=100", 1);
            int commits = calculate(
                "https://api.github.com/repos/" + githubRepo + "/commits?page={page}&per_page=100", 1);
```

```java
                int forks = calculate("https://api.github.com/repos/" + githubRepo
+ "/forks?page={page}&per_page=100",
                        1);
                int issues = calculate(
                        "https://api.github.com/repos/" + githubRepo + "/issues?page=
{page}&per_page=100", 1);
                jdbcTemplate.update(
                    "insert into GITHUB_REPO(repo_name, commits, issues, forks,
releases) values ( ?,?,?,?,? )",
                    new PreparedStatementSetter() {
                        @Override
                        public void setValues(PreparedStatement preparedStatement)
throws SQLException {
                            preparedStatement.setString(1, githubRepo);
                            preparedStatement.setInt(2, commits);
                            preparedStatement.setInt(3, issues);
                            preparedStatement.setInt(4, forks);
                            preparedStatement.setInt(5, releases);
                        }
                    });
            });
        };
    }

    private int calculate(String url, int page) {
        String realUrl = url.replace("{page}", String.valueOf(page));
        int realPage = page;
        List<Object> releaseResult = restTemplate.getForObject(realUrl, List.class);
        int num = releaseResult.size(), totalNum = releaseResult.size();
        logger.info("calculate [" + num + "]from [" + realUrl + "]");
        num = releaseResult.size();
        try {
            Thread.sleep(sleepTime);
        } catch (InterruptedException e) {
            // 忽略
            return -1;
```

```
        }
        while (num != 0) {
            realPage++;
            realUrl = url.replace("{page}", String.valueOf(realPage));
            releaseResult = restTemplate.getForObject(realUrl, List.class);
            totalNum = totalNum + releaseResult.size();
            logger.info("calculate [" + realUrl + "]from [" + realUrl + "], totalNum is " + totalNum);
            num = releaseResult.size();
            try {
                Thread.sleep(sleepTime);
            } catch (InterruptedException e) {
                // 忽略
                return -1;
            }
        }
        return totalNum;
    }
}
```

接下来使用 Spring Cloud Data Flow 创建批处理任务。

1. 注册 Task 应用

批处理任务对应的应用是 Task 类型的应用，如图 8-12 所示。

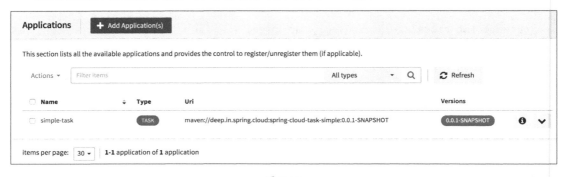

图 8-12

2. 创建 Task 批处理任务

创建 Task 批处理任务，如图 8-13 所示。

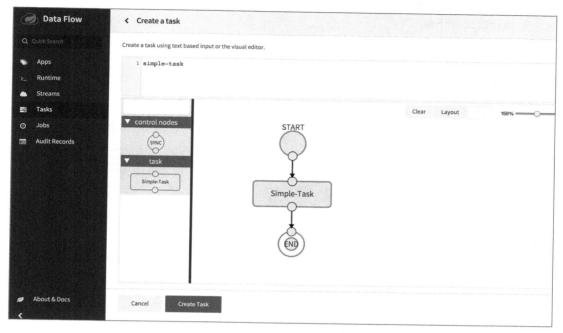

图 8-13

Dashboard 创建批处理任务的方式有以下两种。

- 拖曳界面上的 Task 任务进行编排。
- 在输入框内使用 DSL 描述任务编排情况。

批处理任务添加成功之后。Tasks 批处理列表页会显示刚才添加的批处理任务（这时批处理任务还未开始执行，状态 Status 显示为 UNKNOWN），如图 8-14 所示。

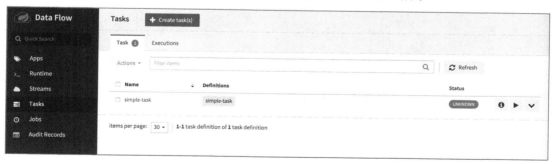

图 8-14

3. 执行批处理任务

单击 Tasks 列表页上对应批处理的运行按钮运行任务，如图 8-15 所示。

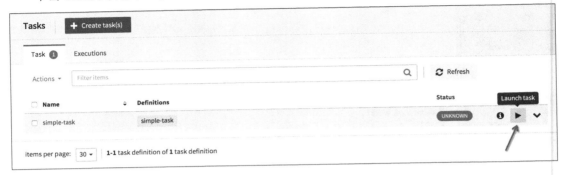

图 8-15

运行时需要填写一些配置（sleep time 为 60000ms，防止被 GitHub 限流），如图 8-16 所示。

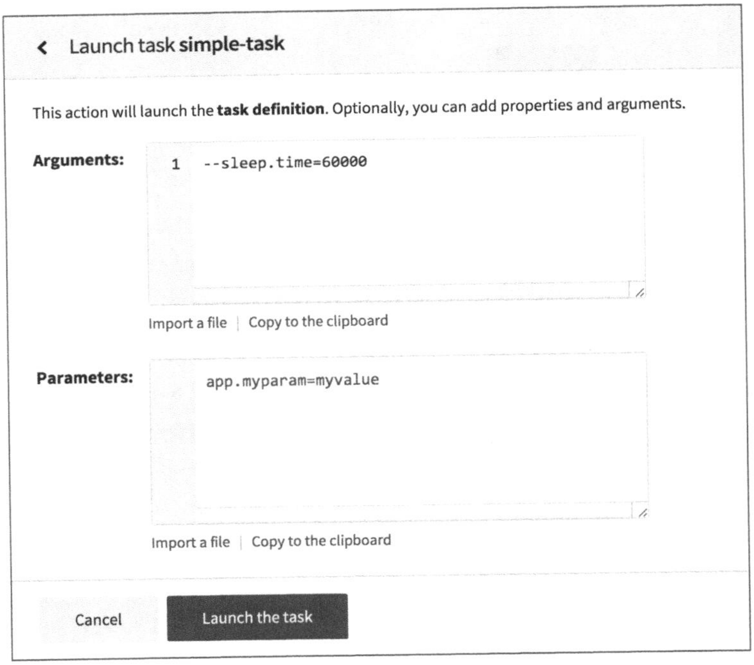

图 8-16

填写完毕并确认没问题后，单击 Launch the task 按钮运行任务。

4. 查看批处理任务执行状态

运行任务后，Tasks 任务列表页的 Executions 页会显示任务的执行情况，如图 8-17 所示。

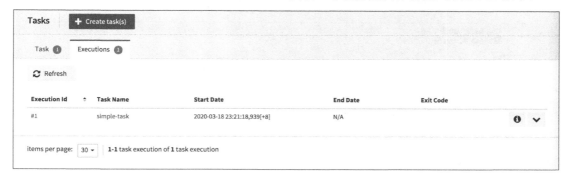

图 8-17

列表页上会显示任务的执行 ID、任务名、开始日期、结束日期和退出码（Exit Code）。

单击 Show details 按钮可以查看任务的详细信息，比如退出消息（Exit Message）、任务执行时间、参数等内容，如图 8-18 所示。

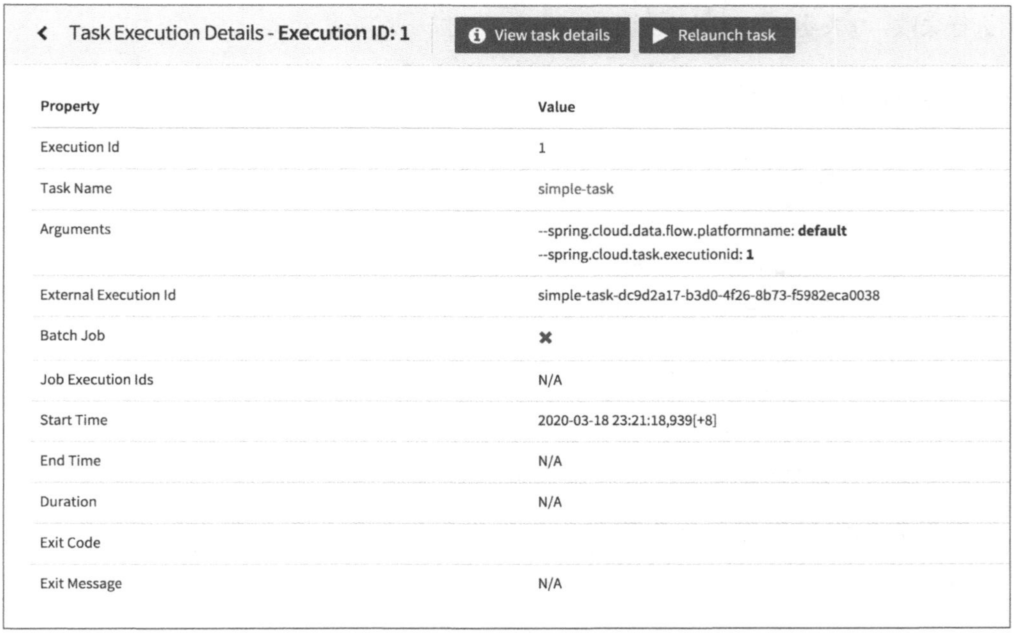

图 8-18

当 Status 为 RUNNING 时，表示任务正在执行，当 Status 为 ERROR 时，表示任务执行失败，如图 8-19 所示。

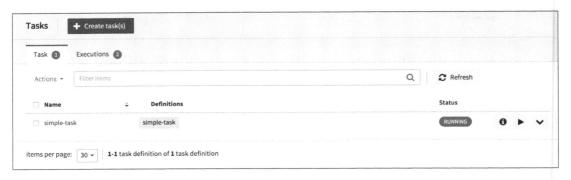

图 8-19

任务执行失败时，可以进入"java.io.tmpdir"系统参数对应的临时目录查看任务的执行日志，该临时目录内部会有一个任务名开头的文件夹，文件夹内部存储着每次执行的日志信息。如图 8-20 所示，Status 为 COMPLETE，表示任务执行成功，退出码（Exit Code）为 0（非 0 表示执行失败，0 表示执行成功）。

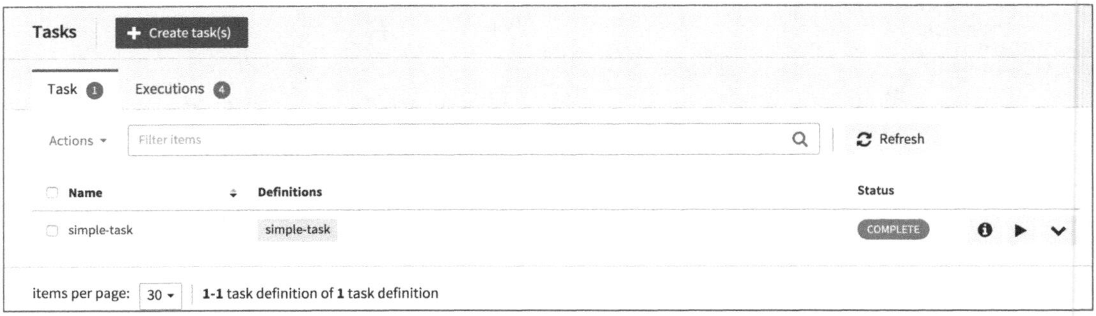

图 8-20

运行成功的日志在 Execution Detail 页面会显示任务运行的日志。

5. 查看数据库验证任务执行是否成功

查看数据库验证任务执行是否成功，如图 8-21 所示。

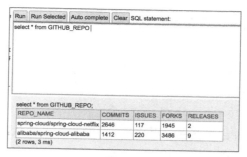

图 8-21

8.4　Spring Cloud Data Flow 批处理任务组合

　　Spring Cloud Data Flow 批处理支持多个任务的组合。如图 8-22 所示，这是一个组合任务，表示 Task01 执行完毕后，Task02 和 Task03 异步同时执行，Task02 执行后开始执行 Task04，Task03 执行后开始执行 Task05，Task04 和 Task05 执行完毕后执行 Task06。

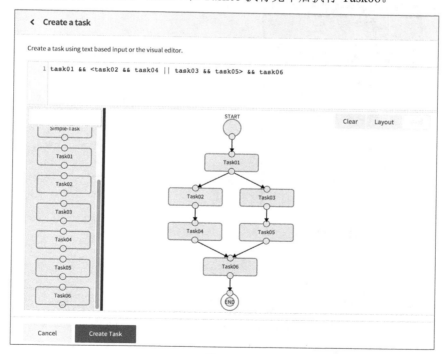

图 8-22

> **提示**：运行组合任务的前提是注册 Spring Cloud Data Flow 提供的 composedtaskrunner-task 这个 Task 类型的应用，可以通过 Data Flow Shell 添加：

```
dataflow:>app register --name composed-task-runner --type task --uri
maven://org.springframework.cloud.task.app:composedtaskrunner-task:{version}
```

在这个过程中，如果有错误发生，后续的任务将不再继续执行（如图 8-23 所示，Task-Error 是一个会发生错误的任务，后续 Task02 就不会执行）。

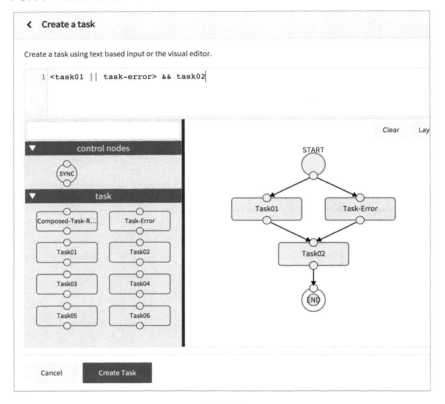

图 8-23

> **提示**：创建批处理任务时可以对任务设置应用级别的参数。比如，task create --definition "task01 --sleep.time=10000 && task02 --sleep.time=20000" --name test-task 脚本对于 task01 任务而言，设置 sleep time 为 10000ms，对于 task02 任务，设置 sleep time 为 20000ms。

可以针对不同的 Exit Message 做不同的处理（如图 8-24 所示，Task-Error 任务的 Exit Message 若为 1，则执行 Task01 任务，Exit Message 为 0，则执行 Task02 任务，否则结束整个流程任务。Task01 和 Task02 任务执行后也会结束整个流程任务）。

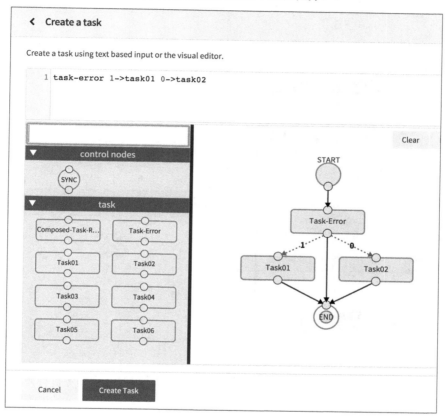

图 8-24

Exit Message 也可以使用通配符来进行匹配（如图 8-25 所示，Task-Error 任务的 Exit Message 为 0，则执行 Task01 任务，否则其他所有的 Exit Message 执行 Task02 任务）。

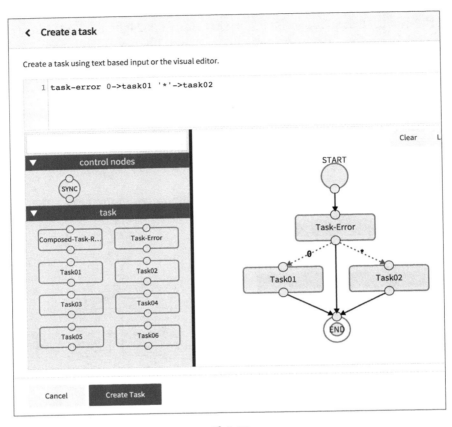

图 8-25

如图 8-26 所示,Task01、Task02 和 Task03 异步同时执行中,3 个任务执行完毕后,Task04 和 Task05 异步同时执行。

 提示:创建组合任务时可以对不同的子任务设置不同的参数。比如,在 Parameters 中输入 `app.two-task.task01.sleep.time=10000, app.two-task.task02.sleep.time=20000`,表示组合任务 two-task 里的 task01 子任务的配置为 sleep.time=10000,task02 子任务的配置为 sleep.time=20000。

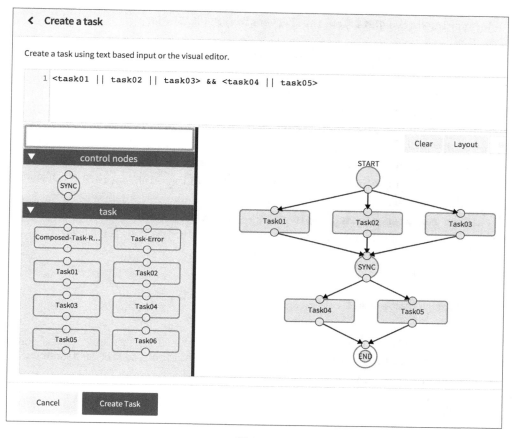

图 8-26

8.5　Spring Cloud Data Flow Shell

Spring Cloud Data Flow Server 对外提供了 UI 界面用于相关操作。Spring Cloud Data Flow Shell 是另外一种使用方式，使用脚本代替 UI 界面操作。

不论是 Spring Cloud Data Flow 的 UI 界面还是 Shell 脚本，底层都是调用 Data Flow Server 对外暴露的 Rest API。这些 Rest API 包括审计日志、Streams 流任务定义、日志、部署；Task 批任务定义、日志、部署等内容。

若要使用 Spring Cloud Data Flow Shell，启动一个 JAR 包即可：

```
java -jar spring-cloud-dataflow-shell-{version}.RELEASE.jar
```

启动后执行对应的命令,即可完成对应的操作,比如,创建应用。

```
# 创建 source 类型的应用
dataflow:>app register --name creditcard-source --type source --uri maven://deep.in.spring.cloud:creditcard-source:jar:0.0.1-SNAPSHOT
# 创建 sink 类型的应用
dataflow:>app register --name creditcard-fraud-detection-sink --type sink --uri maven://deep.in.spring.cloud:creditcard-fraud-detection-sink:jar:0.0.1-SNAPSHOT
# 创建 task 类型的应用
dataflow:>app register --name simple-task --type task --uri maven://deep.in.spring.cloud:spring-cloud-task-simple:jar:0.0.1-SNAPSHOT
# 查看应用列表
dataflow:>app list
```

流处理任务的相关脚本如下:

```
# 创建 Stream 流处理任务
dataflow:>stream create --name credit-stream --definition 'creditcard-source | creditcard-fraud-detection-sink'
# 查看当前 Stream 流处理任务列表
dataflow:>stream list
# 执行流处理任务
dataflow:>stream deploy --name credit-stream
```

批处理任务的相关脚本如下:

```
# 创建 Task 批处理任务
dataflow:>task create --name simple-task --definition "simple-task"
# 查看当前 Task 批处理任务列表
dataflow:>task list
# 运行批处理任务
dataflow:>task launch simple-task --arguments "--sleep.time=30000"
# 查看任务执行情况列表
dataflow:>task execution list
```

8.6 Spring Cloud Skipper

Spring Cloud Skipper 是一个可以在多个云平台（本地、Cloud Foundry、Kubernetes）上发现 Spring Boot 应用并管理其生命周期的工具。Spring Cloud Data Flow 依赖 Skipper 完成应用相关的注册、部署等操作，Skipper 也可以单独使用或者与 CI 工具（比如 Jenkins）结合完成符合自身条件的部署形态。

Spring Cloud Skipper 定义了以下 5 个概念：

- Platforms：部署平台，也就是应用在哪里运行的平台。Spring Cloud Skipper 使用 Spring Cloud Deployer 对应用进行部署，目前支持本地（Local）、Cloud Foundry 和 Kubernetes 平台。一个 Skipper 可以支持多个 Platform 部署平台。
- Packages：安装包，用于描述应用在 Platform 部署平台上的元数据信息。一个安装包可以定义一个应用或者一组应用，这些应用包含 Spring Boot JAR 包的位置（maven 仓库、docker 仓库、本地文件目录或 HTTP 路径）、部署属性等元数据信息。一个安装包是一组 yaml 文件的集合，这些文件会被打包到 ZIP 压缩包内，ZIP 文件名的格式为 name-version.zip（比如 myapp-1.0.0.zip）。
- Repositories：托管安装包的仓库。支持本地和远程仓库，本地仓库由 Skipper Server 管理，远程仓库必须支持 HTTP 协议且遵循一些约定俗成的内容。
- Releases：安装、更新、回滚或删除安装包的时候会产生一个 Release 发布。
- Release Workflows：更新或回滚应用具体的流程。

接下来通过 Spring Cloud Skipper 模拟应用的发布、升级、回滚过程，让大家对 Skipper 有一个直观的了解。

（1）编写 Spring Boot 应用代码。

非常简单的一个 Spring Boot 应用，对外暴露了 /hello 端点模拟当前应用的版本号：

```
@SpringBootApplication
public class HelloSkipperApplication {

    public static void main(String[] args) {
        SpringApplication.run(HelloSkipperApplication.class, args);
    }
```

```
@RestController
class HelloController {
    @GetMapping("/hello")
    String hello() {
        return "hello skipper: v1";
    }
}
```

（2）打包应用到 ZIP 压缩包内，按照 Package Metadata 的标准定义 yaml 文件。

```
├── package.yml
├── templates
│   └── helloworld.yml
└── values.yml
```

zip 压缩包名字为 helloskipper-1.0.0.zip。

（3）启动 Skipper Server 和 Skipper Shell 组件。

要使用 Spring Cloud Skipper，需要启动服务器端和客户端。服务器端 Skipper Server 对外提供了 Open API，但没有提供 UI 界面，可以通过 Open API 或 Skipper Shell 进行操作。

（4）上传 Package 到 Repository。

Skipper Server 启动后，Repository 列表会有一个 name 为 local 的仓库，由 Skipper Server 管理。

通过 Skipper Shell 上传 Package，并查看上传的 package 信息：

```
skipper:>package upload --repo-name local --path /YourPath/helloskipper-1.0.0.zip
Package uploaded successfully:[helloskipper:1.0.0]
skipper:>package list
```

Name	Version	Description
helloskipper	1.0.0	Hello Skipper v1

> **注意**：上传的 ZIP 包必须有一个与 ZIP 压缩包名一样的父目录。比如，helloskipper-1.0.0.zip 压缩包内部的文件内容如下：

```
helloskipper-1.0.0/
helloskipper-1.0.0/package.yml
helloskipper-1.0.0/templates/
helloskipper-1.0.0/templates/helloskipper.yml
helloskipper-1.0.0/values.yml
```

（5）安装 Package，产生 Release 发布。

```
skipper:>package install --package-name helloskipper --release-name first-release --properties spec.applicationProperties.server.port=8080
Released first-release. Now at version v1.
```

访问 http://localhost:8080/hello，验证是否部署成功（返回的内容为 hello skipper: v1）。

（6）应用版本升级。

first-release 对应的 Release 发布对应的 Package 版本为 1.0.0。

Package 在升级成 1.0.1 版本之前，需要上传新的 Package 到 Repository 里。

```
skipper:>package upload --repo-name local --path /YourPath/helloskipper-1.0.1.zip
Package uploaded successfully:[helloskipper:1.0.1]
skipper:>package list
```

Name	Version	Description
helloskipper	1.0.0	Hello Skipper v1
helloskipper	1.0.1	Hello Skipper v1.0.1

```
skipper:>release history --release-name first-release
```

Version	Last updated	Status	Package Name	Package Version	Description
2	Sun Mar 22 18:30:53 MYT 2020	DEPLOYED	helloskipper	1.0.1	Upgrade complete
1	Sun Mar 22 18:08:22 MYT 2020	DELETED	helloskipper	1.0.0	Delete complete

Package 升级老的 Release 发布到 1.0.1 版本，并将应用端口改成 9090：

```
skipper:>release upgrade --release-name first-release --package-name helloskipper -
-package-version 1.0.1  --properties spec.applicationProperties.server.port=9090
first-release has been upgraded. Now at version v2.
```

Release 升级之后，first-release 对应的 Release 发布对应的 Package 版本为 1.0.1。

访问 http://localhost:9090/hello，验证部署是否成功（返回的内容为 hello skipper: v1.0.1）。这时访问 http://localhost:8080/hello，出现 Connection Refused，应用进程已经不存在。

（7）应用版本再次升级，修改部署配置。

上传新的 Package 到 Repository 里。

```
skipper:>package upload --repo-name local --path /YourPath/helloskipper-1.0.2.zip
Package uploaded successfully:[helloskipper:1.0.2]
skipper:>package list
```

Name	Version	Description
helloskipper	1.0.0	Hello Skipper v1
helloskipper	1.0.1	Hello Skipper v1.0.1
helloskipper	1.0.2	Hello Skipper v1.0.2

Package 升级 1.0.1 到 1.0.2：

```
skipper:>release upgrade --release-name first-release --package-name helloskipper -
-package-version 1.0.2  --file /YourPath/helloskipper-upgrade.yml
```

helloskipper-upgrade.yml 的内容如下：

```yaml
spec:
  applicationProperties:
    server.port: 9090
  deploymentProperties:
    spring.cloud.deployer.memory: 2048m
```

Release 升级之后，first-release 对应的 Release 发布对应的 Package 版本为 1.0.2。

访问 http://localhost:9090/hello，验证部署是否成功（返回的内容为 hello skipper: v1.0.2）。

查看 Java 进程，启动参数有 `-Xmx2048m` 配置，部署参数生效。

（8）应用版本降级。

查看 Release 部署历史信息，回滚到 Skipper 上版本号为 2 的 Release 部署：

```
skipper:>release history --release-name first-release

╔═════════╤══════════════════════════════╤══════════╤══════════════╤════════════════╤══════════════════╗
║Version │         Last updated         │  Status  │ Package Name │ Package Version│   Description    ║
╠═════════╪══════════════════════════════╪══════════╪══════════════╪════════════════╪══════════════════╣
║3       │Sun Mar 22 18:50:36 MYT 2020  │DEPLOYED  │helloskipper  │1.0.2           │Upgrade complete  ║
║2       │Sun Mar 22 18:30:53 MYT 2020  │DELETED   │helloskipper  │1.0.1           │Delete complete   ║
║1       │Sun Mar 22 18:08:22 MYT 2020  │DELETED   │helloskipper  │1.0.0           │Delete complete   ║
╚═════════╧══════════════════════════════╧══════════╧══════════════╧════════════════╧══════════════════╝

skipper:>release rollback --release-name first-release --release-version 2
first-release has been rolled back. Now at version v4.
```

回滚成功后访问 http://localhost:9090/hello，验证部署是否成功（返回的内容为 hello skipper: v1.0.1）。

8.7 Spring Cloud Deployer

Spring Cloud Deployer 组件是 Spring Cloud 提供用于定义应用部署 SPI 的项目。

Spring Cloud Skipper 内部关于应用的部署、扩容、缩容、内存参数设置，以及 CPU 参数设置等操作都是基于 Spring Cloud Deployer 完成的。

8.7.1 TaskLauncher 接口

TaskLauncher SPI 接口（这里的 Task 代表的是一个容器或者是一个应用的运行任务）是 Spring Cloud Deployer 定义的核心接口之一，这个接口运行和管理这些任务的生命周期。该接口的定义如下：

```java
public interface TaskLauncher {

    // 运行一个任务并得到任务 ID。AppDeploymentRequest 表示应用部署任务所需的一些参数
    String launch(AppDeploymentRequest request);

    // 根据任务 ID 取消任务
    void cancel(String id);

    // 根据任务 ID 查询当前任务的状态
    TaskStatus status(String id);

    // 清理任务 ID 关联的任务所关联的资源。如果没有清理动作，方法内部不需要做任何事
    void cleanup(String id);

    // 根据应用名尝试销毁任务，销毁时也会清理任务关联的资源。AppName 从
    // AppDeploymentRequest#AppDefinition#name 中获取
    void destroy(String appName);

    // 任务运行时的环境信息。比如 Java 版本、Spring 版本、Spring Boot 版本等信息
    RuntimeEnvironmentInfo environmentInfo();

    // 最大并发任务执行数
    default int getMaximumConcurrentTasks() {
        throw new UnsupportedOperationException("'getMaximumConcurrentTasks' is not implemented.");
    };

    // 正在执行的任务数
    default int getRunningTaskExecutionCount() {
```

```
        throw new UnsupportedOperationException("'getRunningTaskExecutionCount' is
not implemented.");
    }

    // 根据任务 ID 获取任务运行的日志内容
    default String getLog(String id) {
        throw new UnsupportedOperationException("'getLog' is not implemented.");
    }
}
```

TaskLauncher 接口内部的类或接口都与部署有关系。

AppDeploymentRequest 表示应用部署任务需要的参数信息，由以下 4 部分组成：

- definition: AppDefinition：应用定义，包括 String name 和 Map<String, String> properties，分别表示应用名和应用参数。
- resource: Resource：应用对应的资源文件，可以是一个 maven 坐标，也可以是一个本地或远程 JAR 包。
- deploymentProperties: Map<String, String>：部署参数。
- commandlineArguments: List<String>：启动参数。

RuntimeEnvironmentInfo 表示任务运行时的环境信息，包括以下内容：

- spiVersion：Spring Cloud Deployer SPI 版本。比如 2.2.0.RELEASE。
- implementationName：具体的 SPI 实现类。比如，LocalTaskLauncher，表示 Spring Cloud Deployer Local。
- implementationVersion：具体的 SPI 实现类版本。比如 2.2.0.RELEASE。
- platformType：具体的 SPI 实现类型。比如 Local。
- platformApiVersion：具体的 SPI 实现类型对应的 API 版本。比如 Mac OS X 10.12.6。
- platformHostVersion：具体的 SPI 实现类型对应的版本信息。比如 10.12.6。
- javaVersion：Java 版本。比如 1.8.0_141。
- springVersion：Spring Framework 版本。比如 5.2.1.RELEASE。
- springBootVersion：Spring Boot 版本。比如 2.2.1.RELEASE。
- platformSpecificInfo：具体的 SPI 实现类型特殊的属性。

TaskStatus 表示 TaskLauncher 运行的一个任务对应的任务状态，包括以下 3 个属性：

- id: String：任务 ID。
- state: LaunchState：任务运行状态。
- attributes: Map<String, String>：任务运行需要的一些属性。比如 working.dir、stdout、stderr 等内容。

LaunchState 包含以下状态：

- launching：任务执行刚请求，正在处理中。
- running：任务成功运行，但还没完成。
- cancelled：任务被取消。
- complete：任务运行完成。
- failed：任务运行失败。
- error：任务运行时发生的系统错误。
- unknown：未知。

8.7.2　AppDeployer 接口

AppDeployer SPI 接口也是 Spring Cloud Deployer 定义的核心接口之一，该接口与 TaskLauncher 的区别在于它用于运行和管理应用的生命周期，该接口的定义如下：

```
public interface AppDeployer {

    // Spring Cloud Deployer 配置前缀
    static final String PREFIX = "spring.cloud.deployer.";

    // 应用部署个数
    static final String COUNT_PROPERTY_KEY = PREFIX + "count";

    // 应用所属分组
    static final String GROUP_PROPERTY_KEY = PREFIX + "group";

    // 应用索引值
    static final String INDEXED_PROPERTY_KEY = PREFIX + "indexed";

    // 应用索引值，一般从环境变量获取
```

```java
static final String INSTANCE_INDEX_PROPERTY_KEY = "INSTANCE_INDEX";

// 应用内存大小
static final String MEMORY_PROPERTY_KEY = PREFIX + "memory";

// 应用磁盘空间大小
static final String DISK_PROPERTY_KEY = PREFIX + "disk";

// 应用 CPU 核数
static final String CPU_PROPERTY_KEY = PREFIX + "cpu";

// 部署一个应用，得到部署 ID。AppDeploymentRequest 表示应用部署所需的一些参数
String deploy(AppDeploymentRequest request);

// 基于部署 ID 停止应用的部署
void undeploy(String id);

// 基于部署 ID 获取应用状态
AppStatus status(String id);

// 应用部署时的环境信息。比如 Java 版本、Spring 版本、Spring Boot 版本等信息
RuntimeEnvironmentInfo environmentInfo();

// 基于部署 ID 获取应用日志
default String getLog(String id) {
    throw new UnsupportedOperationException("'getLog' is not implemented.");
}

// 应用扩容，AppScaleRequest 内部的 count 参数可以指定扩容的实例个数
default void scale(AppScaleRequest appScaleRequest) {
    throw new UnsupportedOperationException("'scale' is not implemented.");
}
}
```

AppDeployer 接口是一个规范，接口内部定义的这些参数（比如，应用部署个数、内存、CPU）会被这个接口的具体实现类所使用，比如，Spring Cloud Deployer Local 的 JavaCommandBuilder 就使用了内存参数：

```java
protected void addJavaOptions(List<String> commands, Map<String, String> deploymentProperties,
            LocalDeployerProperties localDeployerProperties) {
    String memory = null;
    // 获取内存配置
    if (deploymentProperties.containsKey(AppDeployer.MEMORY_PROPERTY_KEY)) {
        memory = "-Xmx" +
ByteSizeUtils.parseToMebibytes(deploymentProperties.get(AppDeployer.MEMORY_PROPERTY_KEY)) + "m";
    }
}
```

再如，Spring Cloud Deployer Kubernetes 的 AbstractKubernetesDeployer 使用了 CPU 和内存参数（读取的是 Spring Cloud Deployer Kubernetes 自定义的配置项）：

```java
protected Map<String, Quantity> deduceResourceRequests(AppDeploymentRequest request) {
    String memOverride = PropertyParserUtils.getDeploymentPropertyValue(request.getDeploymentProperties(),
            "spring.cloud.deployer.kubernetes.requests.memory");
    if (memOverride == null) {
        memOverride = properties.getRequests().getMemory();
    }



    String cpuOverride = PropertyParserUtils.getDeploymentPropertyValue(request.getDeploymentProperties(),
            "spring.cloud.deployer.kubernetes.requests.cpu");
    if (cpuOverride == null) {
        cpuOverride = properties.getRequests().getCpu();
    }

    logger.debug("Using requests - cpu: " + cpuOverride + " mem: " + memOverride);
```

```java
Map<String,Quantity> requests = new HashMap<String, Quantity>();
if (memOverride != null) {
    requests.put("memory", new Quantity(memOverride));
}
if (cpuOverride != null) {
    requests.put("cpu", new Quantity(cpuOverride));
}
return requests;
}
```

AppDeployer 接口内部的 AppStatus 表示部署应用后对应的应用状态，包括以下 3 个属性：

- deploymentId: String：部署 ID。
- generalState: DeploymentState：部署状态。
- instances: Map<String, AppInstanceStatus>：一次部署可能包括多个实例，Map 维护着这些实例的状态。

DeploymentState 包含以下状态：

- deploying：正在部署过程中。
- deployed：部署完成（所有的应用实例全部部署完成）。
- undeployed：未部署。
- partial：多个应用，一些部署成功，另一些未部署成功。
- failed：部署失败。
- error：应用部署时发生的系统错误。
- unknown：未知。

AppInstanceStatus 是一个接口，需要以下 3 个方法：

- getId():String 方法：每个实例对应的 ID。
- getState():DeploymentState：实例的部署状态。
- getAttributes():Map<String, String>：实例的一些属性，比如 pid、port、stdout、stderr、working.dir 等内容。

TaskLauncher 和 AppDeployer 的不同点在于双方关注点不一致。TaskLauncher 关注的是任务的生命周期，一个 Java 应用的启动任务就是一个任务的执行（比如 Spring Cloud Deployer

Kubernetes 对应的 TaskLauncher 内部有可能会创建一个 Job）；AppDeployer 关注的是应用的生命周期，一个 Java 应用的部署、扩容就是一个应用生命周期的展示。

8.7.3　LocalAppDeployer

Spring Cloud Deployer SPI 接口是一个规范，目前有 Local、Cloud Foundry、Kubernetes、Yarn 以及 Mesos 实现类。其中，Spring Cloud Deployer Local 表示这些任务/应用的执行在本地完成。

LocalAppDeployer 是 AppDeployer SPI 接口的 Local 实现，其内部会构造 Java Process 进程完成应用的部署。

下面是使用 Spring Cloud Deployer Local 在本地部署一个 Nacos Provider 应用的例子，代码如下：

```java
public class SpringCloudDeployerLocalApplication {

    public static void main(String[] args) {
        LocalAppDeployer deployer = new LocalAppDeployer(new LocalDeployerProperties());
        String deploymentId = deployer.deploy(createAppDeploymentRequest());
        while (true) {
            try {
                Thread.sleep(1000L);
            } catch (InterruptedException e) {
                // 忽略
            }
            AppStatus status = deployer.status(deploymentId);
            System.out.println("app status: " + status);
            if (status.getState() == DeploymentState.deployed) {
                System.out.println("app is deployed");
                break;
            }
        }
        try {
            Thread.sleep(30000L);
            deployer.shutdown();
        } catch (Exception e) {
```

```
            e.printStackTrace();
        }
    }

    private static AppDeploymentRequest createAppDeploymentRequest() {
        MavenResource resource = new MavenResource.Builder()
            .artifactId("spring-cloud-alibaba-nacos-provider")
            .groupId("deep.in.spring.cloud")
            .version("0.0.1-SNAPSHOT")
            .build();
        Map<String, String> properties = new HashMap<>();
        properties.put("server.port", "8080");
        properties.put("spring.application.name", "spring-cloud-deployer-provider");
        properties.put("spring.cloud.nacos.discovery.server-addr", "localhost:8848");
        AppDefinition definition = new AppDefinition("nacos-provider", properties);
        AppDeploymentRequest request = new AppDeploymentRequest(definition, resource);
        return request;
    }
}
```

LocalAppDeployer 的构造需要 LocalDeployerProperties 配置类，这个配置类内部定义了一些新的配置，比如 spring.cloud.deployer.local.debug-port，用于配置远程 debug 端口；spring.cloud.deployer.local.debug-suspend 用于 suspend 的设置。

8.8 Spring Cloud Task

Spring Cloud Task 是一个任务处理框架，内部使用关系型数据库存储任务执行的状态。任务的执行通过 Spring Boot 提供的 CommandLineRunner 或 ApplicationRunner 完成。所有的 CommandLineRunner 或 ApplicationRunner 执行完成后表示一个 TaskExecution 任务顺利执行。

8.8.1 体验 Spring Cloud Task

下面是最简单的一个 Spring Cloud Task 程序，代码如下：

```
@SpringBootApplication
```

```java
@EnableTask
public class MyTaskApplication {

    public static void main(String[] args) {
        SpringApplication.run(MyTaskApplication.class, args);
    }

    @Bean
    public CommandLineRunner runner1() {
        return (args) -> {
            Thread.sleep(60000L);
            System.out.println("Hello Spring Cloud Task runner1.");
        };
    }

    @Bean
    public CommandLineRunner runner2() {
        return (args) -> {
            System.out.println("Hello Spring Cloud Task runner2.");
        };
    }

    @Bean
    public ApplicationRunner runner3() {
        return (args) -> {
            System.out.println("Hello Spring Cloud Task runner3.");
        };
    }
}
```

在任务执行时，SimpleTaskRepository 会在控制台打印日志，内容如下：

```
2020-03-24 23:00:48.663 DEBUG 9992 --- [           main] o.s.c.t.r.support.SimpleTaskRepository   : Creating: TaskExecution{executionId=0, parentExecutionId=null, exitCode=null, taskName='application', startTime=Tue Mar 24 23:00:48 MYT 2020, endTime=null, exitMessage='null', externalExecutionId='null', errorMessage='null', arguments=[-Dparam=1]}
```

每次 TaskExecution 的执行都会记录到 TASK_EXECUTION 表。TASK_EXECUTION 表维护着以下字段：

- TASK_EXECUTION_ID：每次执行任务产生的唯一 ID。
- START_TIME：任务开始时间。
- END_TIME：任务结束时间。
- TASK_NAME：任务名。
- EXIT_CODE：任务结束码。根据 SpringBoot 的 ExitCodeExceptionMapper 接口得到，0 表示正常结束，1 表示非正常结束。
- EXIT_MESSAGE：任务结束消息。可以使用@AfterTask、@BeforeTask 或@FailedTask 注解修饰方法，修改 TaskExecution 内部的 exitMessage 完成；或者实现 TaskExecution-Listener 接口监听 onTaskStartup、onTaskEnd 或 onTaskFailed 方法修改 TaskExecution 内部的 exitMessage 完成。
- LAST_UPDATED：任务更新时间。
- EXTERNAL_EXECUTION_ID：外部设置的任务执行 ID，可以通过 `spring.cloud.task.external-execution-id` 配置修改。
- PARENTEXECUTIONID：父任务 ID，可以通过 `spring.cloud.task.parent-execution-id` 配置修改。

每个 TaskExecution 任务执行都会有输入参数，这些参数存储在 TASKEXECUTIONPARAM 表内。

8.8.2 深入理解 Spring Cloud Task

Spring Cloud Task 的生效需要加上 `org.springframework.cloud:spring-cloud-starter-task` 依赖和@EnableTask 注解，它们会加载一些自动化配置类。比如，SimpleTaskAutoConfiguration 自动化配置类内部会初始化 TaskProperties 配置类、TaskRepositoryInitializer 初始化器、TaskRepository、TaskExplorer、TaskNameResolver 等核心类，这些类也会被注入到应用 ApplicationContext 中。@EnableTask 注解会让 TaskLifecycleConfiguration 配置类生效，该配置类内部会初始化 TaskLifecycleListener。TaskLifecycleListener 会监控任务的开始和完成阶段，监听到 ApplicationFailedEvent 或 ApplicationReadyEvent 事件会结束任务，监听到 ExitCodeEvent 事件会记录 TASKEXECUTION 表的 EXITCODE 字段。TaskLifecycleListener 实现了 SmartLifecycle 接口，在 start 过程中会启动任务。

TaskRepositoryInitializer 内部会初始化 Spring Cloud Task 相关的数据库，初始化的时候会从 org.springframework.cloud:spring-cloud-task-core JAR 包内 org.springframework.cloud.task package 里加载 schema-@@platform@@.sql 文件，这个 platform 会被替换成具体的关系型数据库，比如 db2、mysql、h2、oracle 等。这个 schema-@@platform@@.sql 文件会创建 4 个表，分别是 TASKEXECUTION、TASKEXECUTIONPARAMS、TASKTASKBATCH 和 TASKLOCK。

比如，schema-h2.sql 内容如下：

```sql
CREATE TABLE TASK_EXECUTION  (
    TASK_EXECUTION_ID BIGINT NOT NULL PRIMARY KEY ,
    START_TIME TIMESTAMP DEFAULT NULL ,
    END_TIME TIMESTAMP DEFAULT NULL ,
    TASK_NAME  VARCHAR(100) ,
    EXIT_CODE INTEGER ,
    EXIT_MESSAGE VARCHAR(2500) ,
    ERROR_MESSAGE VARCHAR(2500) ,
    LAST_UPDATED TIMESTAMP,
    EXTERNAL_EXECUTION_ID VARCHAR(255),
    PARENT_EXECUTION_ID BIGINT
);

CREATE TABLE TASK_EXECUTION_PARAMS  (
    TASK_EXECUTION_ID BIGINT NOT NULL ,
    TASK_PARAM VARCHAR(2500) ,
    constraint TASK_EXEC_PARAMS_FK foreign key (TASK_EXECUTION_ID)
    references TASK_EXECUTION(TASK_EXECUTION_ID)
) ;

CREATE TABLE TASK_TASK_BATCH (
  TASK_EXECUTION_ID BIGINT NOT NULL ,
  JOB_EXECUTION_ID BIGINT NOT NULL ,
    constraint TASK_EXEC_BATCH_FK foreign key (TASK_EXECUTION_ID)
    references TASK_EXECUTION(TASK_EXECUTION_ID)
) ;

CREATE SEQUENCE TASK_SEQ ;
```

```
CREATE TABLE TASK_LOCK  (
    LOCK_KEY CHAR(36) NOT NULL,
    REGION VARCHAR(100) NOT NULL,
    CLIENT_ID CHAR(36),
    CREATED_DATE TIMESTAMP NOT NULL,
    constraint LOCK_PK primary key (LOCK_KEY, REGION)
);
```

TASK_EXECUTION 数据库表记录每个任务的执行情况，如表 8-1 所示。

表 8-1

TASKEXECUTIONID	1	EXIT_MESSAGE	null
START_TIME	2020-03-24 22:43:09.734	ERROR_MESSAGE	null
END_TIME	2020-03-24 22:43:31.376	LAST_UPDATED	2020-03-24 22:43:31.394
TASK_NAME	application	EXTERNALEXECUTIONID	888
EXIT_CODE	0	PARENTEXECUTIONID	999

TASKEXECUTIONPARAM 数据库表记录每个任务执行的参数信息，如表 8-2 所示。

表 8-2

TASKEXECUTIONID	TASK_PARAM
1	-Dparam=1

TASK_LOCK 数据库表用于判断任务名相同的任务是否可以同时执行，如表 8-3 所示。该表的使用需要配合 Spring-Integration 项目。

表 8-3

LOCK_KEY	REGION	CLIENT_ID	CREATED_DATE
3676d55f-8449-3cbe-adfc-614c1b1b62fc	DEFAULT	1	2020-03-24 23:01:10.416

TaskRepository 用于关联任务与数据库之间的操作，默认会构造 SimpleTaskRepository 这个实现类。对外提供了如 completeTaskExecution（完成任务执行）、createTaskExecutio（创建任务执行）、startTaskExecution（开始任务执行）、validateCompletedTaskExitInformation（验证任务是否执行完毕）等方法。

TaskExplorer 用于查询任务的执行情况，默认会构造 SimpleTaskExplorer 类。对外提供了如 getTaskExecution（获取任务执行信息）、getTaskExecutionCount（当前执行的任务数）、findTaskExecutionsByName（String taskName、Pageable pageable）（根据任务名和分页信息查

询任务执行情况)等方法。

TaskNameResolver 是任务名解析器接口,默认会构造 SimpleTaskNameResolver 这个实现类。SimpleTaskNameResolver 内部会使用 `spring.cloud.task.name` 配置项作为任务名。

如表 8-4 所示,这是 TaskProperties 支持任务相关的配置项。

表 8-4

配 置 项	作 用	默 认 值
spring.cloud.task.external-execution-id	外部自行设置的任务执行 ID(Spring Cloud Task 自身的任务执行 ID 是一个从 1 开始递增的 ID,Spring Cloud Data Flow 内部的批处理任务就使用了自行设置的任务执行 ID)	null
spring.cloud.task.executionid	任务执行 ID。若不设置,默认从 1 开始递增,设置后会读取 TASK_EXECUTION 表对应的数据,如不存在数据,则会报错	从 1 开始递增
spring.cloud.task.parent-execution-id	父任务执行 ID。用于父子关联任务的设置	null
spring.cloud.task.table-prefix	数据库表前缀	TASK_
spring.cloud.task.closecontext-enabled	任务完成后是否关闭 ApplicationContext	false
spring.cloud.task.single-instance-enabled	同一时间是否只允许相同 name 的任务一起执行(依赖 spring-integration 项目),锁存储在 TABLE_LOCK 表内	false
spring.cloud.task.single-instance-lock-ttl	同一时间相同 name 的任务上锁的 ttl 最大等待时间(需要打开 single-instance-enabled 配置)	Integer.MAX_VALUE
spring.cloud.task.single-instance-lock-check-interval	同一时间相同 name 的任务上锁后每次 sleep 的时间,sleep 后需要验证(需要打开 single-instance-enabled 配置)	Integer.500 ms

8.8.3 Spring Cloud Task Batch

Spring Cloud Task 可以与 Spring Batch 整合组成 Spring Cloud Task Batch。Spring Cloud

Task Batch 内部的 TaskBatchAutoConfiguration 自动化配置类会构造一个 TaskBatchExecution-ListenerBeanPostProcessor，这个 Processor 会为每一个 Spring Batch 里的 Job 注册一个 TaskBatchExecutionListener 监听器，该监听器内部会关联 TaskExecution 和 JobExecution，并把它们的 ID 存储数据到 TASKTASKBATCH 关联表中。TASKTASKBATCH 数据库表内容如表 8-5 所示。

表 8-5

TASKEXECUTIONID	JOBEXECUTIONID
1	1

Spring Cloud Task Batch 可以与 Spring Cloud Stream 整合，会在 Spring Batch 里注册 ChunkListener、ItemProcessListener、ItemReadListener、ItemWriteListener、JobExecutionListener、SkipListener、StepExecutionListener 类型的监听器，监听器内部把每个过程需要记录的数据通过 Spring Cloud Stream 发送到消息中间件中：

```java
public class EventEmittingJobExecutionListener implements JobExecutionListener, Ordered {

    private MessagePublisher<JobExecutionEvent> messagePublisher;

    private int order = Ordered.LOWEST_PRECEDENCE;

    public EventEmittingJobExecutionListener(MessageChannel output) {
        Assert.notNull(output, "An output channel is required");
        this.messagePublisher = new MessagePublisher<>(output);
    }

    public EventEmittingJobExecutionListener(MessageChannel output, int order) {
        this(output);
        this.order = order;
    }

    @Override
    public void beforeJob(JobExecution jobExecution) {
        this.messagePublisher.publish(new JobExecutionEvent(jobExecution));
```

```
    }

    @Override
    public void afterJob(JobExecution jobExecution) {
        this.messagePublisher.publish(new JobExecutionEvent(jobExecution));
    }

    @Override
    public int getOrder() {
        return this.order;
    }
}
```

EventEmittingJobExecutionListener 的构造在 BatchEventAutoConfiguration 自动化配置类中完成:

```
// BatchEventAutoConfiguration.java
@Bean
@Lazy
@ConditionalOnProperty(prefix = "spring.cloud.task.batch.events.job-execution",
        name = "enabled", havingValue = "true", matchIfMissing = true)
// @checkstyle:on
public JobExecutionListener jobExecutionEventsListener() {
    return new EventEmittingJobExecutionListener(
            this.listenerChannels.jobExecutionEvents(),
            this.taskEventProperties.getJobExecutionOrder());
}
```

这个 this.listenerChannels.jobExecutionEvents() 得到的是通过 Spring Cloud Stream 构造的 Output Binding:

```
// BatchEventAutoConfiguration$BatchEventsChannels.java
@Output(JOB_EXECUTION_EVENTS)
MessageChannel jobExecutionEvents();
```

这些通过 Spring Cloud Stream 记录数据的操作可以通过 spring.cloud.task.batch.events.

`<type>.enabled=false` 配置取消。

Spring Cloud Task Batch 与 Spring Cloud Stream 整合还提供了以下两个功能：

① 提供@EnableTaskLauncher 注解，通过 Spring Cloud Stream 接收消息（Binding name 为 input），消息的 Payload 类型为 TaskLaunchRequest。接收到消息之后会根据 TaskLaunchRequest 和 Spring Cloud Deployer 提供的 TaskLauncher 完成任务的运行。TaskLaunchRequest 属性如下：

- uri:String：要运行任务的 resource，会被构造成 Resource。
- applicationName:String：任务名。
- commandLineArguments:List<String>：任务需要的启动参数。
- environmentProperties:Map<String, String>：任务需要的环境变量。
- deploymentProperties:Map<String, String>：任务需要的部署参数。

```java
// TaskLauncherSink.java
@ServiceActivator(inputChannel = Sink.INPUT)
public void taskLauncherSink(TaskLaunchRequest taskLaunchRequest) throws Exception {
    launchTask(taskLaunchRequest);
}

private void launchTask(TaskLaunchRequest taskLaunchRequest) {
    Assert.notNull(this.taskLauncher, "TaskLauncher has not been initialized");
    logger.info("Launching Task for the following uri " + taskLaunchRequest.getUri());
    Resource resource = this.resourceLoader.getResource(taskLaunchRequest.getUri());
    AppDefinition definition = new AppDefinition(
            taskLaunchRequest.getApplicationName(),
            taskLaunchRequest.getEnvironmentProperties());
    AppDeploymentRequest request = new AppDeploymentRequest(definition, resource,
            taskLaunchRequest.getDeploymentProperties(),
            taskLaunchRequest.getCommandlineArguments());
    this.taskLauncher.launch(request);
}
```

② 使用 TaskExecutionListener 监听器监听任务执行的过程，将监听过程产生的一些数据发送到 MQ。

TaskExecutionListener 监听器在每个过程执行到的方法的返回值或参数被包装成消息，并通过 Spring Cloud Stream 发送到 Binding name 为 task-events 对应的 Topic 中：

```java
// TaskEventAutoConfiguration.java
public interface TaskEventChannels {

    String TASK_EVENTS = "task-events";

    @Output(TASK_EVENTS)
    MessageChannel taskEvents();

}

@Configuration
@EnableBinding(TaskEventChannels.class)
public static class ListenerConfiguration {

    @Bean
    public GatewayProxyFactoryBean taskEventListener() {
        GatewayProxyFactoryBean factoryBean = new GatewayProxyFactoryBean(
                TaskExecutionListener.class);

        factoryBean.setDefaultRequestChannelName(TaskEventChannels.TASK_EVENTS);

        return factoryBean;
    }

}
```

8.9 Spring Batch

Spring Batch 是 Spring 生态体系的一个批处理框架，该框架并不是一个调度框架（比如 Quartz），而是一个可以和调度框架一起工作的批处理框架。所有的批处理状态与 Spring Cloud Task 一样记录在关系型数据库中。

8.9.1 Spring Batch 核心组件

Spring Batch 的生效需要加上 `org.springframework.boot:spring-cloud-starter-batch` 依赖

和@EnableBatchProcessing 注解，它们会加载一些自动化配置类。比如，BatchAutoConfiguration 自动化配置类内部会初始化 BatchProperties 配置类，SimpleBatchConfiguration 配置类，以及 JobRepository、JobLauncher、JobRegistry、JobExplorer 等核心类也会被注入到应用 ApplicationContext 中。

Spring Batch 批处理框架不仅有批处理功能，还包括以下功能：

- 日志/链路追踪。
- 事务管理。
- Job 运行统计信息。
- 支持 Job 重跑（restart）、重试（retry）、跳过（skip）。
- 提供常用第三方组件的整合：Kafka、Avro、DataBase、File 等。
- 多种执行策略：正常的批处理过程、批次窗口内的批后处理、统一时间并行运行多个批次、分区（超多数量的 Job 数）。

Spring Batch 核心模块的架构如图 8-27 所示。下面是各个核心组件的介绍：

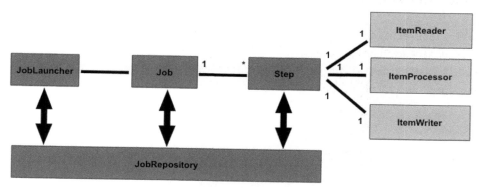

图 8-27

① Job：批次作业。Job 批次作业内部包含 3 个概念：JobInstance、JobParameters 和 JobExecution。比如，便利店有每天凌晨计算前一天总营业额账单的批次作业。这个账单批次作业就是 Job，每天执行的 Job 作业表示一个 JobInstance（一年就有 365 个 JobInstance）作业实例；每个 JobInstance 执行都需要一些参数（执行日期、销售商品的价格和数量），这些参数就是 JobParameters 作业参数；JobExecution 表示每个 JobInstance 的执行情况，包含作业的创建、开始、结束时间，以及退出状态、JobInstance id、JobParameters 等内容。

每个 Job 拥有一个 name（名称）、多个 Steps（执行步骤）以及是否支持重启（isRestartable）

这 3 个重要配置。

② Step：Job 作业内的一个执行步骤，一个 Job 可以有多个 Step，一个 Step 内部可以有多个 StepExecution（步骤执行过程）。

③ ExecutionContext：作业执行上下文，内存存储着一组 k/v 结构的数据，这些数据可以被 StepExecution、JobExecution、ItemReader 等组件所读取/修改。

④ JobRepository：提供上述组件的 CRUD 操作。比如，createJobInstance（创建作业实例）、createJobExecution（创建作业执行情况）、update(JobExecution jobExecution)（更新作业执行情况）、add(StepExecution stepExecution)（添加步骤执行情况）、updateExecutionContext（更新执行上下文）等方法。

⑤ JobLauncher：运行作业 Job 的组件，运行时可以加上 JobParameters 作业参数，运行后得到 JobExecution 作业执行情况。JobLauncher 接口定义如下：

```java
public interface JobLauncher {

    public JobExecution run(Job job, JobParameters jobParameters) throws JobExecutionAlreadyRunningException,
            JobRestartException, JobInstanceAlreadyCompleteException, JobParametersInvalidException;

}
```

⑥ ItemReader：每个步骤的输入数据抽象，当读取完所有的数据库后会返回 null。

```java
public interface ItemReader<T> {

    @Nullable
    T read() throws Exception, UnexpectedInputException, ParseException, NonTransientResourceException;

}
```

⑦ ItemWriter：每个步骤的输出抽象。

```
public interface ItemWriter<T> {

    void write(List<? extends T> items) throws Exception;

}
```

⑧ ItemProcessor：ItemReader 和 ItemWriter 之间的组件，用于数据的转换。

```
public interface ItemProcessor<I, O> {

    @Nullable
    O process(I item) throws Exception;

}
```

Spring Batch 针对上述核心组件都有对应的监听器：

⑨ JobExecutionListener：JobExecution 作业执行前后的回调接口。

⑩ StepExecutionListener：StepExecution 步骤执行前后的回调接口。

⑪ ChunkListener：每个提交批次的前后以及错误发生情况下的回调接口。

⑫ ItemReadListener：每个步骤读取数据前后以及读取错误情况下的回调接口。

⑬ ItemProcessListener：每个步骤转换数据前后以及转换错误情况下的回调接口。

⑭ ItemWriteListener：每个步骤写入数据前后以及写入错误情况下的回调接口。

⑮ SkipListener：过滤每个步骤读取数据、转换数据和写入数据情况下的回调接口。

8.9.2　案例：使用 Spring Batch 完成便利店每日账单统计

理解了前面介绍的核心组件后，下面介绍使用 Spring Batch 完成便利店每日总营业额账单数据的生成任务。

ItemReader、ItemProcessor 和 ItemWriter 的构造代码如下：

```
@Bean
public JsonItemReader<Goods> jsonItemReader() {
    ObjectMapper objectMapper = new ObjectMapper();
```

```java
        JacksonJsonObjectReader<Goods> jsonObjectReader =
            new JacksonJsonObjectReader<>(Goods.class);
        jsonObjectReader.setMapper(objectMapper);

        return new JsonItemReaderBuilder<Goods>()
            .jsonObjectReader(jsonObjectReader)
            .resource(goodsResource)
            .name("GoodsJsonItemReader")
            .build();
    }

    @Bean
    ItemProcessor itemProcessor() {
        return new BillItemProcessor();
    }

    @Bean
    public ItemWriter<Bill> jdbcBillWriter(DataSource dataSource) {
        JdbcBatchItemWriter<Bill> writer = new JdbcBatchItemWriterBuilder<Bill>()
            .beanMapped()
            .dataSource(dataSource)
            .sql("INSERT INTO BILLS (name, amount) VALUES " +
                "(:name, :amount)")
            .build();
        return writer;
    }
```

BillItemProcessor 这个 ItemProcessor 会根据 ItemReader 读取的 Goods 对象转换成 Bill 对象：

```java
class BillItemProcessor implements ItemProcessor<Goods, Bill> {

    @Override
    public Bill process(Goods item) throws Exception {
        return new Bill(item.getName(), item.getCount() * item.getPrice());
    }
}
```

最后构造 Job 作业，代码如下：

```
@Bean
public Job bullJob(ItemReader<Goods> reader,
    ItemProcessor<Goods, Bill> itemProcessor, ItemWriter<Bill> writer) {
    Step step = stepBuilderFactory.get("BillProcessing")
        .<Goods, Bill>chunk(2)
        .reader(reader)
        .processor(itemProcessor)
        .writer(writer)
        .build();

    return jobBuilderFactory.get("BillJob")
        .incrementer(new RunIdIncrementer())
        .start(step)
        .build();
}
```

在 Spring Cloud Data Flow 上使用 Spring Batch 创建批处理任务与 Spring Cloud Task 唯一的区别是：Spring Batch 执行的结果会展示在 Jobs 菜单中，如图 8-28 所示。

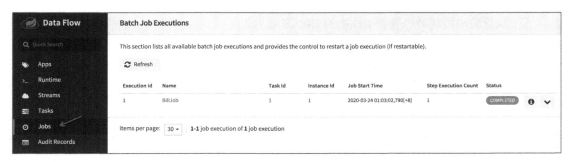

图 8-28

Jobs 作业详情页会显示读取的数据量、写入的数据量以及提交的次数，如图 8-29 所示。

图 8-29

从图 8-29 可知，Reads 表示一共读取 5 条数据，Writes 表示一共写入 5 条数据，Commits 表示一共提交 3 次（chunk 设置为 2，每 2 条提交一次，一共 5 条，提交 3 次），Rollbacks 表示回滚的次数，批处理执行成功的，就没有回滚。

Job 执行完毕后就可以查看 DB 中的 BILLS 数据。

在这个简单的批任务背后，Spring Batch 产生了如下 DB 数据。

① BATCHJOBINSTANCE 数据库表如表 8-6 所示。

表 8-6

JOBINSTANCEID	VERSION	JOB_NAME	JOB_KEY
1	0	BillJob	853d3449e311f40366811cbefb3d93d7

② BATCHJOBEXECUTION 数据库表如表 8-7 所示。

表 8-7

JOBEXECUTIONID	1	STATUS	COMPLETED
VERSION	1	EXIT_CODE	COMPLETED
JOBINSTANCEID	1	EXIT_MESSAGE	

续表

CREATE_TIME	2020-03-24 01:13:00.806	LAST_UPDATED	2020-03-24 01:13:00.951
START_TIME	2020-03-24 01:13:00.828	JOBCONFIGURATIONLOCATION	null
END_TIME	2020-03-24 01:13:00.951		

③ BATCHJOBEXECUTION_PARAMS 数据库表如表 8-8 所示。

表 8-8

JOBEXECUTIONID	TYPE_CD	KEY_NAME	STRING_VAL	DATE_VAL	LONG_VAL	DOUBLE_VAL	IDENTIFYING
1	STRING	-spring.cloud.data.flow.platformname	default	1970-01-01 07:30:00	0	0.0	N
1	STRING	-spring.cloud.task.executionid	1	1970-01-01 07:30:00	0	0.0	N
1	LONG	run.id		1970-01-01 07:30:00	1	0.0	Y

④ BATCHSTEPEXECUTION 数据库表如表 8-9 所示。

表 8-9

STEPEXECUTIONID	1	STATUS	COMPLETED	WRITESKIPCOUNT	0
VERSION	5	COMMIT_COUNT	3	PROCESSSKIPCOUNT	0
STEP_NAME	BillProcessing	READ_COUNT	5	ROLLBACK_COUNT	0
JOBEXECUTIONID	1	FILTER_COUNT	0	EXIT_CODE	COMPLETED
START_TIME	2020-03-24 01:13:00.846	WRITE_COUNT	5	EXIT_MESSAGE	
END_TIME	2020-03-24 01:13:00.947	READSKIPCOUNT	0	LAST_UPDATED	2020-03-24 01:13:00.947

第 9 章

网关

本章将介绍 Spring Cloud 体系内的网关 Spring Cloud Gateway 和 Netflix Zuul。网关（Gateway）是微服务体系中非常重要的一个模块，它提供了鉴权、路由、限流等功能。

由于 Netflix Zuul 已经不再是 Spring Cloud 官方推荐的网关，所以本章将对 Spring Cloud Gateway 的架构以及内部的 Route（路由）、Predicate（断言）、Filter（拦截器）等内容进行详细讲解。

最后使用 Spring Cloud Gateway 路由转发案例体验 Spring Cloud Gateway 的作用。

9.1 API 网关概述

网关在微服务体系中属于比较重要的一个模块。如果没有网关存在，单体应用改造成微服务后服务调用的链路就会像图 9-1 所示的这样。

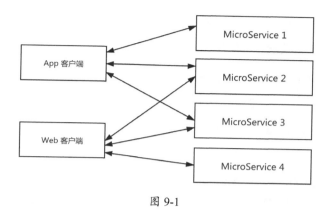

图 9-1

图 9-1 的模型存在以下问题：

- 如果添加鉴权功能，需要对每一个服务进行改造。
- 跨域问题需要对每一个服务进行改造。
- 流量控制需要对每一个服务进行改造。
- 灰度发布、动态路由需要对每一个服务进行改造。
- 收集用户日志访问记录需要聚合各个微服务里的数据。
- 存在安全问题。每个微服务暴露的 Endpoint 是固定的，客户端访问需要清楚各个微服务真实的 Endpoint。

上述问题的核心其实就是所有的微服务都是一个入口应用，这时引入网关来解决这个问题，如图 9-2 所示。

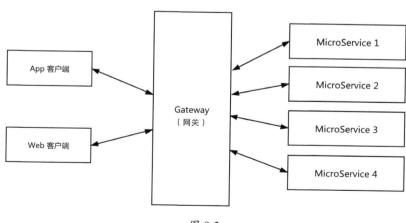

图 9-2

有了网关之后，客户端所有的流量全部先经过网关，网关再转发到对应的微服务，然后微服务返回结果给网关，网关最终将结果展示给客户。

目前业界比较有名的网关有：Nginx、Apache、Kong、Netflix Zuul、Alibaba Tengine 和 Spring Cloud Gateway。其中，Spring Cloud Gateway 是本章要讨论的网关。

由于 Spring Cloud Gateway 是 Spring 生态体系的一个子项目，它必然会整合 Spring 生态体系的其他项目：

- 整合服务注册/发现。在 Spring Cloud Gateway 中转发请求到注册中心提供的服务。
- 整合分布式配置。在 Spring Cloud Gateway 中可以依赖分布式配置动态修改一些路由规则。
- 使用 Circuit Breaker 提供容错机制。

9.2 Netflix Zuul

Spring Cloud Gateway 并不是 Spring Cloud 一开始推荐的网关实现，Netflix Zuul 1.x 才是一开始被 Spring Cloud 推荐的网关组件。之后 Netflix 开发了 Zuul 2.x。Spring Cloud Netflix GitHub 项目上有人提出是否有计划集成 Zuul 2.x，结果事与愿违，Zuul 2.x 在开发时延期比较严重。Spring Cloud 最终可能由于某种原因导致自己开发了 Spring Cloud Gateway 网关组件，用于代替 Zuul 网关。

虽然官方并没有声明已经停止维护 Netflix Zuul 1.x，但是 Spring Cloud 已经不再对其进行集成。目前 Zuul 对应的 Spring Cloud 集成在 Spring Cloud Netflix GitHub 项目上也已经不复存在。

如图 9-3 所示，Netflix Zuul 1.x 采用 Java Servlet 技术，在 ZuulServlet 技术上封装了各个 route 过程（运行不同类型的 ZuulFilter）。

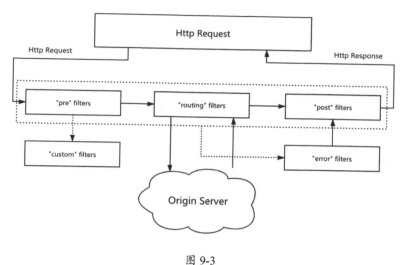

图 9-3

ZuulServlet 对请求的处理代码如下：

```
@Override
public void service(javax.servlet.ServletRequest servletRequest,
javax.servlet.ServletResponse servletResponse) throws ServletException, IOException
{
    try {
        init((HttpServletRequest) servletRequest, (HttpServletResponse) servletResponse);

        RequestContext context = RequestContext.getCurrentContext(); //①
        context.setZuulEngineRan();

        try {
            preRoute();      // ②
        } catch (ZuulException e) {
            error(e);        // ③
            postRoute();     // ④
            return;
        }
        try {
```

```
            route();        // ⑤
        } catch (ZuulException e) {
            error(e);       // ⑥
            postRoute();    // ⑦
            return;
        }
        try {
            postRoute();    // ⑧
        } catch (ZuulException e) {
            error(e);       // ⑨
            return;
        }

    } catch (Throwable e) {
        error(new ZuulException(e, 500, "UNHANDLED_EXCEPTION_" +
e.getClass().getName()));    // ⑩
    } finally {
        RequestContext.getCurrentContext().unset();    // ⑪
    }
}
```

上述代码中：

① 通过 ThreadLocal 保存 RequestContext，RequestContext 用于保存/透传每次请求需要的内容和发生异常后的异常信息。

② 进行 preRoute 操作，这个过程会运行 Zuul 里定义的所有 pre 类型的 ZuulFilter。

③ preRoute 过程发生错误后，运行 Zuul 里定义的所有 error 类型的 ZuulFilter。

④ preRoute 过程发生错误后依旧执行 postRoute 操作，这个过程会运行 Zuul 里定义的所有 post 类型的 ZuulFilter。

⑤ 若 preRoute 执行没问题，就开始执行 route 操作，这个过程会运行 Zuul 里定义的所有 route 类型的 ZuulFilter。

⑥ route 过程发生错误后，运行 Zuul 里定义的所有 error 类型的 ZuulFilter。

⑦ route 过程发生错误后，依旧运行 Zuul 里定义的所有 post 类型的 ZuulFilter。

⑧ 若 preRoute 执行没问题，就开始执行 post 操作，这个过程会运行 Zuul 里定义的所有 post 类型的 ZuulFilter。

⑨ postRoute 过程发生错误后，运行 Zuul 里定义的所有 error 类型的 ZuulFilter。

⑩ 若整个过程发生了没有被捕获住的异常，就运行 Zuul 里定义的所有 error 类型的 ZuulFilter。

⑪ ThreadLocal 还原，避免内存泄漏。

Spring Cloud Gateway 项目完全采用响应式（reactive）架构，依赖 Reactor 和 Spring WebFlux 项目。Spring WebFlux 是 Spring Framework 5.x（Spring Boot 2.x）新推出的基于 Reactive 实现的响应式 Web 服务器端框架。Spring Cloud Finchley 版本是第一个支持 Spring Boot 2.x 的 Spring Cloud 版本（这是 Spring Cloud Gateway 最低的 Spring Cloud 版本要求）。

9.3 非阻塞式的 Spring Cloud Gateway

Spring Cloud Gateway 基于 reactive 响应式模型如图 9-4 所示，这是一个非阻塞式的网关实现。

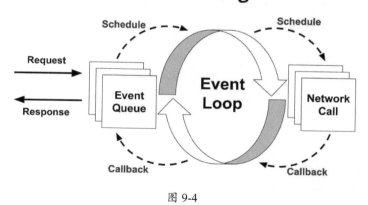

图 9-4

Spring Cloud Gateway 定义了 RoutePredicateHandlerMapping 这个 HandlerMapping 接口实现类（该接口是 WebFlux 定义的一个处理流量请求的处理器），用于处理请求的映射关系，得到的结果被 HandlerAdapter 处理。下面这段代码是 RoutePredicateHandlerMapping 对请求的处理过程：

```java
// RoutePredicateHandlerMapping.java
@Override
protected Mono<?> getHandlerInternal(ServerWebExchange exchange) {
    if (this.managementPortType == DIFFERENT && this.managementPort != null
            && exchange.getRequest().getURI().getPort() == this.managementPort) {    // ①
        return Mono.empty();
    }
    exchange.getAttributes().put(GATEWAY_HANDLER_MAPPER_ATTR, getSimpleName());

    return lookupRoute(exchange)    // ②
            .flatMap((Function<Route, Mono<?>>) r -> {    // ③
                exchange.getAttributes().remove(GATEWAY_PREDICATE_ROUTE_ATTR);
                if (logger.isDebugEnabled()) {
                    logger.debug(
                            "Mapping [" + getExchangeDesc(exchange) + "] to " + r);
                }

                exchange.getAttributes().put(GATEWAY_ROUTE_ATTR, r);
                return Mono.just(webHandler);    // ④
            }).switchIfEmpty(Mono.empty().then(Mono.fromRunnable(() -> {
                exchange.getAttributes().remove(GATEWAY_PREDICATE_ROUTE_ATTR);
                if (logger.isTraceEnabled()) {
                    logger.trace("No RouteDefinition found for ["
                            + getExchangeDesc(exchange) + "]");
                }
            })));
}

protected Mono<Route> lookupRoute(ServerWebExchange exchange) {
    return this.routeLocator.getRoutes()    // ⑤
            .concatMap(route -> Mono.just(route).filterWhen(r -> {
                exchange.getAttributes().put(GATEWAY_PREDICATE_ROUTE_ATTR, r.getId());
                return r.getPredicate().apply(exchange);    // ⑥
            })
            .doOnError(e -> logger.error(
                    "Error applying predicate for route: " + route.getId(),
```

```
                e))
            .onErrorResume(e -> Mono.empty()))
            .next()
            .map(route -> {
                if (logger.isDebugEnabled()) {
                    logger.debug("Route matched: " + route.getId());
                }
                validateRoute(route, exchange);    // ⑦
                return route;
            });
}
```

上述代码中：

① 应用开启了 management 功能（Spring Boot Actuator），若端口与应用不一致，则不对请求进行处理。

② 根据请求信息，通过 lookup 方法找出合适的 Route 信息。

③ 通过 flatMap 将 Route 信息存储在请求属性中，为后续 FilteringWebHandler 的执行准备数据。

④ HandlerMapping 返回的结果是一个 FilteringWebHandler。这个 WebHandler 会在 GatewayAutoConfiguration 自动化配置类中被构造。

⑤ lookupRoute 方法寻找合适的路由信息，首先根据 RouteLocator 找出所有定义的路由信息。

⑥ 通过 Route 里的 Predicate 和请求信息判断路由是否满足条件。

⑦ 验证 Route 信息是否合法。

RoutePredicateHandlerMapping 返回的 WebHandler 是 FilteringWebHandler，它会被 HandlerAdapter 接口处理。Spring Cloud Gateway 会被 SimpleHandlerAdapter 这个 HandlerAdapter 实现类（如果是 WebFlux，则会被 RequestMappingHandlerAdapter 处理）处理。

下面是 SimpleHandlerAdapter 处理 FilteringWebHandler 的过程，内部直接使用 FilteringWebHandler 对请求进行处理：

```java
public class SimpleHandlerAdapter implements HandlerAdapter {

    @Override
    public boolean supports(Object handler) {
        return WebHandler.class.isAssignableFrom(handler.getClass());
    }

    @Override
    public Mono<HandlerResult> handle(ServerWebExchange exchange, Object handler) {
        WebHandler webHandler = (WebHandler) handler;
        Mono<Void> mono = webHandler.handle(exchange);
        return mono.then(Mono.empty());
    }
}
```

下面是 FilteringWebHandler 处理请求的过程：

```java
// FilteringWebHandler.java
@Override
public Mono<Void> handle(ServerWebExchange exchange) {
    Route route = exchange.getRequiredAttribute(GATEWAY_ROUTE_ATTR);
    List<GatewayFilter> gatewayFilters = route.getFilters();

    List<GatewayFilter> combined = new ArrayList<>(this.globalFilters);
    combined.addAll(gatewayFilters);
    AnnotationAwareOrderComparator.sort(combined);

    if (logger.isDebugEnabled()) {
        logger.debug("Sorted gatewayFilterFactories: " + combined);
    }

    return new DefaultGatewayFilterChain(combined).filter(exchange);
}
```

DefaultGatewayFilterChain 是一个过滤器链，其内部存储着各种过滤器（Filter），过滤器链处理请求的时候会遍历这些过滤器并得到最终结果。这里的 globalFilters 属性是构造

FilteringWebHandler 的过程中注入的 GlobalFilter 拦截器集合。

如果在 Route 里定义了路由对应的 GatewayFilter，也会加入到这个整合的 Filter 集合内。

DefaultGatewayFilterChain 处理过程如下：

```java
// FilteringWebHandler$DefaultGatewayFilterChain.java
@Override
public Mono<Void> filter(ServerWebExchange exchange) {
    return Mono.defer(() -> {
        if (this.index < filters.size()) {
            GatewayFilter filter = filters.get(this.index);
            DefaultGatewayFilterChain chain = new DefaultGatewayFilterChain(this,
                    this.index + 1);
            return filter.filter(exchange, chain);
        }
        else {
            return Mono.empty(); // complete
        }
    });
}
```

Spring Cloud Gateway 整个请求处理的流程如图 9-5 所示。

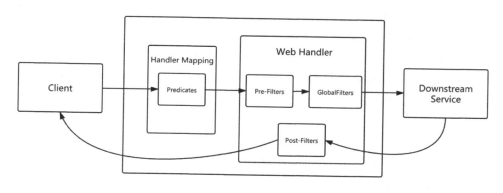

图 9-5

图 9-5 中的 Pre-Filters 和 Post-Filters 是逻辑上的概念。Spring Cloud Gateway 内的 GatewayFilter 并没有具体的属性或方法表示这是一个 Pre 或 Post 类型的拦截器，但是从作用上可以区分为

Pre-Filters 或 Post-Filters。

总体来说，如果读者对 Spring Web 或 Spring WebFlux 项目比较熟悉，理解 Spring Cloud Gateway 的架构还是非常容易的。

① 请求进来后，被 RoutePredicateHandlerMapping 这个 HandlerMapping 处理。

② 在 HandlerMapping 处理的过程中，会通过所有的 Route 信息列表中每个路由信息里的 Predicate 找出合适的 Route（路由）。

③ RoutePredicateHandlerMapping 处理后得到的结果是一个 FilteringWebHandler。

④ FilteringWebHandler 使用内部的 GlobalFilter 对请求进行处理。如果路由信息也配置了 GatewayFilter，这些 GatewayFilter 会和 GlobalFilter 一起被执行（GlobalFilter 对全局生效，GatewayFilter 对部分（需要配置）生效）。

⑤ 一些路由 Filter 会调用下游的服务，最终将结果返回给客户端。

9.4　Route 路由信息

Spring Cloud Gateway 在 RoutePredicateHandlerMapping 映射请求的过程中，首先找出所有的路由信息，这个路由信息对应的是 Route 类。Route 类存在以下 6 个属性：

- id:String：路由的 ID。
- uri:URI：路由的 URI 信息。每个 URI 有对应的 protocol 协议和 path（路径）。Spring Cloud Gateway 会根据 URI 里的属性完成不同的事情。比如，lb 协议的 URI 会根据 ReactiveLoadBalancerClientFilter 从注册中心获取实例；又如，HTTP 或 HTTPS 协议的 URI 会根据 WebClientHttpRoutingFilter 通过 WebClient 发起 HTTP 请求；再如，WS 或 WSS 协议的 URI 会根据 WebsocketRoutingFilter 通过 WebSocketClient 发起 websocket 请求。
- order:int：路由的优先级。Route 实现了 Spring 的 Ordered 接口，该接口对应的 getOrder 方法的返回值就是该属性。
- predicate:AsyncPredicate<ServerWebExchange>：AsyncPredicate 是支持 Reactive 的 Predicate 实现，内部的 Predicate 通过 RoutePredicateFactory 构造。比如，PathRoutePredicateFactory 会通过 pattern 解析器解析路径是否匹配。

- gatewayFilters:List<GatewayFilter>：GatewayFilter 集合。这里是 GatewayFilter，并不是 GlobalFilter。Spring Cloud Gateway 默认提供了一些 GatewayFilter，比如 StripPrefix-GatewayFilterFactory 创建过滤前缀信息、SetResponseHeaderGatewayFilterFactory 扩展 Response Header 内容、SetRequestHeaderGatewayFilterFactory 扩展 Request Header 内容等信息。这些 GatewayFilter 都可以被添加到路由里。
- metadata:Map<String, Object>：路由信息的元数据信息。

predicate 和 gatewayFilters 属性的构造是 Spring Cloud Gateway 的核心。

9.5 Predicate 机制

Predicate 是 Spring Cloud Gateway 的核心组件之一，其作用是在请求映射的过程中匹配用户的请求。比如，PathRoutePredicateFactory 构造的 Predicate 会用于路径的匹配，HeaderRoutePredicateFactory 构造的 Predicate 用于 Header 的匹配。

9.5.1 PredicateDefinition 和 AsyncPredicate

Spring Cloud Gateway 路由配置对应的 predicate 信息通过对象 PredicateDefinition 构造。

PredicateDefinition 定义如下：

```
@Validated
public class PredicateDefinition {

    @NotNull
    private String name;

    private Map<String, String> args = new LinkedHashMap<>();

    public PredicateDefinition() {
    }

    public PredicateDefinition(String text) {
        int eqIdx = text.indexOf('=');
        if (eqIdx <= 0) {
```

```
            throw new ValidationException("Unable to parse PredicateDefinition text '"
                    + text + "'" + ", must be of the form name=value");
        }
        setName(text.substring(0, eqIdx));

        String[] args = tokenizeToStringArray(text.substring(eqIdx + 1), ",");

        for (int i = 0; i < args.length; i++) {
            this.args.put(NameUtils.generateName(i), args[i]);
        }
    }
    ...
}
```

从这个带有 String 参数的构造函数可知，PredicateDefinition 对应的配置信息 Path=/httpbin/** 其实就是根据这个构造函数完成构造的。

那么 PredicateDefinition 最后会转换成什么呢？它们会被转换成 AsyncPredicate（支持 Reactive 的 Predicate），这个动作在 RouteDefinitionRouteLocator 中通过 RoutePredicateFactory 完成。

RouteDefinitionRouteLocator 在构造时会使用通过 Spring ApplicationContext 注入的 predicates 集合来给它的内部属性 predicates 赋值，这个 predicates 属性是 Map<String, RoutePredicateFactory> 类型的，对应的 key 是 RoutePredicateFactory 的 name 方法返回值。name 方法默认的实现是返回类名的缩写，比如 PathRoutePredicateFactory 对应的 name 方法返回值为 Path；HostRoutePredicateFactory 对应的 name 方法返回值为 Host。

9.5.2　RoutePredicateFactory

每个 RoutePredicateFactory 接口的实现有对应的配置 Config，表示这个 Predicate 需要的配置信息，每个 Predicate 的配置都有自己单独的属性。比如 PathRoutePredicateFactory 对应的配置 Config 维护一个 String 集合的 patterns，用于路径匹配；HostRoutePredicateFactory 对应的配置 Config 维护一个 String 集合的 patterns，用于 Host 匹配。

下列代码是 RouteDefinitionRouteLocator 构造函数，内部调用 initFactories 方法初始化 predicates 属性：

```java
public RouteDefinitionRouteLocator(RouteDefinitionLocator routeDefinitionLocator,
        List<RoutePredicateFactory> predicates,
        List<GatewayFilterFactory> gatewayFilterFactories,
        GatewayProperties gatewayProperties,
        ConfigurationService configurationService) {
    this.routeDefinitionLocator = routeDefinitionLocator;
    this.configurationService = configurationService;
    initFactories(predicates);
    gatewayFilterFactories.forEach(
            factory -> this.gatewayFilterFactories.put(factory.name(), factory));
    this.gatewayProperties = gatewayProperties;
}

private void initFactories(List<RoutePredicateFactory> predicates) {
    predicates.forEach(factory -> {
        String key = factory.name();
        if (this.predicates.containsKey(key)) {
            this.logger.warn("A RoutePredicateFactory named " + key
                    + " already exists, class: " + this.predicates.get(key)
                    + ". It will be overwritten.");
        }
        this.predicates.put(key, factory);
        if (logger.isInfoEnabled()) {
            logger.info("Loaded RoutePredicateFactory [" + key + "]");
        }
    });
}
```

这些注入的 RoutePredicateFactory 集合在 GatewayAutoConfiguration 自动化配置类中完成构造。

RouteDefinitionRouteLocator 获取路由信息的时候会根据配置的 PredicateDefinition 在 predicates 属性内寻找对应的 RoutePredicateFactory：

```java
private AsyncPredicate<ServerWebExchange> combinePredicates(
        RouteDefinition routeDefinition) {
```

```java
    List<PredicateDefinition> predicates = routeDefinition.getPredicates();
    AsyncPredicate<ServerWebExchange> predicate = lookup(routeDefinition,
            predicates.get(0));

    for (PredicateDefinition andPredicate : predicates.subList(1,
            predicates.size())) {
        AsyncPredicate<ServerWebExchange> found = lookup(routeDefinition,
                andPredicate);
        predicate = predicate.and(found);
    }

    return predicate;
}
```

lookup 方法内部不仅会找到对应的 RoutePredicateFactory，还会根据 PredicateDefinition 的 args 属性构造对应的 Config 参数作为 RoutePredicateFactory 的 apply 方法里的参数：

```java
private AsyncPredicate<ServerWebExchange> lookup(RouteDefinition route,
        PredicateDefinition predicate) {
    RoutePredicateFactory<Object> factory = this.predicates.get(predicate.getName());
    if (factory == null) {
        throw new IllegalArgumentException(
                "Unable to find RoutePredicateFactory with name "
                        + predicate.getName());
    }
    if (logger.isDebugEnabled()) {
        logger.debug("RouteDefinition " + route.getId() + " applying "
                + predicate.getArgs() + " to " + predicate.getName());
    }

    Object config = this.configurationService.with(factory)
            .name(predicate.getName())
            .properties(predicate.getArgs())
            .eventFunction((bound, properties) -> new PredicateArgsEvent(
                    RouteDefinitionRouteLocator.this, route.getId(), properties))
```

```
            .bind();

    return factory.applyAsync(config);
}
```

比如，Path=/httpbin/**配置对应的 PredicateDefinition 类中的 name 属性为 Path，args 属性只有一个 k-v 项，key 为 genkey0，value 为/httpbin/。这个 PredicateDefinition 对应的 RoutePredicate-Factory 为 PathRoutePredicateFactory（PathRoutePredicateFactory 的 name 方法返回值为 Path），对应的 Config 参数类型为 PathRoutePredicateFactory.Config，且内部的 List 类型的 patterns 内部有一个/httpbin/\的值。

9.5.3 内置 RoutePredicateFactory

Spring Cloud Gateway 提供了一些内置的 RoutePredicateFactory，解释如下：

- PathRoutePredicateFactory：RouteDefinitionRouteLocator 中的 predicates 属性里对应的 key 为 Path，根据请求路径的 pattern 确认是否满足路由条件。对应的 Config 配置内部维护 List<String> 类型的 patterns，用于请求路径的判断。
- HostRoutePredicateFactory：predicates 属性里对应的 key 为 Host，根据 Request Header 里的 Host 属性确认是否满足路由条件。对应的 Config 配置内部维护 List<String> 类型的 patterns，用于 Host Header 的判断。
- MethodRoutePredicateFactory：predicates 属性里对应的 key 为 Method，根据 Request HTTP Method 属性确认是否满足路由条件。对应的 Config 配置内部维护 HttpMethod 类型的 method 属性，用于 Request HTTP Method 的判断。
- HeaderRoutePredicateFactory：predicates 属性里对应的 key 为 Header，根据 Request Header 属性是否满足正则表达式确认是否满足路由条件。对应的 Config 配置内部维护 header（header 的 key）和 regexp（正则表达式）属性，用于从 Request Header 中获取对应的信息，并判断是否满足正则表达式。
- QueryRoutePredicateFactory：predicates 属性里对应的 key 为 Query，根据 Request Query Parameter 参数是否满足正则表达式确认是否满足路由条件。对应的 Config 配置内部维护 param（Request Query Parameter 的 key）和 regexp（正则表达式）属性，用于从 Request Query Parameter 中获取对应的信息，并判断是否满足正则表达式。
- CookieRoutePredicateFactory：predicates 属性里对应的 key 为 Cookie，根据 Cookie 属性是否满足正则表达式确认是否满足路由条件。对应的 Config 配置内部维护 name

（cookie 的 key）和 regexp（正则表达式）属性，用于从 Cookie 中获取对应的信息，并判断是否满足正则表达式。

- AfterRoutePredicateFactory：predicates 属性里对应的 key 为 After，根据当前时间是否晚于配置的时间确认是否满足路由条件。对应的 Config 配置内部维护 datetime 属性用于与当前时间进行比较。
- BeforeRoutePredicateFactory：predicates 属性里对应的 key 为 Before，根据当前时间是否早于配置的时间确认是否满足路由条件。对应的 Config 配置内部维护 datetime 属性，用于与当前时间进行比较。
- BetweenRoutePredicateFactory：predicates 属性里对应的 key 为 Between，根据当前时间是否处于配置的时间区间确认是否满足路由条件。对应的 Config 配置内部维护 datetime1 和 datetime2 属性，用于判断当前时间是否处于这两个时间之内。
- RemoteAddrRoutePredicateFactory：predicates 属性里对应的 key 为 RemoteAddr，根据请求的 IP 是否处于配置的 IP 段内确认是否满足路由条件。对应的 Config 配置内部维护 RemoteAddressResolver 接口和 String 集合的 source 属性。RemoteAddressResolver 根据请求解析 IP，source 用于生产 Netty 的 IpSubnetFilterRule 规则进行 IP 段的判断。

如果这些 RoutePredicateFactory 不满足条件，我们也可以自定义 RoutePredicateFactory。

9.6 Filter 机制

Filter 是 Spring Cloud Gateway 的另一个核心组件，其作用是处理请求。

9.6.1 FilterDefinition 和 GatewayFilter

Spring Cloud Gateway 路由配置对应的 Filter 信息通过对象 FilterDefinition 构造。

FilterDefinition 定义如下：

```
@Validated
public class FilterDefinition {

    @NotNull
    private String name;
```

```java
    private Map<String, String> args = new LinkedHashMap<>();

    public FilterDefinition() {
    }

    public FilterDefinition(String text) {
        int eqIdx = text.indexOf('=');
        if (eqIdx <= 0) {
            setName(text);
            return;
        }
        setName(text.substring(0, eqIdx));

        String[] args = tokenizeToStringArray(text.substring(eqIdx + 1), ",");

        for (int i = 0; i < args.length; i++) {
            this.args.put(NameUtils.generateName(i), args[i]);
        }
    }
    …
}
```

FilterDefinition 的解析与 PredicateDefinition 基本一致。这个带有 String 参数的构造函数根据配置信息解析成对应的属性。FilterDefinition 最终会被转换成 GatewayFilter，这个动作在 RouteDefinitionRouteLocator 中通过 GatewayFilterFactory 完成。

这些操作也都是在 RouteDefinitionRouteLocator 内完成的，convertToRoute 方法内部不但会根据 PredicateDefinition 集合构造 AsyncPredicate，还会根据 FilterDefinition 集合构造 GatewayFilter，并把这些信息设置到 Route 属性内，代码如下：

```java
// RouteDefinitionRouteLocator.java
private Route convertToRoute(RouteDefinition routeDefinition) {
    AsyncPredicate<ServerWebExchange> predicate = combinePredicates
(routeDefinition);    // ①
```

```java
    List<GatewayFilter> gatewayFilters = getFilters(routeDefinition);     // ②

    return Route.async(routeDefinition).asyncPredicate(predicate)
            .replaceFilters(gatewayFilters).build();
}

private List<GatewayFilter> getFilters(RouteDefinition routeDefinition) {
    List<GatewayFilter> filters = new ArrayList<>();

    if (!this.gatewayProperties.getDefaultFilters().isEmpty()) {
        filters.addAll(loadGatewayFilters(DEFAULT_FILTERS,
                this.gatewayProperties.getDefaultFilters()));     // ③
    }

    if (!routeDefinition.getFilters().isEmpty()) {
        filters.addAll(loadGatewayFilters(routeDefinition.getId(),
                routeDefinition.getFilters()));     // ④
    }

    AnnotationAwareOrderComparator.sort(filters);
    return filters;
}
```

上述代码中：

① 基于定义的路由信息内部的 predicate 配置构造 AsyncPredicate。

② 基于定义的路由信息内部的 Filter 配置构造 GatewayFilter。

③ 如果存在 spring.cloud.gateway.defaultFilters 配置（默认的 GatewayFilter，所有的路由都会生效），则添加到拦截器集合内。

④ 如果配置的拦截器信息存在，则添加到拦截器集合内。

loadGatewayFilters 方法内部会根据 gatewayFilterFactories 属性和 FilterDefinition 的 name 属性加载 GatewayFilter。

这个 gatewayFilterFactories 属性与 predicates 属性的构造过程相同，在 RouteDefinition-RouteLocator 的构造方法内注入 Spring ApplicationContext 内的 GatewayFilterFactory 集合，用

于赋值这个属性。这些 Spring ApplicationContext 内的 GatewayFilterFactory 全部都在 GatewayAutoConfiguration 自动化配置类中初始化完成。

gatewayFilterFactories 的属性类型与 predicates 的属性类型相同，都是 Map 类型。Map 中对应的 value 是 GatewayFilterFactory，key 是 gatewayFilterFactories 的 name 方法返回值，name 方法默认的实现是返回类名的缩写，比如 RedirectToGatewayFilterFactory 对应的 name 方法返回值为 RedirectTo；StripPrefixGatewayFilterFactory 对应的 name 方法返回值为 StripPrefix。

loadGatewayFilters 根据配置的 FilterDefinition 里的内容基于 GatewayFilterFactory 构造 GatewayFilter 集合，代码如下：

```java
List<GatewayFilter> loadGatewayFilters(String id,
        List<FilterDefinition> filterDefinitions) {
    ArrayList<GatewayFilter> ordered = new ArrayList<>(filterDefinitions.size());
    for (int i = 0; i < filterDefinitions.size(); i++) {
        FilterDefinition definition = filterDefinitions.get(i);
        GatewayFilterFactory factory = this.gatewayFilterFactories
                .get(definition.getName());
        if (factory == null) {
            throw new IllegalArgumentException(
                    "Unable to find GatewayFilterFactory with name "
                            + definition.getName());
        }
        if (logger.isDebugEnabled()) {
            logger.debug("RouteDefinition " + id + " applying filter "
                    + definition.getArgs() + " to " + definition.getName());
        }

        Object configuration = this.configurationService.with(factory)
                .name(definition.getName())
                .properties(definition.getArgs())
                .eventFunction((bound, properties) -> new FilterArgsEvent(
                        RouteDefinitionRouteLocator.this, id, (Map<String, Object>) properties))
                .bind();
```

```
        if (configuration instanceof HasRouteId) {
            HasRouteId hasRouteId = (HasRouteId) configuration;
            hasRouteId.setRouteId(id);
        }

        GatewayFilter gatewayFilter = factory.apply(configuration);
        if (gatewayFilter instanceof Ordered) {
            ordered.add(gatewayFilter);
        }
        else {
            ordered.add(new OrderedGatewayFilter(gatewayFilter, i + 1));
        }
    }

    return ordered;
}
```

每个 GatewayFilterFactory 接口的实现也有对应的配置 Config，表示这个 GatewayFilter 需要的配置信息，每个 GatewayFilter 的配置都有自己单独的属性。比如 StripPrefixGatewayFilterFactory 对应的配置 Config 维护一个 int 类型的 parts，用于请求路径的跳过；RemoveRequestHeaderGatewayFilterFactory 对应的配置 Config 维护一个 String 类型的 name，用于删除 Request Header 中对应的 key。

9.6.2 GlobalFilter

Route 配置内部保存的 Filter 类型是 GatewayFilter。Spring Cloud Gateway 还提供了另外一种 GlobalFilter 的拦截器类型。

GlobalFilter 类型的拦截器会让所有的路由都生效，GatewayFilter 类型的拦截器只会让配置的路由生效。

GlobalFilter 拦截器让所有的路由都生效的原因是：在 FilteringWebHandler 对请求做处理的时候会自动加上所有的 GlobalFilter 拦截器。这些 GlobalFilter 拦截器全部通过 Spring ApplicationContext 注入，它们都是在 GatewayAutoConfiguration 自动化配置内初始化完成的。

FilteringWebHandler 内自动添加所有的 GlobalFilter，代码如下：

```java
// FilteringWebHandler.java
@Override
public Mono<Void> handle(ServerWebExchange exchange) {
    Route route = exchange.getRequiredAttribute(GATEWAY_ROUTE_ATTR);
    List<GatewayFilter> gatewayFilters = route.getFilters();

    List<GatewayFilter> combined = new ArrayList<>(this.globalFilters);
    combined.addAll(gatewayFilters);
    AnnotationAwareOrderComparator.sort(combined);

    if (logger.isDebugEnabled()) {
        logger.debug("Sorted gatewayFilterFactories: " + combined);
    }

    return new DefaultGatewayFilterChain(combined).filter(exchange);
}
```

9.6.3 内置 GatewayFilterFactory

Spring Cloud Gateway 还提供了一些内置的 GatewayFilterFactory，解释如下：

- AddRequestHeaderGatewayFilterFactory：RouteDefinitionRouteLocator 中的 gatewayFilterFactories 属性对应的 key 为 AddRequestHeader，作用是为 Request 请求添加新的 Header 内容。对应的 Config 配置内部维护 name 和 value 属性，用于表示 Request Header 中要新添加的 key 和 value。
- MapRequestHeaderGatewayFilterFactory：RouteDefinitionRouteLocator 中的 gatewayFilterFactories 属性对应的 key 为 MapRequestHeader，作用是将 Request Header 中对应的 key 值转移到另外一个 key 上。对应的 Config 配置内部维护 fromHeader 和 toHeader 属性，用于将 Header 中 fromHeader 对应的值转移到 toHeader 这个 key 上。
- AddRequestParameterGatewayFilterFactory：RouteDefinitionRouteLocator 中的 gatewayFilterFactories 属性对应的 key 为 AddRequestParameter，作用是为 Request 请求添加新的 Query Parameter 参数和内容。对应的 Config 配置内部维护 name 和 value 属性，用于表示 Query Parameter 中要新添加的 key 和 value。
- AddResponseHeaderGatewayFilterFactory：RouteDefinitionRouteLocator 中的 gatewayFilterFactories 属性对应的 key 为 AddResponseHeader，作用是为 Response 请求添加新的

Header 内容。对应的 Config 配置内部维护 name 和 value 属性，用于表示 Response Header 中要新添加的 key 和 value。

- ModifyRequestBodyGatewayFilterFactory：RouteDefinitionRouteLocator 中的 gateway-FilterFactories 属性对应的 key 为 ModifyRequestBody，作用是修改 Request Body 内容。对应的 Config 配置内部维护 inClass、outClass 和 RewriteFunction 等属性，用于将原先 Request Body 中对应的内容转换成另外的内容。

- DedupeResponseHeaderGatewayFilterFactory：RouteDefinitionRouteLocator 中的 gateway-FilterFactories 属性对应的 key 为 DedupeResponseHeader，作用是修改 Response Header 内容，对应的 Config 配置内部维护 strategy 和 name 属性。strategy 表示保留策略，分为 3 种：只保留第一个、只保留最后一个、删除重复值，对应的值分别为：RETAIN_FIRST、RETAIN_LAST 和 RETAIN_UNIQUE。name 则是要修改的 HEADER key 值。

- ModifyResponseBodyGatewayFilterFactory：RouteDefinitionRouteLocator 中的 gateway-FilterFactories 属性对应的 key 为 ModifyResponseBody，作用是修改 Response Body 内容，对应的 Config 配置内部维护 inClass、outClass 和 RewriteFunction 等属性，用于将原先 Response Body 中对应的内容转换成另外的内容。

- PrefixPathGatewayFilterFactory：RouteDefinitionRouteLocator 中的 gatewayFilterFactories 属性对应的 key 为 PrefixPath，作用是为 Request Path 添加前缀，对应的 Config 配置内部维护 prefix 属性，表示要添加的前缀。

- PreserveHostHeaderGatewayFilterFactory：RouteDefinitionRouteLocator 中的 gateway-FilterFactories 属性对应的 key 为 PreserveHostHeader，作用是为 Request 保留 Header 中 key 为 Host 的内容。没有 Config 配置信息，如果添加了 Filter，则会保存 Host Header，否则会删除。

- RedirectToGatewayFilterFactory：RouteDefinitionRouteLocator 中的 gatewayFilterFactories 属性对应的 key 为 RedirectTo，作用是对请求进行重定向处理，对应的 Config 配置内部维护 status 和 url 属性，其中，status 表示 Response Code，必须为 3xx 状态码，url 表示重定向的地址。

- RemoveRequestHeaderGatewayFilterFactory：RouteDefinitionRouteLocator 中的 gateway-FilterFactories 属性对应的 key 为 RemoveRequestHeader，作用是为 Request 请求删除 Header 内容，对应的 Config 配置内部维护 name 属性，用于表示 Request Header 中要新删除的 key。

- RemoveRequestParameterGatewayFilterFactory：RouteDefinitionRouteLocator 中的 gateway-

FilterFactories 属性对应的 key 为 RemoveRequestParameter，作用是为 Request 请求删除 Query Parameter 内容，对应的 Config 配置内部维护 name 属性，用于表示 Query Parameter 中要删除的 key。

- RemoveResponseHeaderGatewayFilterFactory：RouteDefinitionRouteLocator 中的 gateway-FilterFactories 属性对应的 key 为 RemoveResponseHeader，作用是为 Response 删除 Header 内容，对应的 Config 配置内部维护 name 属性，用于表示 Response Header 中要删除的 key。

- RewritePathGatewayFilterFactory：RouteDefinitionRouteLocator 中的 gatewayFilter-Factories 属性对应的 key 为 RewritePath，作用是基于正则表达式修改 Request 的请求 URL，对应的 Config 配置内部维护 regexp 和 replacement 属性，regexp 表示匹配的正则表达式，replacement 表示替换内容。

- RetryGatewayFilterFactory：RouteDefinitionRouteLocator 中的 gatewayFilterFactories 属性对应的 key 为 Retry，作用是基于配置信息判断是否需要重试，对应的 Config 配置包括 Response Code 的类型（1xx、2xx、3xx、4xx 或 5xx）、具体的 Code 状态码和 HTTP Method 等信息。

- SetPathGatewayFilterFactory：RouteDefinitionRouteLocator 中的 gatewayFilterFactories 属性对应的 key 为 SetPath，作用是基于 UriTemplate 覆盖请求路径，对应的 Config 配置包括 template 信息、根据请求路径的参数信息和 template 信息覆盖原先的请求路径。

- SecureHeadersGatewayFilterFactory：RouteDefinitionRouteLocator 中的 gatewayFilter-Factories 属性对应的 key 为 SecureHeaders，作用是为请求信息添加一些安全相关的 header。没有 Config 配置信息。

- SetRequestHeaderGatewayFilterFactory：RouteDefinitionRouteLocator 中的 gateway-FilterFactories 属性里对应的 key 为 SetRequestHeader，作用是为 Request 请求修改 Header 内容，对应的 Config 配置内部维护 name 和 value 属性，用于表示 Request Header 中要修改的 key 和 value。

- SetResponseHeaderGatewayFilterFactory：RouteDefinitionRouteLocator 中的 gateway-FilterFactories 属性对应的 key 为 SetResponseHeader，作用是为 Response 请求修改 Header 内容，对应的 Config 配置内部维护 name 和 value 属性，用于表示 Response Header 中要修改的 key 和 value。

- RewriteResponseHeaderGatewayFilterFactory：RouteDefinitionRouteLocator 中的 gateway-FilterFactories 属性对应的 key 为 RewriteResponseHeader，作用是为 Response 请求修

改 Header 内容，对应的 Config 配置内部维护 regexp 和 replacement 属性，regexp 表示匹配的正则表达式，replacement 表示替换内容。跟其他修改 Response 的 Filter 相比，RewriteResponseHeaderGatewayFilterFactory 基于正则完成。

- SetStatusGatewayFilterFactory：RouteDefinitionRouteLocator 中的 gatewayFilterFactories 属性对应的 key 为 SetStatus，作用是修改 Response Status 状态码，对应的 Config 内部维护 status 属性，表示修改后的状态码。
- SaveSessionGatewayFilterFactory：RouteDefinitionRouteLocator 中的 gatewayFilterFactories 属性对应的 key 为 SaveSession，作用是基于 Spring Web 内的 WebSession 接口保存 Session 内容。没有 Config 配置信息。
- StripPrefixGatewayFilterFactory：RouteDefinitionRouteLocator 中的 gatewayFilterFactories 属性对应的 key 为 StripPrefix，作用是跳过请求部分的路径，对应的 Config 内部维护 parts 属性，表示跳过的路径个数。
- RequestHeaderToRequestUriGatewayFilterFactory：RouteDefinitionRouteLocator 中的 gatewayFilterFactories 属性对应的 key 为 RequestHeaderToRequestUri，作用是将请求路径修改成 Request Header 内的信息，对应的 Config 内部维护 name 属性，表示 Request Header 中的 key。
- RequestSizeGatewayFilterFactory：RouteDefinitionRouteLocator 中的 gatewayFilterFactories 属性对应的 key 为 RequestSize，用于阻止传递内容超过限制值的请求，对应的 Config 内部维护 maxSize 属性，表示请求内容的大小限制值，默认是 5 MB。
- RequestHeaderSizeGatewayFilterFactory：RouteDefinitionRouteLocator 中的 gatewayFilter-Factories 属性里对应的 key 为 RequestHeaderSize，用于阻止当请求 Header 内容超过限制值的请求，对应的 Config 内部维护 maxSize 属性，表示请求 Header 的大小限制值，默认是 16 KB。

9.6.4 网关内置的 GlobalFilter

Spring Cloud Gateway 提供了一些内置的 GlobalFilter，解释如下：

- ForwardPathFilter：处理 schema 为 forward 的请求，会对请求的路径进行解析。配合 ForwardRoutingFilter 完成请求处理。
- ForwardRoutingFilter：处理 schema 为 forward 的请求，配合 ForwardPathFilter 完成请求的转发。内部会根据 DispatcherHandler 处理请求。
- GatewayMetricsFilter：基于 MicroMeter 完成 Spring Cloud Gateway Metrics 信息的统计。

- **LoadBalancerClientFilter**：处理 schema 为 lb 的请求。基于 Spring Cloud LoadBalancer 和 Spring Cloud 服务注册/发现的编程模型完成服务名对应实例的查询。
- **NettyRoutingFilter**：处理 schema 为 http 或 https 的请求，内部使用 Netty HttpClient 客户端完成请求调用。
- **NettyWriteResponseFilter**：通过 Netty HttpClient 写回 Response 信息到客户端。
- **ReactiveLoadBalancerClientFilter**：LoadBalancerClientFilter 的 reactive 实现。
- **SentinelGatewayFilter**：基于 Sentinel 完成限流/降级。
- **WebClientHttpRoutingFilter**：处理 schema 为 http 或 https 的请求，内部使用 Spring WebClient 完成请求调用。默认情况下，使用 NettyRoutingFilter，可通过相关配置修改成 WebClientHttpRoutingFilter。
- **WebClientWriteResponseFilter**：通过 Spring WebClient 写回 Response 信息到客户端。
- **WebsocketRoutingFilter**：代理 WebSocket 请求。

9.7 整合注册中心和配置中心

试想这样一个场景：如果注册中心里有几十个甚至上百个服务，那么是不是要配置几十个甚至上百个路由信息？

Spring Cloud Gateway 提供了一个参数用于根据请求路径自动识别服务名，并进行服务调用的功能。比如，注册中心有一个服务为 nacos-provider，直接访问 http://localhost:8080/nacos-provider/echo，即可访问 nacos-provider 服务对外暴露的 echo 端点。这个配置项是 `spring.cloud.gateway.discovery.locator.enabled`，需要设置为 true。

该配置数据完成配置后，会触发 GatewayDiscoveryClientAutoConfiguration$ReactiveDiscoveryClientRouteDefinitionLocatorConfiguration 自动化配置类，自动化配置类内部会构造 DiscoveryClientRouteDefinitionLocator 这个 RouteDefinitionLocator 接口的实现类（默认情况下，构造的是 PropertiesRouteDefinitionLocator 这个 RouteDefinitionLocator 接口的实现类），代码如下：

```
@Configuration(proxyBeanMethods = false)
@ConditionalOnProperty(value = "spring.cloud.discovery.reactive.enabled",
        matchIfMissing = true)
public static class ReactiveDiscoveryClientRouteDefinitionLocatorConfiguration {
```

```java
    @Bean
    @ConditionalOnProperty(name = "spring.cloud.gateway.discovery.locator.enabled")
    public DiscoveryClientRouteDefinitionLocator discoveryClientRouteDefinitionLocator(
            ReactiveDiscoveryClient discoveryClient,
            DiscoveryLocatorProperties properties) {
        return new DiscoveryClientRouteDefinitionLocator(discoveryClient, properties);
    }

}
```

DiscoveryClientRouteDefinitionLocator 实现的 getRouteDefinitions 方法内部会根据 ReactiveDiscoveryClient 找出所有的服务，并构造对应的路由信息。

DiscoveryClientRouteDefinitionLocator#getRouteDefinitions 源码如下：

```java
// DiscoveryClientRouteDefinitionLocator.java
public DiscoveryClientRouteDefinitionLocator(ReactiveDiscoveryClient discoveryClient,
        DiscoveryLocatorProperties properties) {
    this(discoveryClient.getClass().getSimpleName(), properties);
    serviceInstances = discoveryClient.getServices()
            .flatMap(service ->
    discoveryClient.getInstances(service).collectList());    // ①
}

@Override
public Flux<RouteDefinition> getRouteDefinitions() {

    SpelExpressionParser parser = new SpelExpressionParser();
    Expression includeExpr = parser
            .parseExpression(properties.getIncludeExpression());
    Expression urlExpr = parser.parseExpression(properties.getUrlExpression());

    Predicate<ServiceInstance> includePredicate;
    if (properties.getIncludeExpression() == null    // ②
            || "true".equalsIgnoreCase(properties.getIncludeExpression())) {
```

```
            includePredicate = instance -> true;
        }
        else {
            includePredicate = instance -> {
                Boolean include = includeExpr.getValue(evalCtxt, instance, Boolean.class);
                if (include == null) {
                    return false;
                }
                return include;
            };
        }

        return serviceInstances.filter(instances -> !instances.isEmpty())    // ③
                .map(instances -> instances.get(0)).filter(includePredicate)
                .map(instance -> {
                    String serviceId = instance.getServiceId();

                    RouteDefinition routeDefinition = new RouteDefinition();    // ④
                    routeDefinition.setId(this.routeIdPrefix + serviceId);
                    String uri = urlExpr.getValue(evalCtxt, instance, String.class);
                    routeDefinition.setUri(URI.create(uri));

                    final ServiceInstance instanceForEval = new DelegatingServiceInstance(
                            instance, properties);

                    for (PredicateDefinition original : this.properties.getPredicates()) {
                        PredicateDefinition predicate = new PredicateDefinition(); // ⑤
                        predicate.setName(original.getName());
                        for (Map.Entry<String, String> entry : original.getArgs()
                                .entrySet()) {
                            String value = getValueFromExpr(evalCtxt, parser,
                                    instanceForEval, entry);
                            predicate.addArg(entry.getKey(), value);
                        }
                        routeDefinition.getPredicates().add(predicate);
                    }
```

```java
            for (FilterDefinition original : this.properties.getFilters()) {
                FilterDefinition filter = new FilterDefinition();      // ⑥
                filter.setName(original.getName());
                for (Map.Entry<String, String> entry : original.getArgs()
                        .entrySet()) {
                    String value = getValueFromExpr(evalCtxt, parser,
                            instanceForEval, entry);
                    filter.addArg(entry.getKey(), value);
                }
                routeDefinition.getFilters().add(filter);
            }

            return routeDefinition;
        });
}
```

上述代码中：

① DiscoveryClientRouteDefinitionLocator 构造函数会根据 ReactiveDiscoveryClient 获取所有的服务名和这些服务名对应的实例列表。

② 基于 SPEL 表达式和 `spring.cloud.gateway.discovery.locator.includeExpression` 配置项，构造 ServiceInstance 的 Predicate。

③ 根据代码②处的 Predicate 过滤一些 ServiceInstance。

④ 根据 ServiceInstance 信息构造 RouteDefinition。

⑤ 根据 `spring.cloud.gateway.discovery.locator.predicates` 前缀对应的 predicate 配置构造 FilterDefinition。

⑥ 根据 `spring.cloud.gateway.discovery.locator.filters` 前缀对应的 Filter 配置构造 PredicateDefinition。

Spring Cloud Gateway 内部提供 RouteRefreshListener 这个 ApplicationListener。当接收到 ContextRefreshedEvent、RefreshScopeRefreshedEvent 或 InstanceRegisteredEvent 事件的时候，会发送 RefreshRoutesEvent 事件，当发生 HeartbeatEvent 事件的时候，会判断是否需要发送 RefreshRoutesEvent 事件，其代码如下：

```java
public class RouteRefreshListener implements ApplicationListener<ApplicationEvent> {

    private final ApplicationEventPublisher publisher;

    private HeartbeatMonitor monitor = new HeartbeatMonitor();

    public RouteRefreshListener(ApplicationEventPublisher publisher) {
        Assert.notNull(publisher, "publisher may not be null");
        this.publisher = publisher;
    }

    @Override
    public void onApplicationEvent(ApplicationEvent event) {
        if (event instanceof ContextRefreshedEvent
                || event instanceof RefreshScopeRefreshedEvent
                || event instanceof InstanceRegisteredEvent) {
            reset();
        }
        else if (event instanceof ParentHeartbeatEvent) {
            ParentHeartbeatEvent e = (ParentHeartbeatEvent) event;
            resetIfNeeded(e.getValue());
        }
        else if (event instanceof HeartbeatEvent) {
            HeartbeatEvent e = (HeartbeatEvent) event;
            resetIfNeeded(e.getValue());
        }
    }

    private void resetIfNeeded(Object value) {
        if (this.monitor.update(value)) {
            reset();
        }
    }

    private void reset() {
```

```
        this.publisher.publishEvent(new RefreshRoutesEvent(this));
    }
}
```

ContextRefreshedEvent 和 RefreshScopeRefreshedEvent 事件是 Spring Cloud 配置动态改变相关的事件，InstanceRegisteredEvent 事件表示服务实例注册成功的事件，Spring Cloud Gateway 自身对应的实例在注册中心注册成功后，也会调用 reset 方法。

reset 方法内部会发送 RefreshRoutesEvent 事件。CachingRouteLocator 内部会接收 RefreshRoutesEvent 事件，并刷新路由信息，代码如下：

```java
public class CachingRouteLocator
        implements RouteLocator, ApplicationListener<RefreshRoutesEvent> {

    private final RouteLocator delegate;

    private final Flux<Route> routes;

    private final Map<String, List> cache = new HashMap<>();

    public CachingRouteLocator(RouteLocator delegate) {
        this.delegate = delegate;
        routes = CacheFlux.lookup(cache, "routes", Route.class)
                .onCacheMissResume(() -> this.delegate.getRoutes()
                        .sort(AnnotationAwareOrderComparator.INSTANCE));
    }

    @Override
    public Flux<Route> getRoutes() {
        return this.routes;
    }

    public Flux<Route> refresh() {
        this.cache.clear();
        return this.routes;
```

```java
    }

    @Override
    public void onApplicationEvent(RefreshRoutesEvent event) {
        refresh();
    }

    @Deprecated
    void handleRefresh() {
        refresh();
    }
}
```

CachingRouteLocator 使用了装饰者设计模式装饰原先的 RouteLocator。

9.8　GatewayControllerEndpoint

GatewayControllerEndpoint 由 Spring Cloud Gateway 提供，用于暴露 Gateway 内部的一些信息。GatewayControllerEndpoint 对应的 ID 为 gateway，提供如下路径已经对应的功能：

- GET/routes：返回定义的全部路由信息。
- GET /routes/{id}：基于 ID 返回路由信息。

```json
{
    "predicate": "Paths: [/nacos/**], match trailing slash: true",
    "route_id": "nacos",
    "filters": ["[[StripPrefix parts = 1], order = 1]"],
    "uri": "lb://nacos-provider",
    "order": 0
}
```

- POST /refresh：发送 RefreshRoutesEvent 事件强制刷新路由信息。
- GET /globalfilters：返回所有的 GlobalFilter 信息。
- GET /routefilters：返回所有的 GatewayFilterFactory 信息。
- GET /routepredicates：返回所有的 RoutePredicateFactory 信息。

- POST /routes/{id}：基于 ID 修改对应的路由信息。
- DELETE /routes/{id}：基于 ID 删除对应的路由信息。
- GET /routes/{id}/combinedfilters：基于 ID 返回 Route 路由信息。与 GET /routes/{id}的区别在于，该方法返回的是 Object 的 toString 方法，而另外一个转成了 json 数据。

```
{
    "Route{id='nacos', uri=lb://nacos-provider, order=0, predicate=Paths:
[/nacos/**], match trailing slash: true, gatewayFilters=[[[StripPrefix parts = 1],
order = 1]], metadata={}}": 0
}
```

9.9 案例：使用 Spring Cloud Gateway 进行路由转发

本例的依赖文件需要加上 Spring Cloud Gateway 对应的依赖信息：`org.springframework.cloud:spring-cloud-starter-gateway`。

使用 Spring Cloud Gateway 的核心步骤是定义路由信息，路由信息需要配置对应的 Predicates，确保请求地址可以根据路由信息列表找到合适的路由信息。

配置文件里路由信息的定义如下：

```
spring:
  cloud:
    nacos:
      discovery:
        server-addr: localhost:8848    // ①
    gateway:
      routes:    // ②
        - id: http    // ③
          uri: http://httpbin.org    // ④
          predicates:    // ⑤
            - Path=/httpbin/**
          filters:    // ⑥
            - StripPrefix=1
        - id: nacos    // ⑦
```

```
uri: lb://nacos-provider    // ⑧
predicates:
- Path=/nacos/**
filters:
- StripPrefix=1
```

上述代码中：

① Spring Cloud Gateway 本身已经与 Spring Cloud 服务注册/发现机制打通，需要配置注册中心的地址。

② 路由信息的定义通过 spring.cloud.gateway.routes 前缀配置，这是一个 List 集合，对应的配置类为 RouteDefinition。

③ 路由信息的 ID 信息，若不配置，默认是一个 UUID。这个 ID 没有实质性的作用。

④ 路由信息的 URI 信息，这是一个 Java URI 对象。

⑤ 路由的 predicate 信息，这是一个 PredicateDefinition 集合对象。PredicateDefinition 对象内部有 String name 和 Map<String, String> args 属性。Path=/httpbin/** 表示一个 name 属性为 Path，args 属性只有一个元素，key 为 genkey0，value 为 /httpbin/** 的 PredicateDefinition 对象。关于 PredicateDefinition 如何转换成对应的 Predicate，请参考 9.5 节的内容。

⑥ 路由信息的 Filter 信息，这是 GatewayFilter 类型的 Filter，是一个 FilterDefinition 集合对象。FilterDefinition 对象内部有 String name 和 Map<String, String> args 属性。StripPrefix=1 表示一个 name 属性为 StripPrefix，args 属性只有一对 k-v（key 为 genkey0，value 为 1）的 FilterDefinition 对象。关于 FilterDefinition 如何转换成对应的 GatewayFilter，请参考 9.6 节的内容。

⑦ 另外一个 ID 为 nacos 的路由信息。这个配置总共有两个路由信息。

⑧ 这个路由信息的 URI 的 protocol 协议是 lb，表示是一个负载均衡协议，会从注册中心查询 nacos-provider 服务名对应的实例。

Spring Cloud Gateway 路由配置信息中的两段路由配置完成的效果如下：

- 访问 contextPath 为 /httpbin/** 的所有请求，路由到 http://httpbin.org 下。比如，访问 http://ip:port/httpbin/status/200，会路由到 http://httpbin.org/status/200 地址。

- 访问 contextPath 为 /nacos/** 的所有请求，使用负载均衡路由到注册中心里服务名为 nacos-provider 对应的实例。比如，访问 http://ip:port/nacos/echo，会路由到 http://192.168.0.1:8080/echo 地址（注册中心里服务名为 nacos-provider 的实例只有一个，IP 为 192.168.0.1，port 为 8080）。

Spring Cloud Gateway 启动代码与一个普通的 Spring Boot 应用启动代码一致：

```
@SpringBootApplication
public class SpringCloudGatewayApplication {

    public static void main(String[] args) {
        SpringApplication.run(SpringCloudGatewayApplication.class, args);
    }

}
```

启动完成后使用 curl 进行验证：

```
$ curl -i -XGET 'http://localhost:8080/httpbin/status/400'
HTTP/1.1 400 Bad Request
Date: Fri, 03 Apr 2020 15:19:01 GMT
Content-Type: text/html; charset=utf-8
Content-Length: 0
Server: gunicorn/19.9.0
Access-Control-Allow-Origin: *
Access-Control-Allow-Credentials: true


$ curl -i -XGET 'http://localhost:8080/nacos/echo?name=jim'
HTTP/1.1 200 OK
Content-Type: text/plain;charset=UTF-8
Content-Length: 9
Date: Fri, 03 Apr 2020 15:19:09 GMT

echo: jim
```

第 10 章

Spring Cloud 与 Serverless

本章将首先介绍 Serverless 的概念，Serverless 目前主要分为 BaaS 和 FaaS。Spring Cloud 生态内的 Spring Cloud Function 涉及 FaaS 领域。Spring Cloud Function 出现的目的是为了统一目前业界的一些 FaaS 产品，比如 AWS Lambda、Azure Functions、Google Cloud Functions。开发者并不需要关心底层使用的是哪一套 FaaS 模型，只要了解 Spring Cloud Function 的使用即可，它会屏蔽这些 FaaS 产品的细节。然后讲解 Spring Cloud Function 与 Spring 生态内的其他组件的深度集成。最后介绍在 GCP 上用 Cloud Function 体验 Spring Cloud Function。

10.1 Serverless

近几年，Serverless 的概念非常火，各个云厂商和开源社区也开始拥抱 Serverless，比如，最早提出 Serverless 概念的 Amazon 推出的 AWS Lambda，或者是 Google Cloud Functions、Azure Functions、Alibaba Cloud Function Compute（阿里云函数计算）、SAE（Serverless App Engine，即阿里云 Serverless 应用引擎），或者是开源框架如 Apache Openwhisk、Spring Cloud

Function、Knative、Riff 等，都是 Serverless 火爆的体现。

那么什么是 Serverless？为什么近几年热度这么高？

其实 Serverless 目前还没有一个明确的定义，也没有事实的标准（因此，出现了如此多的开源框架和商业化产品）。

维基百科上对 Serverless 的定义为：无服务器运算（Serverless Computing）又被称为函数即服务（Function-as-a-Service，缩写为 FaaS），是云计算的一种模型。以平台即服务（PaaS）为基础，无服务器运算提供一个微型的架构，终端客户不需要部署、配置或管理服务器服务，代码运行所需要的服务器服务皆由云端平台来提供。

对上述定义提取重点如下：

- Serverless 属于云计算的一种模型。
- 终端客户不需要关注 IaaS（Infrastructure-as-a-Service）层资源，只需要提供代码（函数）。

Martin Fowler 官网上也有一篇关于 Serverless Architectures 的文章介绍 Serverless 架构。该文章把 Serverless 应用分为 BaaS 和 FaaS 两种类型。

- BaaS：后端即服务，这里的 Backend 可以理解为任何第三方提供的应用和服务。比如，各个云厂商对外提供的服务——阿里云 SAE（Serverless 应用引擎）（免运维 IaaS、按需使用、按量计费）、Google Firebase、LeanCloud 等。
- FaaS：函数即服务。应用以函数的形式存在，并由第三方云平台托管运行，比如，各个云厂商对外提供的 FaaS 服务——AWS Lambda、Google Cloud Functions、阿里云函数计算等。

如图 10-1 所示，FaaS 这套 Serverless 出现后，微服务架构的形态逐渐演变成函数形态。

无论是 FaaS 还是 BaaS，使用 Serverless 时，开发者只需要关注应用业务逻辑的代码开发。

下面来看看 Spring 对 Serverless 的支持——Spring Cloud Function，在介绍它之前，先了解一下 Java Function。

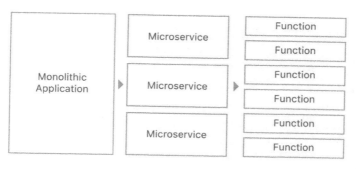

图 10-1

10.2 Java Function

说起 Spring Cloud Function,我们需要先了解 Java 语言中的函数概念。

JDK 1.8 推出了新特性 Lambda 表达式(函数可以作为方法的参数),java.util.function 包中提供了很多函数,其中有以下 3 个函数尤为重要。

(1) java.util.function.Function:需要一个参数,然后根据这个参数得到另一个结果。

```
@FunctionalInterface
public interface Function<T, R> {

    R apply(T t);

}
```

比如,利用 Stream 里的 map 方法,可以通过 Function 把一个字母从小写变成大写。

```
Stream.of("a", "b", "c").map(String::toUpperCase);
```

map 方法需要一个 Function 参数:

```
<R> Stream<R> map(Function<? super T, ? extends R> mapper);
```

比如,方法内部需要一个 Function 参数的使用方法:

```
func(s -> s.toUpperCase(), "a");
Function<String, String> func = s -> s.toUpperCase();
func(func, "b");
func(func, "c");

String func(Function<String, String> func, String origin) {
    return func.apply(origin);
}
```

（2）java.util.function.Consumer：直接对一个参数进行操作，无返回值。

```
@FunctionalInterface
public interface Consumer<T> {

    void accept(T t);

}
```

比如，通过 Stream 里的 forEach 方法遍历每个元素，做对应的业务逻辑处理：

```
RestTemplate restTemplate = new RestTemplate();
Stream.of("200", "201", "202").forEach(code -> {
    ResponseEntity<String> responseEntity =
        restTemplate.getForEntity("http://httpbin.org/status/" + code, String.class);
    System.out.println(responseEntity.getStatusCode());
});
```

（3）java.util.function.Supplier：得到一个结果，无输入参数。

```
@FunctionalInterface
public interface Supplier<T> {

    T get();

}
```

比如，自定义 Supplier 可以返回随机数：

```
Random random = new Random();

Supplier supplier100 = () -> random.nextInt(100);
Supplier supplier1000 = () -> random.nextInt(1000);

System.out.println(supplier100.get());
System.out.println(supplier1000.get());
```

Spring Cloud Function 的出现可以管控这些函数，让每个函数都可以被 Spring ApplicationContext 所管理，成为内部的一个 Bean：

```
@SpringBootApplication
public class SpringCloudFunctionApplication {

    public static void main(String[] args) {
        ApplicationContext applicationContext =
            SpringApplication.run(SpringCloudFunctionApplication.class, args);
        applicationContext.getBean("upperCase", Function.class).apply("a"); //a 通
过 uppercase 函数变成了 A
    }

    @Bean
    public Function<String, String> upperCase() {
        return s -> s.toUpperCase();
    }
}
```

upperCase 方法被@Bean 注解修饰，且返回值是一个 Function，会被解析成 ApplicationContext 下的一个 Spring Bean。通过 ApplicationContext 的 getBean 方法获取 Function，再调用 apply 操作。

10.3 Spring Cloud Function

Spring Cloud Function 出现的目的是为了统一目前业界的一些 FaaS 产品，比如 AWS

Lambda、Azure Functions、Google Cloud Functions、OpenWhisk 的编程模型,这与 Spring Cloud Stream 为不同消息中间件提供统一的上层 API 是一致的。开发者并不需要关心底层使用哪一套 FaaS 模型,只需了解 Spring Cloud Function 的使用即可,它会屏蔽掉这些 FaaS 产品的细节。

在 Spring Cloud Function 没有出现之前,Function 也可以作为一个 Bean 被注册到 ApplicationContext 内。以下代码在未使用 Spring Cloud Function 相关依赖的时候也可以正常运行:

```java
@SpringBootApplication
public class JavaFunctionApplication {

    public static void main(String[] args) {
        ApplicationContext applicationContext =
            SpringApplication.run(JavaFunctionApplication.class, args);
        System.out.println(applicationContext.getBean("upperCase", Function.class));
    }

    @Bean
    public Function<String, String> upperCase() {
        return s -> s.toUpperCase();
    }

    @Bean
    public Function<String, String> lowerCase() {
        return s -> s.toLowerCase();
    }

    @Bean
    public Consumer<String> print() {
        return s -> System.out.println(s);
    }
}
```

Spring Cloud Function 模块主要集中在以下 3 点内容。

- 对 Function 的注册/使用做了一些增强。比如，spring.cloud.function.scan.packages 用于进行自定义 package 函数的注册；FunctionCatalog 用于 ApplicationContext 内函数的获取；FunctionRegistry 用于动态注册函数 Bean；Functional-SpringApplication 专注于函数应用。
- 新增一些函数，比如 FluxFunction、FluxConsumer、FluxSupplier 等函数支持 Reactor，IsolatedFunction、IsolatedConsumer、IsolatedSupplier 支持 ClassLoader 隔离的 Function 等。
- 跟 Spring 生态里的一些组件进行了集成，比如 Spring Web、Spring WebFlux、Spring Cloud Stream 等。

ContextFunctionCatalogAutoConfiguration 是 spring-cloud-function-context 模块内部的自动化配置类，内部会初始化关键/功能组件：

- FunctionRegistry：函数目录接口。
- RoutingFunction：充当函数路由的角色，Function 接口的实现类，会根据参数得到另一个 Function，然后发起一个函数调用。

FunctionCatalog 函数目录接口的定义如下：

```java
public interface FunctionCatalog {

    // 基于函数定义和 MIME（Spring Web 或 Spring Cloud Stream 会用到）类型返回具体的函数
    default <T> T lookup(String functionDefinition, String... acceptedOutputMimeTypes) {
        throw new UnsupportedOperationException("This instance of FunctionCatalog does not support this operation");
    }

    // 基于函数定义返回具体的函数
    default <T> T lookup(String functionDefinition) {
        return this.lookup(null, functionDefinition);
    }

    // 基于函数定义和函数类型返回具体的函数
    <T> T lookup(Class<?> type, String functionDefinition);
```

```
// 返回函数目录中目前对于类型的函数集合
Set<String> getNames(Class<?> type);

// 返回函数目录中所有的函数个数
default int size() {
    throw new UnsupportedOperationException("This instance of FunctionCatalog does not support this operation");
}
}
```

FunctionCatalog 默认有两种实现：BeanFactoryAwareFunctionRegistry 和 AbstractComposableFunctionRegistry（实现类有 InMemoryFunctionCatalog）。其中，BeanFactoryAwareFunctionRegistry 对函数的操作基于 Spring BeanFactory 中的 Bean 函数。

```
// BeanFactoryAwareFunctionRegistry.java
...
@Override
public <T> T lookup(Class<?> type, String definition) {
    return this.lookup(definition, new String[] {});
}

@Override
public int size() {
    return this.applicationContext.getBeanNamesForType(Supplier.class).length +
            this.applicationContext.getBeanNamesForType(Function.class).length +
            this.applicationContext.getBeanNamesForType(Consumer.class).length;
}

@SuppressWarnings("unchecked")
public <T> T lookup(String definition, String... acceptedOutputTypes) {
    Object function = this.proxyInvokerIfNecessary((FunctionInvocationWrapper)
this.compose(null, definition, acceptedOutputTypes));
    return (T) function;
}
...
```

AbstractComposableFunctionRegistry 则是内部维护着一个类型为 Map<String, Object>的 function 属性，接口对应的 lookup 操作在 function 属性中查找。

```java
...
@Override
public <T> T lookup(Class<?> type, String name) {
    String functionDefinitionName = !StringUtils.hasText(name)
            && this.environment.containsProperty("spring.cloud.function.definition")
                    ? this.environment.getProperty("spring.cloud.function.definition")
                    : name;
    return (T) this.doLookup(type, functionDefinitionName);
}

@SuppressWarnings("serial")
@Override
public Set<String> getNames(Class<?> type) {
    if (type == null) {
        return new HashSet<String>(getSupplierNames()) {
            {
                addAll(getFunctionNames());
            }
        };
    }
    if (Supplier.class.isAssignableFrom(type)) {
        return this.getSupplierNames();
    }
    if (Function.class.isAssignableFrom(type)) {
        return this.getFunctionNames();
    }
    return Collections.emptySet();
}

@Override
public int size() {
    return this.functions.size();
}
```

```java
public Set<String> getFunctionNames() {
    return this.functions.entrySet().stream()
            .filter(entry -> !(entry.getValue() instanceof Supplier))
            .map(entry -> entry.getKey())
            .collect(Collectors.toSet());
}
...
```

默认情况下，ContextFunctionCatalogAutoConfiguration 内部构造的 FunctionCatalog 是 BeanFactoryAwareFunctionRegistry。

如果在配置文件中开启了 spring.functional.enabled=true 配置，ContextFunctionCatalog-Initializer 初始化器内部会读取 spring.cloud.function.scan.packages 配置项（未配置时，默认读取 packages）的值，并将该值当成一个包路径读取函数类型的 Bean，然后会注册 InMemoryFunctionCatalog（AbstractComposableFunctionRegistry 实现类）到 ApplicationContext 内，InMemoryFunctionCatalog 内部的函数属性则是 package 内扫描到的函数 Bean。

FunctionRegistry 继承 FunctionCatalog，在除拥有函数目录所有的功能外，还提供函数动态注册的功能：

```java
public interface FunctionRegistry extends FunctionCatalog {

    // 基于 FunctionRegistration 注册函数
    <T> void register(FunctionRegistration<T> registration);

}
```

使用 FunctionRegistry 注册函数：

```java
Consumer<String> consumer = code -> {
    RestTemplate restTemplate = new RestTemplate();
    ResponseEntity<String> responseEntity =
        restTemplate.getForEntity("http://httpbin.org/status/" + code,
String.class);
    System.out.println(responseEntity.getStatusCode());
};
```

```
registry.register(new FunctionRegistration(consumer, "httpFunc").type(FunctionType.of
(Consumer.class)));
Consumer<String> httpFunc = registry.lookup(Consumer.class, "httpFunc");
httpFunc.accept("200");
```

ContextFunctionCatalogAutoConfiguration 自动化配置类里会构造一个类型为 RoutingFunction 的函数。RoutingFunction 函数的作用是 Function Routing 函数路由会把真正的函数调用通过某种方式路由到另外一个。比如，通过 Spring Cloud Stream 与 Spring Cloud Function 整合完成消息的消费，可以通过 Header 中 spring.cloud.function.definition 或 spring.cloud.function.routing-expression 配置决定消息通过函数路由被哪些函数处理。比如，Spring Boot 的作用是通过 spring.cloud.function.definition=upperCase 配置项决定路由到哪个函数调用。

Spring Cloud Function 支持 Function Composition 函数组合功能，可以将多个函数组合起来使用，类似 UNIX 管道机制 history | grep，Spring Cloud Function 也支持"|"字符串让多个函数组合使用：

```
Function composite = registry.lookup("upperCase|print");
composite.apply("hello function"); // HELLO FUNCTION
```

10.4　Spring Cloud Function 与 Spring 生态的整合

Spring Cloud Function 作为 Spring 生态内的一个项目，会跟 Spring 生态内部原有的组件进行深度集成，如图 10-2 所示。

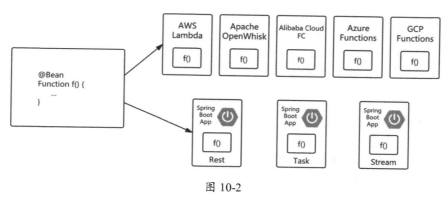

图 10-2

- Spring Web/Spring WebFlux：一次 HTTP 请求是一次函数调用。
- Spring Cloud Task：一次任务执行是一次函数调用。
- Spring Cloud Stream：一次消息消费、生产、转换是一次函数调用。

10.4.1 Spring Cloud Function 与 Spring Web/WebFlux

Spring Cloud Function 与 Spring Web/WebFlux 整合需要使用 org.springframework.cloud: spring-cloud-starter-function-web(spring-cloud-starter-function-webflux)模块实现。

双方整合的示例代码如下：

```java
@SpringBootApplication
public class SpringCloudFunctionWebApplication {

    private Random random = new Random();

    private RestTemplate restTemplate = new RestTemplate();

    public static void main(String[] args) {
        SpringApplication.run(SpringCloudFunctionWebApplication.class, args);
    }

    @Bean
    public Function<String, String> upperCase() {
        return s -> s.toUpperCase();
    }

    @Bean
    public Function<String, String> lowerCase() {
        return s -> s.toUpperCase();
    }

    @Bean
    public Supplier<Integer> random() {
        return () -> random.nextInt(1000);
    }
```

```java
    @Bean
    public Consumer<String> consumer() {
        return code -> {
            ResponseEntity<String> responseEntity =
                restTemplate.getForEntity("http://httpbin.org/status/" + code, String.class);
            System.out.println(responseEntity.getStatusCode().toString());
        };
    }
}
```

调用相关的请求，查看返回数据，内容如下：

```
$ curl -XPOST -H "Content-Type: text/plain" localhost:8080/upperCase -d hello
HELLO
$ curl -XPOST -H "Content-Type: text/plain" localhost:8080/lowerCase -d HELLO
hello
$ curl -XGET -H "Content-Type: text/plain" localhost:8080/random
198
$ curl -XGET -H "Content-Type: text/plain" localhost:8080/random
$ curl -XPOST -H "Content-Type: text/plain" localhost:8080/consumer -d 200
276
$ curl -XGET -H "Content-Type: text/plain" localhost:8080/upperCase?s=hello
{"timestamp":"2020-03-30T04:15:07.851+0000","status":500,"error":"Internal Server Error","message":"No message available","path":"/upperCase"}
```

从上述示例可以看到，定义的所有 Function Bean 都有对应的 URI，HTTP 调用直接选择对应的 URI 即可。upperCase 和 lowerCase 作为 Function 读取的参数内容是 Request Body，返回的结果是 Response Body。

如果使用 Query Parameter 参数，比如发起 `curl -XGET -H "Content-Type: text/plain" localhost:8080/upperCase\?s=hello` 调用，后端会触发 NPE 异常，因为 Function 读取的参数为 Null。

Spring Cloud Function 与 Spring Web 整合过程会涉及以下核心处理类：

- FunctionHandlerMapping：HandlerMapping 请求处理映射接口的实现类，继承 RequestMappingHandlerMapping，通过 ReactorAutoConfiguration 自动化配置类构造。
- FunctionController：接收所有请求的 Controller，内部使用 RequestProcessor 对请求做解析，最后调用对应的函数，通过 ReactorAutoConfiguration 自动化配置类构造。
- RequestProcessor：请求处理器，针对 Request 和 Response 做 Function 特性的处理。

下列代码展示了 FunctionHandlerMapping 处理请求的步骤：

```java
// FunctionHandlerMapping.java
@Override
protected HandlerMethod getHandlerInternal(HttpServletRequest request)
        throws Exception {
    HandlerMethod handler = super.getHandlerInternal(request);    // ①
    if (handler == null) {
        return null;
    }
    String path = (String) request
            .getAttribute(HandlerMapping.PATH_WITHIN_HANDLER_MAPPING_ATTRIBUTE); // ②
    if (path == null) {
        return handler;
    }
    if (StringUtils.hasText(this.prefix) && !path.startsWith(this.prefix)) {
        return null;
    }
    if (path.startsWith(this.prefix)) {    // ③
        path = path.substring(this.prefix.length());
    }

    Object function =
    FunctionWebUtils.findFunction(HttpMethod.resolve(request.getMethod()),
            this.functions, new HttpRequestAttributeDelegate(request), path); // ④
    if (function != null) {
        if (this.logger.isDebugEnabled()) {
            this.logger.debug("Found function for GET: " + path);
        }
        request.setAttribute(WebRequestConstants.HANDLER, function);
```

```java
        return handler;
    }
    return null;
}

// FunctionWebUtils.java
public final class FunctionWebUtils {

    private FunctionWebUtils() {

    }

    public static Object findFunction(HttpMethod method, FunctionCatalog functionCatalog,
                        Map<String, Object> attributes, String path) {
        if (method.equals(HttpMethod.GET) || method.equals(HttpMethod.POST)) { // ⑤
            return doFindFunction(method, functionCatalog, attributes, path);
        }
        else {   // ⑥
            throw new IllegalStateException("HTTP method '" + method + "' is not supported;");
        }
    }

    private static Object doFindFunction(HttpMethod method, FunctionCatalog functionCatalog,
                        Map<String, Object> attributes, String path) {
        path = path.startsWith("/") ? path.substring(1) : path;
        if (method.equals(HttpMethod.GET)) {
            Supplier<Publisher<?>> supplier = functionCatalog.lookup(Supplier.class, path);
            if (supplier != null) {
                attributes.put(WebRequestConstants.SUPPLIER, supplier);   // ⑦
                return supplier;
            }
        }

        StringBuilder builder = new StringBuilder();
        String name = path;
```

```
            String value = null;
            for (String element : path.split("/")) {
                if (builder.length() > 0) {
                    builder.append("/");
                }
                builder.append(element);
                name = builder.toString();
                value = path.length() > name.length() ? path.substring(name.length() + 1)
                        : null;
                Function<Object, Object> function = functionCatalog.lookup(Function.class,
                        name);
                if (function != null) {
                    attributes.put(WebRequestConstants.FUNCTION, function);    // ④
                    if (value != null) {
                        attributes.put(WebRequestConstants.ARGUMENT, value);
                    }
                    return function;
                }
            }
            return null;
        }
    }
```

上述代码中：

① 调用父类 RequestMappingHandlerMapping 的 getHandlerInternal 方法获取 HandlerMethod 请求映射的方法。

② 获取本次请求对应的 path。比如，请求为 localhost:8080/upperCase，那么对应的 path 为 /upperCase。

③ 如果有 spring.cloud.function.web.path 配置项（默认为空），取 substring，得到真正的 path。

④ 通过 FunctionWebUtils 工具类的 findFunction 方法，根据本次请求的 HTTP Method、FunctionCatalog（自动注入到 FunctionHandlerMapping 内）、Request 和 path，找到对应的 Function。

⑤ FunctionWebUtils#findFunction 方法目前只支持 GET 和 POST 方法。

⑥ 非 GET 和 POST 方法，则直接抛出异常。

⑦ 如果是 GET 方法，通过 FunctionCatalog 和 path 找到对应的函数，并设置到 Request 作用域内。

⑧ 如果是 POST 方法，以"/"分割 path，根据路径的组合通过 FunctionCatalog 和 path 找到对应的函数，并设置到 Request 作用域内。

FunctionController 处理对应的请求代码如下：

```java
// FunctionController.java
@PostMapping(path = "/**", consumes = { MediaType.APPLICATION_FORM_URLENCODED_VALUE,
    MediaType.MULTIPART_FORM_DATA_VALUE })    // ①
@ResponseBody
public Mono<ResponseEntity<?>> form(WebRequest request) {
    FunctionWrapper wrapper = wrapper(request);    // ②
    return this.processor.post(wrapper, null, false);    // ③
}

@GetMapping(path = "/**")    // ④
@ResponseBody
public Mono<ResponseEntity<?>> get(WebRequest request) {
    FunctionWrapper wrapper = wrapper(request);
    return this.processor.get(wrapper);
}

private FunctionWrapper wrapper(WebRequest request) {

    Function<Publisher<?>, Publisher<?>> function = (Function<Publisher<?>,
Publisher<?>>) request
            .getAttribute(WebRequestConstants.FUNCTION, WebRequest.SCOPE_REQUEST);

    Consumer<Publisher<?>> consumer = (Consumer<Publisher<?>>) request
            .getAttribute(WebRequestConstants.CONSUMER, WebRequest.SCOPE_REQUEST);

    Supplier<Publisher<?>> supplier = (Supplier<Publisher<?>>) request
```

```java
            .getAttribute(WebRequestConstants.SUPPLIER, WebRequest.SCOPE_REQUEST); // ⑤

    FunctionWrapper wrapper = RequestProcessor.wrapper(function, consumer, supplier);    // ⑥
        for (String key : request.getParameterMap().keySet()) {    // ⑦
            wrapper.params().addAll(key, Arrays.asList(request.getParameterValues(key)));
        }
        for (Iterator<String> keys = request.getHeaderNames(); keys.hasNext();) {    // ⑧
            String key = keys.next();
            wrapper.headers().addAll(key, Arrays.asList(request.getHeaderValues(key)));
        }
        String argument = (String) request.getAttribute(WebRequestConstants.ARGUMENT,
                WebRequest.SCOPE_REQUEST);
        if (argument != null) {
            wrapper.argument(argument);
        }
        return wrapper;
}

// RequestProcessor.java
public Mono<ResponseEntity<?>> get(FunctionWrapper wrapper) {    // ⑨
    if (wrapper.function() != null) {
        return response(wrapper, wrapper.function(), value(wrapper), true, true);
    }
    else {
        Object result = wrapper.supplier().get();
        return response(wrapper, wrapper.supplier(), result instanceof Publisher ?
(Publisher) result : Flux.just(result), null,
                true);
    }

}

@PostMapping(path = "/**")
@ResponseBody
```

```java
public Mono<ResponseEntity<?>> post(WebRequest request,
        @RequestBody(required = false) String body) {    // ⑩
    FunctionWrapper wrapper = wrapper(request);
    return this.processor.post(wrapper, body, false);
}
```

FunctionController 内所有的 Mapping value 都是 /**，表示拦截所有的请求，不同的方法通过 HTTP Method 以及 Content-Type Header 进行区分。

上述代码中：

① 针对 POST 请求且 Content-Type 为表单提交类型。

② 通过 wrapper 包装请求信息，得到 FunctionWrapper。

③ 使用 RequestProcessor 请求处理器处理请求。

④ 针对 GET 请求，不区分 Content-Type 类型。

⑤ 根据 Request 作用域获取 Function、Consumer 和 Supplier 函数。

⑥ 通过 RequestProcessor 请求处理器包装这些函数，得到 FunctionWrapper，后续也会把本次请求的 HEADER 和参数信息放到 FunctionWrapper 内。

⑦ 遍历请求的 Query Parameter 信息，设置到 FunctionWrapper 内。

⑧ 遍历请求的 Header 信息，设置到 FunctionWrapper 内。

⑨ RequestProcessor 请求处理器对 get 方法而言，决定使用 Function 还是 Supplier 处理。如果是 Function 函数，通过 value 方法直接调用 Function 的 apply 方法获得结果，在这个过程中会涉及一些内容转换；如果是 Supplier 函数，则调用 Supplier 函数得到结果。

⑩ RequestProcessor 请求处理器对 post 方法而言，通过 @RequestBody 获取 Request Body 内容作为 Function 参数的参数。

Spring Cloud Function 与 WebFlux 的整合基本也是同样的处理逻辑。org.springframework.cloud.function.web.flux 包下的 FunctionHandlerMapping 和 FunctionController 与 org.springframework.cloud.function.web.mvc 包下的 FunctionHandlerMapping 和 FunctionController 的作用一致，只是针对 Spring Web 和 Spring WebFlux 这两个场景创建不同的类。

10.4.2 Spring Cloud Function 与 Spring Cloud Stream

使用 Spring Cloud Data Flow 进行流处理（Spring Cloud Stream）的过程如图 10-3 所示。

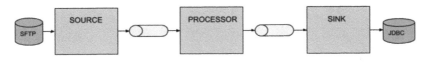

图 10-3

SOURCE、PROCESSOR 和 SINK 对应 Java Function 中的 3 种函数。SOURCE 只读取数据，对应 Supplier 函数，PROCESSOR 读取数据后进行转换并返回，对应 Function 函数，SINK 对最终的数据进行处理，对应 Consumer 函数，如图 10-4 所示。

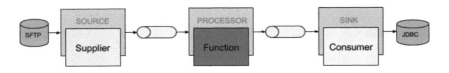

图 10-4

Spring Cloud Stream 自身已经整合了 Spring Cloud Function，对于消息的消费，可以使用 Function 函数进行处理，代码如下：

```java
@SpringBootApplication
@EnableBinding(Processor.class)
public class SpringCloudFunctionStreamApplication {

    public static void main(String[] args) {
        new SpringApplicationBuilder(SpringCloudFunctionStreamApplication.class)
            .web(WebApplicationType.NONE).run(args);
    }

    @Bean
    public Function<String, String> uppercase() {
        return x -> x.toUpperCase();
    }

}
```

配置文件内容如下：

```
spring.cloud.stream.bindings.input.destination=test-input
spring.cloud.stream.bindings.input.group=test-input-function
spring.cloud.stream.bindings.output.destination=upper-test-input
```

读取 test-input 这个 Topic 和 test-input-function group 对应的消息，通过 upper|prefix 函数组合，将消息 Payload 转换成大写，并加上 "prefix-" 前缀，最终写入消息到 upper-test-input 内。

使用 Kafka CLI 工具发送/消费消息，查看是否将内容转换成了大写。

消息发送代码如下：

```
$ ./bin/kafka-console-producer.sh --broker-list localhost:9092 --topic test-input
>hi
>nihao
>spring cloud
>spring cloud stream
>spring cloud function
```

消息消费代码如下：

```
$ ./bin/kafka-console-consumer.sh --bootstrap-server localhost:9092 --topic upper-test-input --from-beginning
HI
NIHAO
SPRING CLOUD
SPRING CLOUD STREAM
SPRING CLOUD FUNCTION
```

以上 Spring Cloud Function 和 Spring Cloud Stream 结合的代码可以进一步简化为如下代码：

```
@SpringBootApplication
public class SpringCloudFunctionStreamApplication {

    public static void main(String[] args) {
```

```
        SpringApplication.run(SpringCloudFunctionStreamApplication.class, args);
    }

    @Bean
    public Function<String, String> uppercase() {
        return x -> x.toUpperCase();
    }
}
```

去掉 @EnableBinding 注解后，Binding name 可以按照以下格式出现：

- input binding - ＋ -in- ＋
- output binding - ＋ -out- ＋

因此，所需的配置文件内容如下：

```
spring.cloud.stream.bindings.uppercase-in-0.destination=test-input
spring.cloud.stream.bindings.uppercase-in-0.group=test-input-function

spring.cloud.stream.bindings.uppercase-out-0.destination=upper-test-input
```

Spring Cloud Stream 可以使用函数组合来对消息内容进行操作：

```
@SpringBootApplication
public class SpringCloudFunctionStreamApplication {

    public static void main(String[] args) {
        SpringApplication.run(SpringCloudFunctionStreamApplication.class, args);
    }

    @Bean
    public Function<String, String> uppercase() {
        return x -> x.toUpperCase();
    }

    @Bean
```

```java
    public Function<String, String> prefix() {
        return x -> "prefix-" + x;
    }
}
```

通过 spring.cloud.function.definition 定义函数组合,Binding name 需要以定义的 definition 为前缀,后续再加上对应的格式:

```
spring.cloud.function.definition=uppercase|prefix

spring.cloud.stream.bindings.uppercase|prefix-in-0.destination=test-input
spring.cloud.stream.bindings.uppercase|prefix-in-0.group=test-input-function
spring.cloud.stream.bindings.uppercase|prefix-out-0.destination=upper-test-input
```

Spring Cloud Stream 支持以 HEADER 为 spring.cloud.function.definition 的值,决定函数路由到哪个符合条件的函数:

```java
@SpringBootApplication
public class SpringCloudFunctionStreamMultiFuncsApplication {

    public static void main(String[] args) {
        SpringApplication.run(SpringCloudFunctionStreamMultiFuncsApplication.class, args);
    }

    private final EmitterProcessor<Message<?>> processor = EmitterProcessor.create();

    @RestController
    class MyController {
        @GetMapping("/go")
        public String go(@RequestParam("func") String func, @RequestParam("payload") String payload) {
            processor.onNext(MessageBuilder.withPayload(payload).
                setHeader("spring.cloud.function.definition", func)
```

```java
                .setHeader("spring.cloud.stream.sendto.destination", "test-input").
build());
        return "ok";
    }
}

@Bean
public Supplier<Flux<Message<?>>> supplier() {
    return () -> processor;
}

@Bean
public Consumer<String> consume1() {
    return x -> System.out.println("consume1: " + x);
}

@Bean
public Consumer<String> consume2() {
    return x -> System.out.println("consume2: " + x);
}

@Bean
public Consumer<String> consume3() {
    return x -> System.out.println("consume3: " + x);
}
}
```

在配置文件中，需要配置 RoutingFunction 对应的 topic 和函数信息：

```
spring.cloud.stream.bindings.functionRouter-in-0.destination=test-input
spring.cloud.stream.bindings.functionRouter-in-0.group=test-input-function

spring.cloud.function.definition=functionRouter;supplier;consume1;consume2;consume3
```

使用 curl 发送消息，查看消费是否正确路由到了对应的函数：

```
curl -XGET http://localhost:8080/go?payload=3&func=consume3
curl -XGET http://localhost:8080/go?payload=1&func=consume1
curl -XGET http://localhost:8080/go?payload=2&func=consume2
```

Spring Cloud Stream 与 Spring Cloud Function 结合的原理是通过 StreamFunctionProperties 配置类解析配置文件里的 Function 内容，再通过 BindableFunctionProxyFactory 构造对应的 BindingTargetFactory 接口实现类（默认情况下，BindableFunctionProxyFactory 的父类 BindableProxyFactory 会对 @Output 和 @Input 注解注释的接口进行代理，生成 BindingTarget-Factory 接口实现类）：

```java
// BindableFunctionProxyFactory.java
private String buildInputNameForIndex(int index) {
    return new StringBuilder(this.functionDefinition.replace("|", ""))
        .append(FunctionConstants.DELIMITER)
        .append(FunctionConstants.DEFAULT_INPUT_SUFFIX)
        .append(FunctionConstants.DELIMITER)
        .append(index)
        .toString();
}

private String buildOutputNameForIndex(int index) {
    return new StringBuilder(this.functionDefinition.replace("|", ""))
            .append(FunctionConstants.DELIMITER)
            .append(FunctionConstants.DEFAULT_OUTPUT_SUFFIX)
            .append(FunctionConstants.DELIMITER)
            .append(index)
            .toString();
}
```

BindableFunctionProxyFactory 会被 FunctionConfiguration 配置类的内部类 FunctionBindingRegistrar 所构造，其中 functionDefinition 属性使用 StreamFunctionProperties 配置类的 definition 属性：

```java
// FunctionConfiguration$FunctionBindingRegistrar.java
@Override
public void afterPropertiesSet() throws Exception {
```

```java
        if (ObjectUtils.isEmpty(applicationContext.getBeanNamesForAnnotation
(EnableBinding.class))
                && this.determineFunctionName(functionCatalog, environment)) {
            BeanDefinitionRegistry registry = (BeanDefinitionRegistry)
applicationContext.getBeanFactory();
            String[] functionDefinitions =
streamFunctionProperties.getDefinition().split(";");
            for (String functionDefinition : functionDefinitions) {
                RootBeanDefinition functionBindableProxyDefinition = new
RootBeanDefinition(BindableFunctionProxyFactory.class);
                FunctionInvocationWrapper function =
functionCatalog.lookup(functionDefinition);
                if (function != null) {
                    Type functionType = function.getFunctionType();
                    if (function.isSupplier()) {
                        this.inputCount = 0;
                        this.outputCount = FunctionTypeUtils.getOutputCount(functionType);
                    }
                    else if (function.isConsumer()) {
                        this.inputCount = FunctionTypeUtils.getInputCount(functionType);
                        this.outputCount = 0;
                    }
                    else {
                        this.inputCount = FunctionTypeUtils.getInputCount(functionType);
                        this.outputCount = FunctionTypeUtils.getOutputCount(functionType);
                    }

                    functionBindableProxyDefinition.getConstructorArgumentValues().
addGenericArgumentValue(functionDefinition);
                    functionBindableProxyDefinition.getConstructorArgumentValues().
addGenericArgumentValue(this.inputCount);
                    functionBindableProxyDefinition.getConstructorArgumentValues().
addGenericArgumentValue(this.outputCount);
                    functionBindableProxyDefinition.getConstructorArgumentValues().
addGenericArgumentValue(streamFunctionProperties);
```

```
            registry.registerBeanDefinition(functionDefinition + "_binding",
functionBindableProxyDefinition);
        }
      }
    }
}
```

StreamFunctionProperties 配置类内部的属性如果没有在配置文件中指定，也会根据其他一些配置项进行初始化：

```
// FunctionConfiguration.java
private boolean determineFunctionName(FunctionCatalog catalog, Environment
environment) {
    String definition = streamFunctionProperties.getDefinition();
    if (!StringUtils.hasText(definition)) {
        definition = environment.getProperty("spring.cloud.function.definition");
    }

    if (StringUtils.hasText(definition)) {
        streamFunctionProperties.setDefinition(definition);
    }
    else if (Boolean.parseBoolean(environment.getProperty
("spring.cloud.stream.function.routing.enabled", "false"))
            || environment.containsProperty("spring.cloud.function.routing-expression")) {
        streamFunctionProperties.setDefinition(RoutingFunction.FUNCTION_NAME);
    }
    else {
        streamFunctionProperties.setDefinition(((FunctionInspector)
functionCatalog).getName(functionCatalog.lookup("")));
    }
    return StringUtils.hasText(streamFunctionProperties.getDefinition());
}
```

10.4.3 Spring Cloud Function 与 Spring Cloud Task

Spring Cloud Function 与 Spring Cloud Task 整合需要使用 org.springframework.cloud:

spring-cloud-function-task 模块。

Spring Cloud Function 与 Spring Cloud Task 整合的示例代码如下：

```java
@SpringBootApplication
public class SpringCloudFunctionTaskApplication {

    public static void main(String[] args) {
        SpringApplication.run(SpringCloudFunctionTaskApplication.class, args);
    }

    @Bean
    public Supplier<List<String>> supplier() {
        return () -> Arrays.asList("200", "201", "202");
    }

    @Bean
    public Function<List<String>, List<String>> function() {
        return (list) ->
            list.stream().map( item -> "prefix-" + item).collect(Collectors.toList());
    }

    @Bean
    public Consumer<List<String>> consumer() {
        return (list) -> {
            list.stream().forEach(System.out::println);
        };
    }
}
```

定义 Supplier 函数获取数据来源，使用 Function 函数对数据来源中的数据进行转换，最后通过 Consumer 函数消费转换后的数据。

配置文件加上函数 Bean name：

```
spring.cloud.function.task.function=function
spring.cloud.function.task.supplier=supplier
spring.cloud.function.task.consumer=consumer
```

Spring Cloud Function 与 Spring Cloud Task 整合过程中定义了任务执行的方式。引入 `spring-cloud-function-task` 模块后，TaskConfiguration 是内部的自动化配置类：

```
@Configuration(proxyBeanMethods = false)
@EnableTask    // ①
@EnableConfigurationProperties(TaskConfigurationProperties.class)    // ②
@ConditionalOnClass({ EnableTask.class })
public class TaskConfiguration {

    @Autowired
    private TaskConfigurationProperties properties;

    @Bean
    public CommandLineRunner commandLineRunner(FunctionCatalog registry) {
        final Supplier<Publisher<Object>> supplier = registry.lookup(Supplier.class,
this.properties.getSupplier());    // ③
        final Function<Publisher<Object>, Publisher<Object>> function =
registry.lookup(Function.class,
                this.properties.getFunction());    // ④
        final Consumer<Publisher<Object>> consumer = consumer(registry);    // ⑤
        CommandLineRunner runner = new CommandLineRunner() {

            @Override
            public void run(String... args) throws Exception {
                consumer.accept(function.apply(supplier.get()));    // ⑥
            }
        };
        return runner;
    }

    private Consumer<Publisher<Object>> consumer(FunctionCatalog registry) {
```

```
        Consumer<Publisher<Object>> consumer = registry.lookup(Consumer.class,
this.properties.getConsumer());
        if (consumer != null) {
            return consumer;
        }
        Function<Publisher<Object>, Publisher<Void>> function =
registry.lookup(Function.class,
            this.properties.getConsumer());
        return flux -> Mono.from(function.apply(flux)).subscribe();
    }
}
```

上述代码中：

① 使用 @EnableTask 注解开启 Spring Cloud Task 功能。

② 初始化 Spring Cloud Function Task 相关的配置信息 TaskConfigurationProperties Bean。

③ 通过函数目录查找配置文件里对应的 Supplier 函数。

④ 通过函数目录查找配置文件里对应的 Function 函数。

⑤ 通过函数目录查找配置文件里对应的 Consumer 函数。

⑥ CommandLineRunner 内部先使用 Supplier 得到输入数据，再通过 Function 转换数据，最后使用 Conusmer 去消费。

TaskConfigurationProperties 配置信息对应的配置前缀如下：

- `spring.cloud.function.task.supplier`。
- `spring.cloud.function.task.function`。
- `spring.cloud.function.task.consumer`。

10.5 案例：使用 GCP Cloud Functions 体验 Spring Cloud Function

Google Cloud Functions 在 2020 年 3 月才支持 Java Function，目前处于 Alpha 状态。使用它需要进入官网申请，如果不申请，控制台将看不到 Java 类型的函数，如图 10-5 所示。

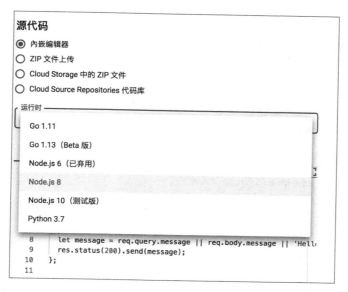

图 10-5

以下 3 个步骤是使用 GCP Cloud Functions 的过程。

（1）编写 Spring Cloud Function 函数代码。

```
@SpringBootApplication
public class SpringCloudFunctionGCPApplication {

    public static void main(String[] args) {
        SpringApplication.run(SpringCloudFunctionGCPApplication.class, args);
    }

    @Bean
    public Function<User, String> user() {
        return s -> s.toString();
    }

    public static class User {
        private String name;

        public User() {
        }
```

```
    public User(String name) {
        this.name = name;
    }

    public String getName() {
        return name;
    }

    public void setName(String name) {
        this.name = name;
    }

    @Override
    public String toString() {
        return "User{" +
            "name='" + name + '\'' +
            '}';
    }
}
```

代码编写完毕后，使用 mvn package 构造 JAR 包并放到 target/deploy 目录下。

（2）创建 GCP 函数。

由于 GCP Java Function 目前还处于 Alpha 状态，需要使用 gcloud alpha 上传，或者在 GCP Console 上直接创建函数，如图 10-6 所示。图 10-6 中的 FunctionInvoker 是 AbstractSpring-FunctionAdapterInitializer 抽象类对应的 GCP 适配器类。

```
$ gcloud alpha functions deploy scf-function \
--entry-point org.springframework.cloud.function.adapter.gcp.FunctionInvoker \
--runtime java11 \
--trigger-http \
--source target/deploy \
--memory 512MB
```

创建成功后，函数列表会出现刚创建的函数。

（3）运行验证。

在控制台进入测试页进行测试，界面如图 10-7 所示，或者访问对外暴露的 endpoint：

```
$ curl https://us-central1-{projectid}.cloudfunctions.net/scf-function/user -d
'{"name":"hello SCF"}'

"User{name\u003d\u0027hello SCF\u0027}"
```

图 10-6　　　　　　　　　　　　　　　　图 10-7